NEUROMETHODS

Series Editor
Wolfgang Walz
University of Saskatchewan
Saskatoon, SK, Canada

For further volumes:
http://www.springer.com/series/7657

Neuromethods publishes cutting-edge methods and protocols in all areas of neuroscience as well as translational neurological and mental research. Each volume in the series offers tested laboratory protocols, step-by-step methods for reproducible lab experiments and addresses methodological controversies and pitfalls in order to aid neuroscientists in experimentation. *Neuromethods* focuses on traditional and emerging topics with wide-ranging implications to brain function, such as electrophysiology, neuroimaging, behavioral analysis, genomics, neurodegeneration, translational research and clinical trials. *Neuromethods* provides investigators and trainees with highly useful compendiums of key strategies and approaches for successful research in animal and human brain function including translational "bench to bedside" approaches to mental and neurological diseases.

Vectorology for Optogenetics and Chemogenetics

Edited by

Mark A. G. Eldridge

NIMH, National Institutes of Health, Behesda, MD, USA

Adriana Galvan

Department of Neurology, School of Medicine, and Emory National Primate Research Center, Emory University, Atlanta, GA, USA

Humana Press

Editors
Mark A. G. Eldridge
NIMH
National Institutes of Health
Behesda, MD, USA

Adriana Galvan
Department of Neurology, School of Medicine, and Emory
National Primate Research Center
Emory University
Atlanta, GA, USA

ISSN 0893-2336 ISSN 1940-6045 (electronic)
Neuromethods
ISBN 978-1-0716-2920-8 ISBN 978-1-0716-2918-5 (eBook)
https://doi.org/10.1007/978-1-0716-2918-5

Cover photo credit: Bing Li, NIMH, USA

This Humana imprint is published by the registered company Springer Science+Business Media, LLC, part of Springer Nature.
The registered company address is: 1 New York Plaza, New York, NY 10004, U.S.A.

Preface to the Series

Under the guidance of its founders Alan Boulton and Glen Baker, the *Neuromethods* series by Humana Press has been very successful since the first volume appeared in 1985. In about 17 years, 37 volumes have been published. In 2006, Springer Science + Business Media made a renewed commitment to this series. The new program will focus on methods that are either unique to the nervous system and excitable cells or which need special consideration to be applied to the neurosciences. The program will strike a balance between recent and exciting developments like those concerning new animal models of disease, imaging, in vivo methods, and more established techniques. These include immunocytochemistry and electrophysiological technologies. New trainees in neurosciences still need a sound footing in these older methods in order to apply a critical approach to their results. The careful application of methods is probably the most important step in the process of scientific inquiry. In the past, new methodologies led the way in developing new disciplines in the biological and medical sciences. For example, Physiology emerged out of Anatomy in the nineteenth century by harnessing new methods based on the newly discovered phenomenon of electricity. Nowadays, the relationships between disciplines and methods are more complex. Methods are now widely shared between disciplines and research areas. New developments in electronic publishing also make it possible for scientists to download chapters or protocols selectively within a very short time of encountering them. This new approach has been taken into account in the design of individual volumes and chapters in this series.

Saskatoon, SK, Canada ***Wolfgang Walz***

About This Book

This volume provides an overview of the viral vectors and associated methods used to apply chemogenetic and optogenetic tools to the study of neural circuits. The chapters in this book are organized into three parts.

Part I covers the viral vectors that have been optimized for neuronal expression and the methods for their production and quality control. Note that, despite being widely used in the field, adeno-associated viruses (AAVs) are not covered here as their capsids are being rapidly evolved and refined, and many are commercially available. Hence we encourage the reader to search for recent review articles if they are considering using AAVs in their research (e.g., Li, C. and R. J. Samulski, *Nat Rev Genet* (2020); Wang, D., et al. *Nat Rev Drug Discov* (2019)) and to purchase virus from commercial sources, such as Addgene (www.addgene.org), the UNC vector core (www.med.unc.edu/genetherapy/vectorcore/), or the Penn vector core (www.gtp.med.upenn.edu/core-laboratories-public/vector-core).

Part II provides information on vector modifications and applications. These include the appropriate use of capsid selection and enhancer/promoter sequences for cell-type selective expression, the use of transcriptomic analysis for neuron classification, methods for pathway-specific targeting, and a description of the development of vectors for systemic delivery.

Part III provides practical advice for the delivery of viral vectors, and for the verification of injection accuracy and monitoring of transgene expression.

Per the *Neuromethods* series style, all chapters include step-by-step detailed protocols that can be easily followed, along with sections detailing tips and tricks for overcoming common problems.

Contents

PART III VECTOR DELIVERY AND VERIFICATION OF EXPRESSION

Contributors

TIPHAINE BELLOIR • *Washington National Primate Research Center, Seattle, WA, USA; Department of Bioengineering, University of Washington, Seattle, WA, USA*

MARTIN BOHLEN • *Department of Biomedical Engineering, Duke University, Durham, NC, USA; Department of Neurobiology, Duke University School of Medicine, Durham, NC, USA*

SHIH-HENG CHEN • *Viral Vector Core, Neurobiology Laboratory, National Institute of Environmental Health Sciences, NIH/DHHS, Research Triangle Park, NC, USA*

GABRIELLE CLOUSE • *Addgene, Watertown, MA, USA*

KIANA DASH • *Laboratory of Neuropsychology, NIMH, NIH, Bethesda, MD, USA*

VIOLETTE DERMINASSIAN • *Laboratory of Neuropsychology, NIMH, NIH, Bethesda, MD, USA*

JORDANE DIMIDSCHSTEIN • *Broad Institute of MIT and Harvard, Cambridge, MA, USA*

MARK A. G. ELDRIDGE • *Laboratory of Neuropsychology, National Institute of Mental Health, Bethesda, MD, USA*

INA ERSING • *Addgene, Watertown, MA, USA*

MELINA FAN • *Addgene, Watertown, MA, USA*

MITCHELL R. FARRELL • *Department of Neurobiology & Behavior, University of California, Irvine, CA, USA*

ALANNA FAVA • *Addgene, Watertown, MA, USA*

ADRIANA GALVAN • *Division of Neuropharmacology and Neurological Disorders, Emory National Primate Research Center, Emory University, Atlanta, GA, USA; Udall Center of Excellence for Parkinson's Disease Research, Emory University, Atlanta, GA, USA; Department of Neurology, School of Medicine, Emory University, Atlanta, GA, USA*

ALON GREENBAUM • *Joint Department of Biomedical Engineering, North Carolina State University and University of North Carolina at Chapel Hill, Raleigh, NC, USA; Comparative Medicine Institute, North Carolina State University, Raleigh, NC, USA*

DEVON GRIGGS • *Department of Electrical and Computer Engineering, University of Washington, Seattle, WA, USA; Washington National Primate Research Center, Seattle, WA, USA*

KAREN GUERIN • *Vedere Bio II, Inc., Cambridge, MA, USA*

KATE S. HARTEN DEMAIO • *Addgene, Watertown, MA, USA*

BO HE • *Viral Vector Core, Neurobiology Laboratory, National Institute of Environmental Health Sciences, NIH/DHHS, Research Triangle Park, NC, USA*

JING HE • *Department of Neurobiology, University of Pittsburgh, Pittsburgh, PA, USA*

KATE S. HEFFERNAN • *Division of Neuropharmacology and Neurological Disorders, Emory National Primate Research Center, Emory University, Atlanta, GA, USA; Udall Center of Excellence for Parkinson's Disease Research, Emory University, Atlanta, GA, USA*

TADASHI ISA • *Department of Neuroscience, Graduate School of Medicine, Kyoto University, Kyoto, Japan; Institute for the Advanced Study of Human Biology (WPI-ASHBi), Kyoto University, Kyoto, Japan*

MOE IWATA • *Laboratory of Cellular Pharmacology, Graduate School of Pharmaceutical Sciences, Nagoya University, Nagoya, Japan*

RIKI KAMAGUCHI • *Laboratory of Cellular Pharmacology, Graduate School of Pharmaceutical Sciences, Nagoya University, Nagoya, Japan*
SHIGEKI KATO • *Department of Molecular Genetics, Institute of Biomedical Sciences, Fukushima Medical University School of Medicine, Fukushima, Japan*
KAZUTO KOBAYASHI • *Department of Molecular Genetics, Institute of Biomedical Sciences, Fukushima Medical University School of Medicine, Fukushima, Japan*
KENTA KOBAYASHI • *Section of Viral Vector Development, National Institute for Physiological Sciences, Okazaki, Japan; SOKENDAI (The Graduate University for Advanced Studies), Hayama, Japan*
WALTER LERCHNER • *Laboratory of Neuropsychology, NIMH, NIH, Bethesda, MD, USA*
BING LI • *Laboratory of Neuropsychology, National Institute of Mental Health, Bethesda, MD, USA*
JESSICA LIN • *Broad Institute of MIT and Harvard, Cambridge, MA, USA*
ALYSSA LUZ-RICCA • *Laboratory of Neuropsychology, NIMH, NIH, Bethesda, MD, USA*
STEPHEN V. MAHLER • *Department of Neurobiology & Behavior, University of California, Irvine, CA, USA*
NEGIN P. MARTIN • *Viral Vector Core, Neurobiology Laboratory, National Institute of Environmental Health Sciences, NIH/DHHS, Research Triangle Park, NC, USA*
MARICELA X. MARTINEZ • *Department of Neurobiology & Behavior, University of California, Irvine, CA, USA*
TAKAFUMI MINAMIMOTO • *Department of Functional Brain Imaging, National Institutes for Quantum Science and Technology, Chiba, Japan*
NANAMI L. MIYAZAKI • *Laboratory of Neuropsychology, National Institute of Mental Health, Bethesda, MD, USA*
YUJI NAGAI • *Department of Functional Brain Imaging, National Institutes for Quantum Science and Technology, Chiba, Japan*
RACHAEL NEVE • *Gene Delivery Technology Core, Massachusetts General Hospital, Cambridge, MA, USA*
FUMITAKA OSAKADA • *Laboratory of Cellular Pharmacology, Graduate School of Pharmaceutical Sciences, Nagoya University, Nagoya, Japan; Laboratory of Neural Information Processing, Institute for Advanced Research, Nagoya University, Nagoya, Japan; PRESTO/CREST, Japan Science and Technology Agency, Kawaguchi, Japan*
MARCELLA PATRICK • *Addgene, Watertown, MA, USA*
MICHELLE PETROZZI • *Addgene, Watertown, MA, USA*
MEGHAN REGO • *Addgene, Watertown, MA, USA*
BARRY J. RICHMOND • *Laboratory of Neuropsychology, NIMH, NIH, Bethesda, MD, USA*
YOLAND SMITH • *Division of Neuropharmacology and Neurological Disorders, Emory National Primate Research Center, Emory University, Atlanta, GA, USA; Udall Center of Excellence for Parkinson's Disease Research, Emory University, Atlanta, GA, USA; Department of Neurology, School of Medicine, Emory University, Atlanta, GA, USA*
WILLIAM R. STAUFFER • *Department of Neurobiology, University of Pittsburgh, Pittsburgh, PA, USA*
SÉBASTIEN TREMBLAY • *Department of Neuroscience, University of Pennsylvania, Philadelphia, PA, USA*

JENNIFER B. TREWEEK • *Department of Biomedical Engineering, University of Southern California, Los Angeles, CA, USA*
WIM VANDUFFEL • *Laboratory for Neuro-and Psychophysiology, Department of Neurosciences, KU Leuven Medical School, Leuven, Belgium; Leuven Brain Institute, KU Leuven, Leuven, Belgium; Harvard Medical School, Boston, MA, USA; Massachusetts General Hospital, Martinos Center for Biomedical Imaging, Charlestown, MA, USA*
CHEN WANG • *Addgene, Watertown, MA, USA*
ZIKAI WANG • *Department of Biomedical Engineering, University of Southern California, Los Angeles, CA, USA*
MASAHIRO YAMAGUCHI • *Laboratory of Cellular Pharmacology, Graduate School of Pharmaceutical Sciences, Nagoya University, Nagoya, Japan*
AZADEH YAZDAN-SHAHMORAD • *Department of Electrical and Computer Engineering, University of Washington, Seattle, WA, USA; Washington National Primate Research Center, Seattle, WA, USA; Department of Bioengineering, University of Washington, Seattle, WA, USA; Graduate Program in Neuroscience, University of Washington, Seattle, WA, USA*
YIJUN ZHANG • *Addgene, Watertown, MA, USA*
JASMINE ZHOU • *Washington National Primate Research Center, Seattle, WA, USA; Department of Bioengineering, University of Washington, Seattle, WA, USA*

Part I

Vectors Used in Systems Neuroscience Research

Chapter 1

Production, Testing, and Verification of Lentivirus for Regional Targeting in the Old-World Monkey Brain

Walter Lerchner, Alyssa Luz-Ricca, Kiana Dash, Violette DerMinassian, and Barry J. Richmond

Abstract

Genetic tools in the monkey brain are primarily applied through the injection of viral vectors in adult animals, as germline modification is logistically challenging. There are several options for viruses and their serotypes to transduce nonreplicating cells. In this chapter, we describe the salient characteristics of lentivirus. The commonly used lentivirus pseudotyped with a vesicular stomatitis virus G (VSV-G) coat protein appears to transduce all cell types equally and can be combined with various promoters/enhancers to achieve specificity in cell type. The packaging capacity, up to 8 kb, is large enough to include regulatory promoter/enhancer regions and/or multiple transgenes in the construct. It is also relatively easy to produce high-titer and high-quality viruses in the large amounts required for treating large brain regions. For retrograde transduction of projection neurons, we find that replacement of a plasmid encoding VSVG with a plasmid encoding FuGE produces a lentivirus that expresses transgenes efficiently (see Chap. 2). Here we detail our methods for producing, testing, and verifying lentivirus gene expression in the monkey brain.

Key words Lentivirus, High-titer, Monkey, Retrograde, Amygdala, Packaging, Brain, Transgenic

1 Introduction

Recent advances in genetic methods for the inducible modulation of specific neuron populations, such as chemogenetics [1], optogenetics [2], or RNAi [3], have the potential to build on a large body of cognitive research in the monkey [4–6]. One of the challenges for use of genetic technologies in old-world monkeys is targeting the desired neuron population [7]. In rodents, accurate targeting is achieved with germline modification, sometimes in combination with viral vectors (most commonly adeno-associated virus (AAV)). For the monkey, introducing genetic material into the germline, while theoretically possible, is impractical because of the long generation times [8]. Instead, nonreplicating viruses are

Mark A. G. Eldridge and Adriana Galvan (eds.), *Vectorology for Optogenetics and Chemogenetics*,
Neuromethods, vol. 195, https://doi.org/10.1007/978-1-0716-2918-5_1,

used to transduce the desired genetic material into cells of the mature brain. AAV is the most commonly used virus. However, AAV at times shows inconsistent expression [9] and has a substantial variety of serotypes, many of which show idiosyncratic selectivity of expression [7], some of which might be caused by preexisting or induced neutralizing antibodies [10]. Another problem is that purification of AAV can be complex and expensive for an individual research laboratory [11]. Some of these challenges have been alleviated by the increasing availability of commercial AAV constructs (e.g., https://www.addgene.org). Nevertheless, many commercially available AAVs have been optimized and/or tested in rodent research and need to be verified for their effectiveness in the monkey brain. Finally, the capacity of AAV is only about 4 kb, which can limit constructs to a single gene with a small regulatory element. To try to ameliorate these problems, we have been using lentivirus vectors to deliver genetic material. When we use the common VSV-G capsid protein, we find that lentivirus vector injections result in a large percentage of targeted neurons expressing the desired gene at consistent expression levels [3, 12, 13]. In contrast to AAV serotypes, lentivirus VSV-G does not seem to show tropism in the brain [7]. The packing size of up to 8 kb permits the insertion of multiple genetic elements [3] and/or longer regulatory promoter regions [7]. Local injection results in dense local expression and no evidence for transport outside the injected region [3]. When we pseudotype lentivirus with the FuG-E capsid protein, we, like others (Kato and Kobayashi 2020, see Chap. 2), find that it results in efficient retrograde transduction of cortical projection neurons. Lentivirus does have some limitations compared to AAV, some of which we detail here. For example, virus particles are approximately fivefold larger than those of AAV, resulting in slower tissue spread and possibly tissue damage when injected too fast or at large volumes per injection. While lentivirus has a higher packaging capacity than AAV, it can become difficult to reach titers required for sufficient gene expression as the size approaches the packaging limit. Here, we present the details of our methods for production, testing, and verification of lentivirus vectors with examples of targeting the old-world monkey amygdala.

2 Lentivirus Production

One of the most important considerations in using lentivirus for targeting old-world monkey brains is to produce a functional virus with high enough infectious units per ml (iu/ml) to achieve sufficient penetrance in the target region without killing neurons through protein overexpression. We find an infectious titer of 2×10^9 iu/mL works well for receptor or reporter proteins expressed by a neuron-specific promoter, such as human synapsin promoter fragment (hSyn). Lentivirus is relatively large, approximately fivefold the size of AAV particles (100 vs. 20 nm), and VSV-G capsid protein results in good binding of viral particles to native cell surface receptors, making the virus "sticky." Both characteristics lead to good transduction near the injection site, but it means we need relatively large volumes of lentivirus to achieve effective diffusion across the region of interest.

To determine the volume of virus needed to cover one demonstrative brain region, the amygdala, we conducted a study in which we compared four 20 μL virus injections into the left amygdala expressing a chemogenetic receptor (Fig. 1a, top panel) with a single injection of 80 μL of the same virus into the right amygdala (Fig. 1a bottom panel). Both injection methods gave similar coverage across a region of approximately 4 mm^3 (Fig. 1b). However, the 80 μL injection resulted in a region in the center of the injection where the NeuN antigen, which we used to identify neurons, was no longer present, but where CFP was still identified in neurons (Fig. 1c, red outline). The loss of NeuN immunoreactivity without complete cell death could indicate damage to neurons in this region, a finding seen before [14]. In addition, there was also a small region of tissue damage that was somewhat larger than that expected just from needle damage, in which neither NeuN nor CFP was detected (Fig. 1c, red outline). Therefore, we now limit single injections to 20 μL but place multiple injections along a track and/or spread across different injection tracks when necessary to increase the injection coverage. To cover the 300 mm^3 of the amygdala in one hemisphere, we estimate that it is necessary to use a minimum of 300 μL of the virus. To arrive at this estimate, we used the coverage that resulted from the four 20 μL injections in our experiment to extrapolate the volume necessary to cover the entirety of the amygdala. Such injections require large volumes of high titer lentivirus. To achieve large volume production at high titer and high purity, we adapted previously published lentivirus production protocols [15, 16]. Our protocol works equally well for lentivirus with the VSV-G capsid protein for anterograde transduction and the FuGE capsid protein for retrograde transduction (Fig. 2). See the Method section 5.2 for details.

3 Lentivirus Titer Determination and Stability Testing

To determine the concentration of lentivirus particles in the virus that we produce, we use at least one of two methods, depending on the capsid protein of the virus. Our goal is to achieve a functional titer for injections of 2×10^9 infectious units per ml (iu/mL) for constructs encoding chemogenetic and optogenetic receptors. If the titer is higher than this, we dilute the virus to 2×10^9 iu/mL before injection. The following is an explanation of our titer determination for VSV-G and FuG-E lentivirus:

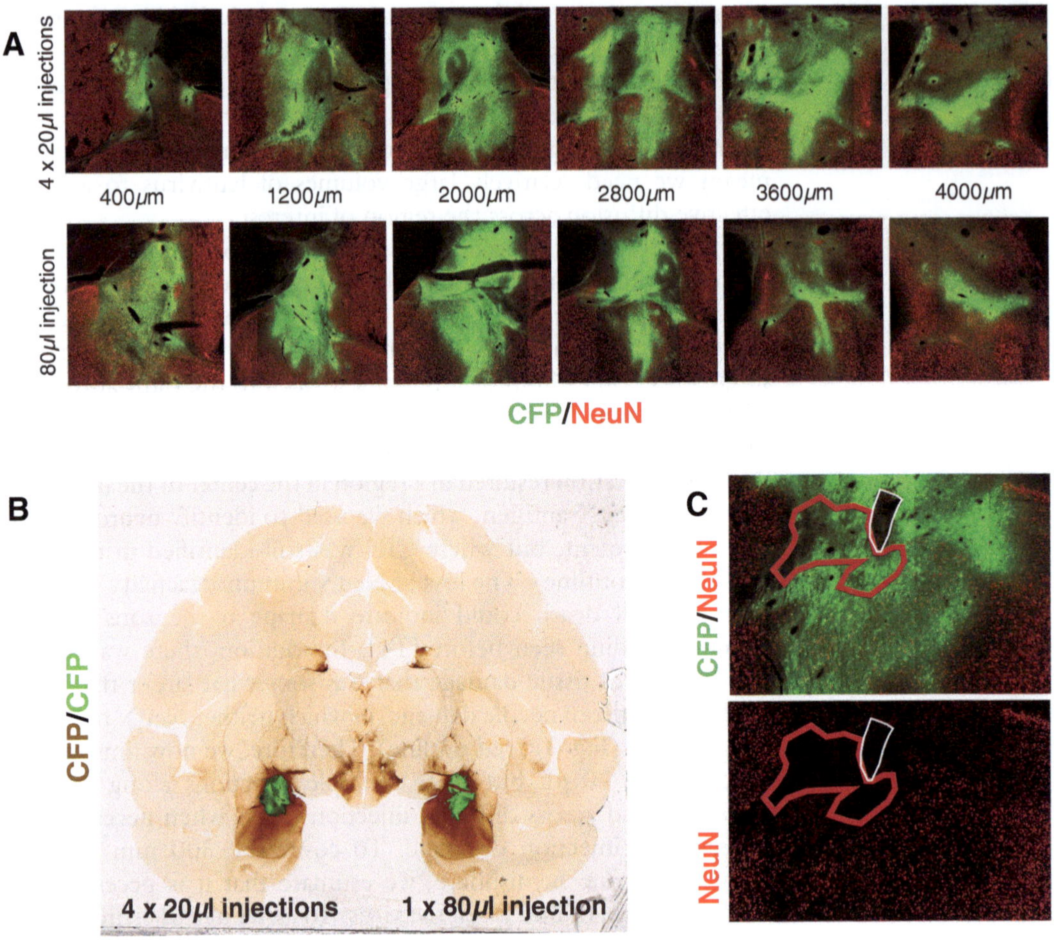

Fig. 1 Establishing targeting parameters for hM4Di_CFP chemogenetic reporter expression in the amygdala. (**a**) Comparative expression regions of either 4 × 20 μL injections in individual tracks at 1.5 mm separation in the anterior-posterior and medial-lateral planes (top panel) or a single 80 μL injection (bottom panel), measurements indicate the position of a section through the anterior to the posterior extent of the expression region. Green: Antibody staining against the CFP portion of the hM4Di_CFP. Red: Antibody staining against the neuronal marker NeuN. (**b**) Virtual overlay of fluorescent staining at 2000 μm on top of an adjacent section stained enzymatically with DAB for CFP detection. Left hemisphere: 4 × 20 μL injection, Right hemisphere: 1 × 80 μL injection. The regions stained outside the fluorescent signal (in brown) indicate staining for axonal fibers emanating from the neurons transduced by the lentivirus. (**c**) NeuN antigen was not detected in the center of the 1 × 80 μL injection outlined in red. Top panel: green signal is the staining for CFP, red staining for NeuN. Bottom panel: NeuN signal is absent in the region outlined in red, even though cells are expressing the CFP reporter gene. The region outlined in white indicates cell damage where neither NeuN nor CFP is detected. (For the method of staining, see Lerchner et al. [7]; for the method of injection, see Lerchner et al. [3])

1. Lentivirus with a VSV-G capsid protein efficiently transduces 293T cells. This makes it possible to test the virus for actual infectivity quantitatively, that is, iu/ml; see detailed protocol below. Theoretically, we can transduce 150,000 cells per well with 2 μL of virus and let the cells grow to 1,000,000 (10^6)

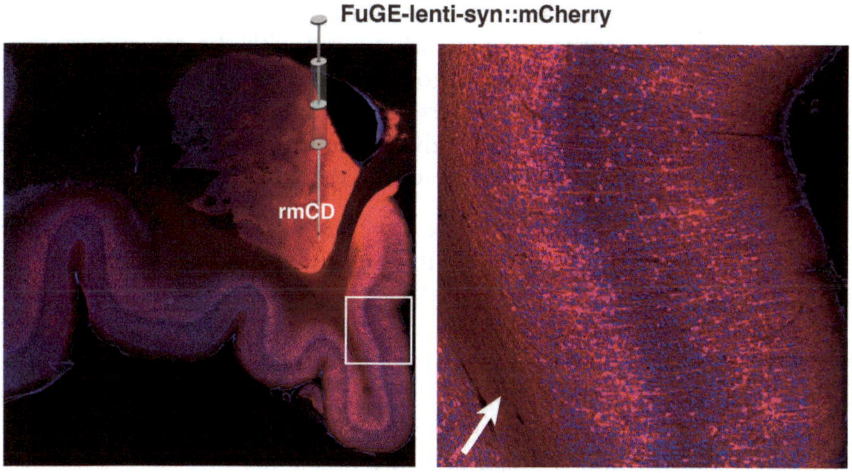

Fig. 2 Retrograde lentivirus expression in cortical neurons projecting to rostromedial Caudate. *Left panel*: FuG-E-lentivirus expressing mCherry under the control of a synapsin promoter was injected into rostromedial Caudate (rmCD). White square indicates the cortical region magnified in the right panel. Red signal is the staining for mCherry, blue signal is the staining for NeuN. *Right panel*: magnified region of ventromedial prefrontal cortex. mCherry is expressed in projection neurons, with a large number of layer five neurons expressing mCherry—white arrow indicates the red fiber bundle of mCherry-expressing cells on the inner cortical surface

cells before harvesting the DNA from these cells and comparing the copy number of a lentiviral gene per cell to the copy number of an endogenous gene (two copies per cell) via their qPCR threshold cycle (CT). If the CT values for virus and endogenous gene are equal, the infectious titer corresponds to 10^6 iu/μL (2×10^6 copies of virus in 10^6 cells, transduced with 2 μL of virus solution), or 10^9 iu/ml, the commonly used way of referring to titer. However, when we used various dilutions from 1:1 to 1:10,000 of lentivirus expressing a green fluorescent protein to visualize cells to count transduced cells and determine infectious units directly, we found that qPCR titering underestimates the titer consistently. To adjust for this underestimation, we now routinely transduce with 3 μL of virus instead of 2 μL of virus and find that qPCR titering and visualized counting corresponded for all dilutions. As methods for titering can vary, we recommend calibrating qPCR titering using a visual method.

2. Lentivirus with a FuG-E capsid protein does not efficiently transduce 293T cells. Therefore, the previously described in vitro cell culture method does not work for titer determination. Instead, we determine the particle titer directly from the concentrated virus using the qPCR Lentivirus Titer Kit (ABM, LV900). To be able to correlate particle titer directly with a

functional titer, we always run samples alongside a VSV-G lentivirus previously titered with the cell culture method. We generally get a good correlation between our functional titers and the numbers calculated from the ABM titer kit control samples, which gives us confidence in the accuracy of the direct titer method, as well as the infectious quality of the virus.

To show that the targeting of lentivirus into the brain is accurate, we mix the MRI contrast agent $MnCl_2$ with the virus and conduct a post-op MRI. To test whether the addition of $MnCl_2$ reduces the infectious titer of lentivirus, we conducted a study in which we loaded virus with or without 10 mM $MnCl_2$ into a Hamilton syringe and expelled aliquots into wells of cultured cells at timepoints from 5 min to 8 h, then used the previously described method to determine the functional titer (Fig. 3). Suspension of lentivirus with 10 mM $MnCl_2$ [17] or addition of the MRI marker gadolinium did not significantly affect titers between 1 and 8 h (Fig. 3). However, we observed increased variability in infectious titer over time with the addition of 10 mM $MnCl_2$ (Fig. 3). This

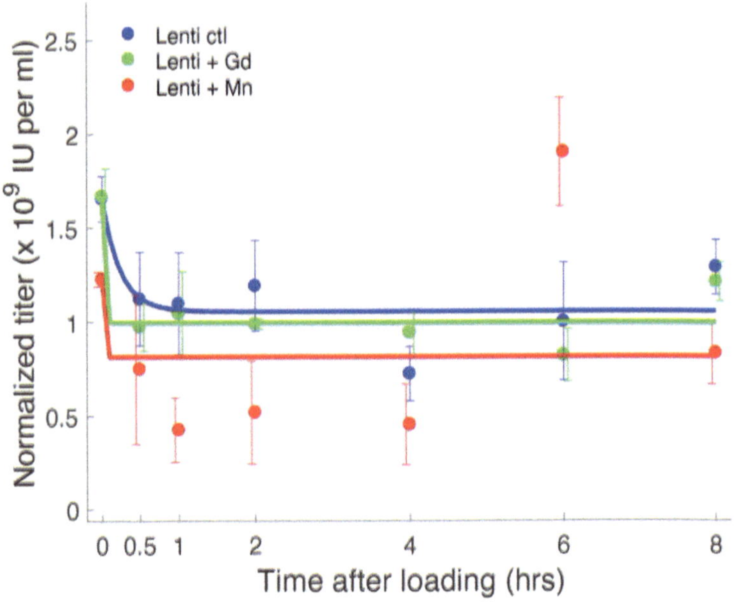

Fig. 3 Lentivirus degradation at room temperature. The virus was loaded into a Hamilton syringe. The viral sample at time 0 was pipetted directly into the culture; all subsequent time points were dispensed from the Hamilton syringe. A virus without the addition of an MRI contrast (blue), a virus with 10 mM $MnCl_2$ (red), and a virus with 100 mM Gadolinium (green). Titers were normalized to virus control samples which were transduced with a freshly thawed virus at an effective titer of 2×10^9 iu/mL. For all samples, the titers stabilized after an initial drop within the first 30 min. Four replicates were conducted for each time point

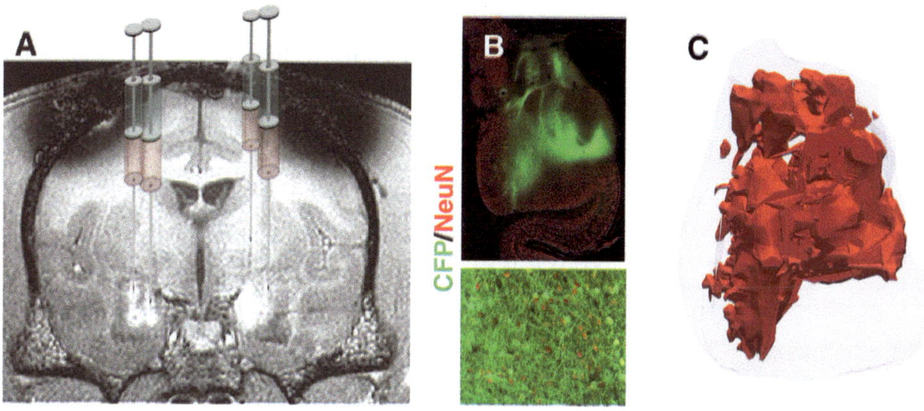

Fig. 4 Representative example of lentivirus injection into the amygdala. (**a**) Lentivirus expressing hM4Di_CFP from a synapsin promoter was injected in multiple tracks with several 20 μL injections per track (a total of 15 × 20 μL injections). Locations of injection regions were visualized via MRI by hypersignal resulting from the addition of 0.1 mM $MnCl_2$ to the virus suspension [12]. (**b**) *Upper panel*. Section through the amygdala, stained for CFP (green) and the neuronal marker NeuN (red). *Lower panel*: Example of cellular staining within a medium penetrance expression region. (**c**) 3D reconstruction of the left amygdala. Red: regions in which at least 50% of neurons expressed CFP. Outlines and reconstruction were performed using the Microbrightfield (MBF) software Neurolucida 360

was possible because the addition caused the $MnCl_2$ to precipitate out of solution, creating needle blockage and more variability in the dispensation of the virus. In current experiments, we avoid this precipitation by adding no more than 1 mM of $MnCl_2$ to the virus mixture and ensuring that the lentivirus is at room temperature before adding the $MnCl_2$. This technique has allowed us to use MRI to confirm successful injections into multiple tracks, for example, required for the injections covering the amygdala (Fig. 4).

4 Materials for Lentivirus Production

- Lenti-X 293T cells (Takara Bio, 632180).
- 75 cm^2 flasks (Thermo Scientific,156499).
- 175 cm^2 flasks (Thermo Scientific, 159910).
- Dulbecco's Modified Eagle's Medium (DMEM) (Lonza, 12-604F).
- Trypsin-EDTA (0.25%) (Corning, 25-053-CI).
- Fetal bovine serum (FBS) (Takara Bio, 631106).
- Sodium pyruvate (Corning, 25-000-CI).
- 0.22 μm filters (Corning, 430513).
- 500 mL sterile receiver bottle (Corning, 430282).
- 0.45 μm filters and 250 mL storage bottle (Corning, 430768).

- Ultra culture serum-free medium (Lonza, 12-725F).
- Sodium butyrate (Sigma-Aldrich, B5887).
- psPAX2 (Addgene, #12260).
- pMD2G (Addgene, #12259) OR pCAGGS-FuG-E (Addgene, #67509).
- pAdVantage vector (Promega, E1711).
- FuGENE 6 transfection reagent (Promega, E2691).
- 5 mL polystyrene round-bottom tubes (Falcon, 352058).
- 15 mL polystyrene conical tubes (Falcon, 352095).
- 50 mL polypropylene conical tubes (Falcon, 352098).
- 1× DPBS without Ca and Mg (Quality Biological, 114-057-101).
- Sucrose (MP Biomedicals, 802536).
- Centrifuge tubes, polypropylene (Beckman Coulter, 358126).
- 24 well plates (Thermo Scientific, 142475).
- qPCR lentivirus titer kit (ABM, LV900).
- Quick-DNA MiniPrep kit (Zymo, D3024).
- iQ SYBR Green Supermix (Bio-Rad, 1708880).
- Lentivirus forward primer.
  - 5′ – AAC CCA CTG CTT AAG CCT CA – 3′.
- Lentivirus reverse primer.
  - 5′ – CGT CGA GAG AGC TCT GGT TT – 3'.
- Control (hVAR1) forward primer.
  - 5′ – GCC CAT GAA CTT GCT CTC TC – 3'.
- Control (hVAR1) reverse primer.
  - 5′ – ACC TGG GAC TTC ACC ACA TC – 3′.

4.1 Solutions

- Cell culture medium (store at 4 °C).
  - Prepare 500 mL of cell culture medium by combining 445 mL of DMEM, 50 mL of FBS, and 5 mL of 100 mM sodium pyruvate; mix well and filter with a 0.22 μm filter.
- Virus production medium (store at 4 °C).
  - Prepare 500 mL of virus production medium by combining 490 mL of Ultraculture serum-free medium, 5 mL of 100 mM sodium pyruvate, and 5 mL of 0.5 M sodium butyrate; mix well and filter with a 0.22 μm filter.
- 20% sucrose solution (store at 4 °C).
  - Add 1× DPBS to 10 g sucrose, to a total volume of 50 mL. Mix thoroughly and filter with a 0.22 μm filter.

5 Methods for Lentivirus Production

5.1 Cell Preparation

Do not exceed 15 passages for cells that will be used in virus production.

Prepare 293T cells (8.4×10^6 cells in 25 mL culture medium/ 175 cm^2 flask × 3 flasks). Grow at 37 °C, 5% CO_2 for 72 h.

Allow cells to grow to nearly 100% confluence. Expand 293T cells to four 175 cm^2 flasks with 25 mL of fresh culture medium per flask. Grow at 37 °C, 5% CO_2 for 24 h.

5.2 Transfection and Virus Collection

Day 1: Transfection

Cells must be close to 100% confluent prior to transfection.

The following protocol is for a single T175 flask. Repeat four times to transfect cells in each T175 flask. See Table 1 for reagents to set up the transfection mix.

Combine the following in a 5 mL polystyrene tube to make the transfection plasmid mixture:

- Lentiviral gene plasmid
- psPAX2 helper plasmid
- pMD2G coat protein plasmid OR pCAGGS-FuG-E coat protein plasmid
- pAdVantage vector

Combine the following in a 15 mL polystyrene tube:

- FuGENE 6 transfection reagent
- DMEM

Table 1
Reagents for transfection

Reagent	Volume/amount
Lentiviral gene plasmid	22 μg
psPAX2 helper plasmid	15 μg
pMD2G coat protein plasmid OR pCAGGS-FuG-E coat protein plasmid	5 μg
pAdVantage vector	2 μg
FuGENE 6 transfection reagent	132 μL
DMEM	Add to 4.5 mL

Incubate the above mixture for 5 min at room temperature.

After 5 min, add the transfection plasmid mixture from the 5 mL tube to the 15 mL tube and mix gently. Incubate the mixture at room temperature for 30 min.

Meanwhile, carefully aspirate the culture medium, without removing cells, from each flask and replace it with 16 mL of fresh culture medium. After the incubation period, add 4.5 mL of transfection mixture to each flask and mix gently, avoiding disturbing the cells.

Incubate at 37 °C for 24 h.

Day 2: Media Replacement

24 h after transfection, carefully aspirate the media and gently replace it with 20 mL of virus production medium. Avoid disturbing the cells. Incubate at 37 °C for 24 h.

Day 3: Virus Collection

Collect virus 46–48 h post-transfection.

Carefully remove the virus production media from each flask and transfer it to a 50 mL conical tube. Note that cells are fragile and not well-adhered at this point. Be careful not to disturb them while collecting the virus production media from each flask. Centrifuge the media at 1000 rpm for 5 min at room temperature. This will pellet any cells that are suspended in the media. Meanwhile, prewet a 0.45 μm filter with 2.5 mL of virus production medium. Make sure the entire membrane is wet. Remove the supernatant from the 50 mL conical tube and filter it through the 0.45 μm filter. Immediately place the filtered supernatant on ice until the next step.

Transfer the supernatant to a 30 mL ultracentrifuge conical tube. Add 2 mL of 20% sucrose to the bottom of the tube by placing the pipette tip at the bottom and very gently pipetting out the solution. Be extremely careful not to introduce any bubbles or disturb the solution. Gently add an additional 7 mL of virus production media to the tube, or enough to ensure that the ultracentrifuge tube is filled nearly to the top, to prevent the tube from collapsing in the ultracentrifuge. Add the media by placing the pipette tip against the tube wall and releasing the media very slowly to avoid disturbing the sucrose solution at the bottom. Prepare and balance an appropriate number of blanks to ensure that each slot in the ultracentrifuge is filled. Ultracentrifuge at 22,000 rpm for 2 h at 4 °C.

After centrifugation, carefully aspirate the supernatant, ensuring that the pellet is not disturbed. Place the tube upside down on a clean kimwipe or paper towel to allow it to dry for several minutes, then remove any remaining droplets of liquid from the walls of the tube using a Pasteur pipette.

Place the tube on ice and add 50 μL of cold 1× PBS. Do not resuspend. Leave the solution on ice for 1 h.

Resuspend the pellet in the cold PBS by gently pipetting the solution up and down. Note: avoid introducing any bubbles in this step. This can be made easier by cutting the pipette tip to create a larger opening. Next, combine resuspensions that contain the same virus and mix thoroughly. Aliquot 10 μL of virus per PCR tube.

Store at −20 °C for 1 h before transferring to −80 °C for long-term storage.

6 Method for Lentivirus Titering

6.1 RT-PCR (for VSV-G or FuG-E Lentivirus)

Prepare 1:10 and 1:100 virus dilutions on ice. Dilute using cold 1× PBS, making sure to mix very thoroughly at each step.

Prepare 1:10 and 1:100 virus dilutions for a control virus of a known titer.

Proceed with qPCR according to the kit specifications (ABM, LV900). Add samples to a 96-well PCR plate on ice. Place an adhesive cover on the plate and centrifuge at 300 × g for 30 s to remove any bubbles.

Compare Ct values between the new virus and the known virus to calculate the titer. If the Ct values of the undiluted virus sample correspond to those of a standard sample of the previously titered virus, then the titers are considered equivalent. Any divergence multiplies or divides the titer by a factor of 2 per cycle as the PCR product doubles with each cycle.

6.2 Transduction (for VSV-G Lentivirus Only)

Prepare 1:10 and 1:100 virus dilutions on ice. Dilute using cold 1× PBS, making sure to mix very thoroughly at each step.

In a 24-well plate, add 150,000 293T cells in 1 mL of cell culture medium per well. Prepare enough wells to have the following:

- 2 wells with the undiluted virus
- 2 wells with the 1:10 virus dilution
- 2 wells with the 1:100 virus dilution
- 2 wells without virus (controls)

Add 3 μL of virus per well. Do so by slowly pipetting the virus into the center of the well, tapping the pipette tip gently against the bottom of the well, and then pipetting up and down to expel all of the viruses from the tip. After adding all virus samples, gently shake the plate back and forth to distribute the virus throughout each well. Incubate at 37 °C for 48 h.

After 48 h, aspirate the supernatant from each well and trypsinize the cells with 200 μL of trypsin. Add 2 mL of cell culture medium to each well and mix, then transfer each mixture to a 15 mL tube. Centrifuge at approximately 500 × g for 5 min.

Table 2
qPCR reaction mixture

Reagent	Volume	Final concentration
Forward primer (20 μM)	0.5 μL	500 nM
Reverse primer (20 μM)	0.5 μL	500 nM
iQ SYBR Green supermix	10 μL	1×
Nuclease free water	7 μL	–
gDNA (50 ng/μL) or NTC	2 μL	100 ng of gDNA
Total volume per reaction	20 μL	

Table 3
qPCR cycling conditions

Step	Temperature	Time	Cycles
Initial denaturation	95 °C	3 min	1
Denaturation Annealing, extension, and read fluorescence	95 °C 62 °C	30 s 30 s	40
Hold (optional)	4 °C		

Aspirate the supernatant and proceed with the gDNA extraction according to the kit instructions (Zymo, D3024).

After gDNA extraction, measure the concentration of each sample. Dilute each sample to a concentration of 50 ng/μL.

Next, perform qPCR to determine the viral titer. See Table 2 for reagents to set up the qPCR reactions.

Each gDNA sample requires two different corresponding PCR reactions: one reaction with lentivirus primers and one reaction with control primers.

Add samples to a 96-well PCR plate on ice. Additionally, include one well that has nuclease-free water instead of gDNA, which will serve as the no template control (NTC). Place an adhesive cover on the plate and centrifuge at 300 × g for 30 s to remove any bubbles. Run the qPCR according to the conditions listed in Table 3.

Compare Ct values between the lentivirus primer samples and the control primer samples. If the Ct values of the undiluted virus sample are identical to corresponding control samples, then the titer is considered 10^9 iu/mL. Any divergence multiplies or divides the titer by a factor of 2 per cycle as the PCR product doubles with each cycle. For example, if the Ct value for the virus sample is one cycle lower than the Ct value for the control primer sample, then the titer for the virus sample is 2×10^9 iu/mL, our preferred value for most applications.

References

1. Sternson SM, Roth BL (2014) Chemogenetic tools to interrogate brain functions. Annu Rev Neurosci 37:387–407. https://doi.org/10.1146/annurev-neuro-071013-014048
2. Reiner A, Isacoff EY (2013) The Brain Prize 2013: the optogenetics revolution. Trends Neurosci 36:557–560. https://doi.org/10.1016/j.tins.2013.08.005
3. Lerchner W et al (2021) RNAi and chemogenetic reporter co-regulation in primate striatal interneurons. Gene Ther 29:69. https://doi.org/10.1038/s41434-021-00260-y
4. Kravitz DJ, Saleem KS, Baker CI, Mishkin M (2011) A new neural framework for visuospatial processing. Nat Rev Neurosci 12:217–230. https://doi.org/10.1038/nrn3008
5. Murray EA, Richmond BJ (2001) Role of perirhinal cortex in object perception, memory, and associations. Curr Opin Neurobiol 11: 188–193. https://doi.org/10.1016/s0959-4388(00)00195-1
6. Richmond BJ, Optican LM (1990) Temporal encoding of two-dimensional patterns by single units in primate primary visual cortex. II. Information transmission. J Neurophysiol 64:370–380. https://doi.org/10.1152/jn.1990.64.2.370
7. Lerchner W, Corgiat B, Der Minassian V, Saunders RC, Richmond BJ (2014) Injection parameters and virus dependent choice of promoters to improve neuron targeting in the nonhuman primate brain. Gene Ther 21:233–241. https://doi.org/10.1038/gt.2013.75
8. Shi L et al (2019) Transgenic rhesus monkeys carrying the human MCPH1 gene copies show human-like neoteny of brain development. Natl Sci Rev 6:480–493. https://doi.org/10.1093/nsr/nwz043
9. Jazayeri M, Lindbloom-Brown Z, Horwitz GD (2012) Saccadic eye movements evoked by optogenetic activation of primate V1. Nat Neurosci 15:1368–1370. https://doi.org/10.1038/nn.3210
10. Verdera HC, Kuranda K, Mingozzi F (2020) AAV vector immunogenicity in humans: a long journey to successful gene transfer. Mol Ther 28:723–746. https://doi.org/10.1016/j.ymthe.2019.12.010
11. Kotin RM (2011) Large-scale recombinant adeno-associated virus production. Hum Mol Genet 20:R2–R6. https://doi.org/10.1093/hmg/ddr141
12. Fredericks JM et al (2020) Methods for mechanical delivery of viral vectors into rhesus monkey brain. J Neurosci Methods 339: 108730. https://doi.org/10.1016/j.jneumeth.2020.108730
13. Nagai Y et al (2016) PET imaging-guided chemogenetic silencing reveals a critical role of primate rostromedial caudate in reward evaluation. Nat Commun 7:13605. https://doi.org/10.1038/ncomms13605
14. Guselnikova VV, Korzhevskiy DE (2015) NeuN as a neuronal nuclear antigen and neuron differentiation marker. Acta Nat 7:42–47
15. Boyden ES, Zhang F, Bamberg E, Nagel G, Deisseroth K (2005) Millisecond-timescale, genetically targeted optical control of neural activity. Nat Neurosci 8:1263–1268. https://doi.org/10.1038/nn1525
16. Han X et al (2009) Millisecond-timescale optical control of neural dynamics in the nonhuman primate brain. Neuron 62:191–198. https://doi.org/10.1016/j.neuron.2009.03.011
17. Simmons JM et al (2008) Mapping prefrontal circuits in vivo with manganese-enhanced magnetic resonance imaging in monkeys. J Neurosci 28:7637–7647. https://doi.org/10.1523/JNEUROSCI.1488-08.2008

Chapter 2

HiRet/NeuRet Vectors: Lentiviral System for Highly Efficient Gene Transfer Through Retrograde Axonal Transport

Kenta Kobayashi, Shigeki Kato, and Kazuto Kobayashi

Abstract

Lentiviral vectors are used for a wide range of research applications in the field of neuroscience. Notably, lentiviral tropism can be manipulated by genetically engineering envelope glycoproteins that are essential for vector transduction. The structural and functional analyses of specific neural pathways are crucial to understanding the neural mechanisms underlying brain functions controlled through complex neural circuits. Viral vectors that produce gene transfer via retrograde axonal transport offer a powerful tool for the analysis of neural pathways. We have succeeded in developing novel types of lentiviral vectors for retrograde gene transfer, named "highly efficient retrograde gene transfer" (HiRet) and "neuron-specific retrograde gene transfer" (NeuRet) vectors, by pseudotyping human immunodeficiency virus type 1 with fusion envelope glycoproteins composed of rabies virus glycoprotein segments and vesicular stomatitis virus glycoprotein segments. HiRet/NeuRet vectors show highly efficient retrograde gene transfer in diverse neural pathways in animal models. These vectors have been harnessed for the analysis of specific neural pathways in combination with various genetic approaches for neuromodulation, including optogenetics and chemogenetics. In this chapter, we describe experimental procedures for producing HiRet/NeuRet vectors and injecting them into brain regions, as well as summarize points important to conduct the experiments smoothly and effectively.

Key words Lentiviral vector, HiRet, NeuRet, Envelope fusion glycoprotein, Retrograde gene transfer, Specific neural pathway

1 Introduction

A lentiviral vector, one of the major viral vectors used for neuroscience research, possesses notable characteristics [1–8]: high infectivity in both dividing and non-dividing cells, long and stable gene expression attributable to integration into host genomes, loading capacity for large-sized transgenes, and low inflammatory potential. Remarkably, lentiviral tropism can be manipulated by genetically engineering envelope glycoproteins, which cover the surface of the virion and are crucial for the virus entry into cells [9–12]. Among

Mark A. G. Eldridge and Adriana Galvan (eds.), *Vectorology for Optogenetics and Chemogenetics*,
Neuromethods, vol. 195, https://doi.org/10.1007/978-1-0716-2918-5_2,

several kinds of lentiviral vectors, a human immunodeficiency virus type 1 (HIV-1)-based lentiviral vector, in particular, has been investigated in detail [13–15].

Brain functions are regulated through complex neural networks. In order to understand the neural mechanisms underlying brain functions, we need to analyze the roles of the specific neural pathways that form the networks. The viral vector system delivering transgenes in the retrograde direction through axons to cell bodies represents a potent tool for the labeling and functional analysis of specific neural pathways. At present, various kinds of efficient retrograde viral vectors have been reported, that include lentiviral vector [10–12], adeno-associated virus (AAV) vector [8, 16], rabies virus vector [8, 17], canine adenovirus type-2 vector [8, 18], pseudorabies virus vector [8, 19], and herpes simplex virus type 1 vector [8, 20]. Here, we focus on our newly developed HIV-1-based retrograde lentiviral vectors, which we have termed highly efficient retrograde gene transfer (HiRet) and neuron-specific retrograde gene transfer (NeuRet) vectors. The HiRet/NeuRet vectors are incorporated at the nerve terminals and then retrogradely transported to cell bodies; next, viral genomes are integrated into host genomes via the reverse transcription reaction, resulting in stable and long-lasting transgene expression (Fig. 1). These vectors are pseudotyped with fusion envelope glycoproteins (FuGs) consisting of rabies virus glycoprotein (RV-G) and vesicular stomatitis virus glycoprotein (VSV-G) segments (*see* Fig. 2 for the structure of FuGs) [10–12, 21–24], displaying a much higher efficiency of retrograde gene transfer than the parent RV-G-pseudotyped vector [10–12, 21–24]. The efficiency of the retrograde gene transfer of HiRet/NeuRet vectors was summarized in our previous review [12]. Particularly, these vectors exhibit markedly high efficiency of retrograde transgene expression in the brains of not only rodents but also non-human primates [24–26]. Each of the HiRet and NeuRet vectors possesses a distinct transduction property; the HiRet vector efficiently transduces dividing cells such as glial and neural stem/progenitor cells, whereas the NeuRet vector scarcely enters these dividing cells [10–12, 21–24]. Therefore, the NeuRet vector is expected to result in less inflammation around the injection site in comparison to the HiRet vector.

HiRet/NeuRet vectors have been applied to the functional analysis of specific neural pathways using optogenetics [27–31] or chemogenetics [29, 32, 33]. The demand for these vectors in the field of neuroscience will continue to expand. Therefore, it is meaningful to present a detailed handling protocol of these vectors to researchers who are trying to use the retrograde lentiviral system for the analysis of neural functions.

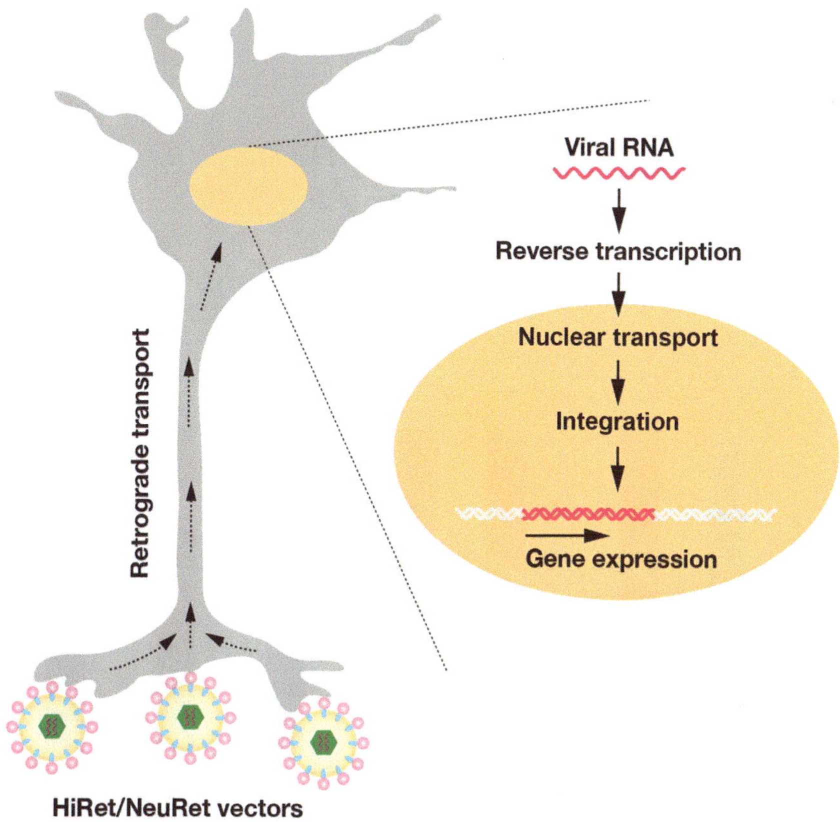

Fig. 1 Retrograde gene delivery into neurons by HiRet/NeuRet vectors. These vectors are internalized into nerve terminals and then retrogradely transported along axons into the cell body. A viral RNA genome is reverse-transcribed into a double-stranded DNA, which is then integrated into the host cell genome, resulting in stable and long-lasting transgene expression

2 Materials

2.1 Production of HiRet/NeuRet Vectors

The materials necessary for vector production are listed below.

- HEK293T cells (ATCC, CRL-3216).
- Primaria™ cell culture dish (Corning, 353803).
- Dulbecco's Modified Eagle's Medium (DMEM) (Sigma-Aldrich, D5796-500ML).
- Fetal bovine serum (FBS) (Nichirei, 174012).

  (Before use, inactivate FBS at 56 °C for 30 min in the water bath. Store at −30 °C).
- Trypsin-EDTA (0.05%) (Gibco, 25300062).
- Penicillin-streptomycin (10,000 U/mL) (Gibco, 5140122).
- Distilled water (Gibco, 15230162).
- Cell culture medium (Store at 4 °C).

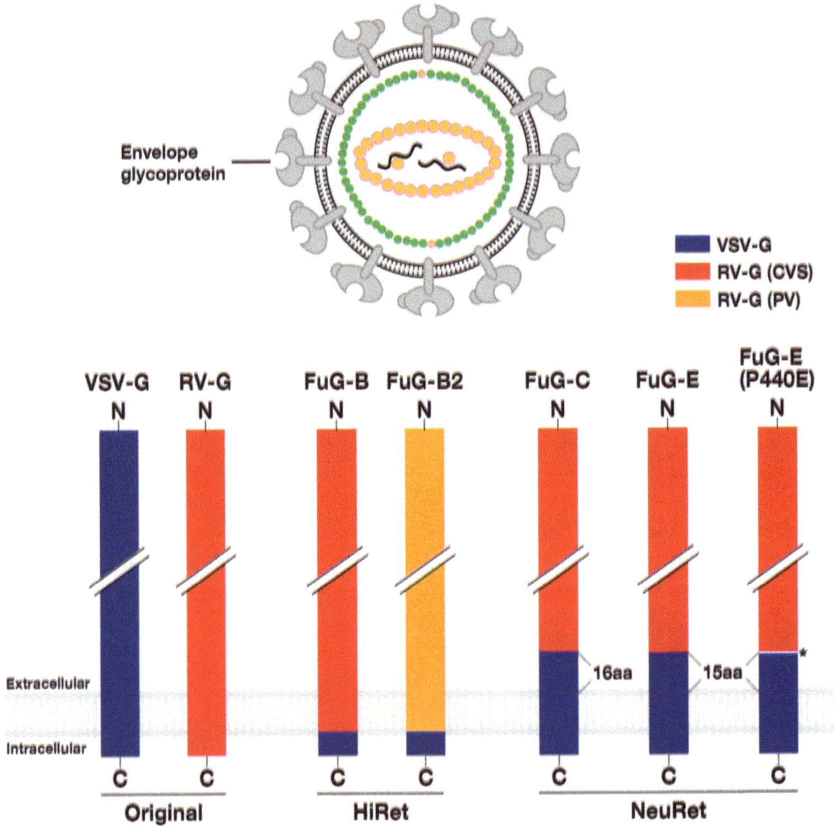

Fig. 2 Schematic representation of the structure of FuGs used for HiRet/NeuRet vectors. The HiRet vector is pseudotyped with FuG-B or FuG-B2. FuG-B is the fusion protein of the extracellular and transmembrane domains of RV-G derived from the challenge virus standard (CVS) and the cytoplasmic domain of VSV-G. FuG-B2 is composed of the extracellular and transmembrane domains of Pasteur virus (PV) strain-derived RV-G and the VSV-G cytoplasmic domain. The NeuRet vector is pseudotyped with FuG-C, FuG-E, or FuG-E (P440E). FuG-C comprises the N-terminal domain of CVS-derived RV-G (439 amino acids) and the C-terminal domain of VSV-G containing the extracellular (16 amino acids), transmembrane and cytoplasmic segments. In FuG-E and FuG-E (P440E), the junction between RV-G and VSV-G is shifted from one amino acid to the C-terminal direction. The 440th proline residue of the original FuG-E is changed to glutamate residue in FuG-E (P440E)

Prepare by mixing 500 mL of DMEM, 56 mL of FBS, and 5 mL of penicillin-streptomycin.

- 10× HBSP stock solution (Store at −30 °C).

NaCl	40.9 g
HEPES	29.8 g
Na_2HPO_4-$12H_2O$	1.35 g
KCl	1.9 g
Glucose	5.4 g

Dilute with distilled water up to 500 mL, and filtrate with 0.45 μm filter (Millipore, SLHV033RB).

- 2× HBSP stock solution (Store at 4 °C).

  Dilute 10× HBSP stock solution with distilled water. Adjust pH to the range from 7.05 to 7.10 with 1 N NaOH.
- 2.5 M $CaCl_2$ solution (Store at −30 °C)

  Dilute 14.7 g of $CaCl_2$-$2H_2O$ with distilled water up to 40 mL, and filtrate with a 0.45 μm filter.
- Sepharose Q FF ion-exchange chromatography column (GE Healthcare, 28-9365-43).
- Polypropylene round-bottom tube (Falcon, 352053).
- Vivaspin 10 K MWCO filter (Sartorius, VS2002).
- NucleoSpin RNA virus kit (Clontech, 631235).
- Lenti-X qRT-PCR titration kit (Clontech, 631235).

2.2 Intracranial Injection of HiRet/NeuRet Vectors

The materials necessary for the vector injection are listed below.

- Set of surgical instruments: scissors, tweezers, forceps, clamp, hemostatic supplies, anesthesia, and suture supplies.
- Stereotaxic instrument (Narishige, SR-5M-HT).
- Surgical stereomicroscope (Leica Microsystems, M320TC12).
- Infusion pump (Eicom, ESP-32).
- Dental drill (Minitor Co., Ltd., C10M).
- Microsyringe (Hamilton, 1801N).
- Glass capillary (Sutter, B100-75-10).
- Connecting tube (Eicom, JT-10-50).
- Mineral oil or Fluorinert (3M, C-3283).
- HiRet/NeuRet vectors (~1.0×10^{12} genome copies/mL).

3 Methods

3.1 Production of HiRet/NeuRet Vectors

We herein describe the production protocol for viral vectors carrying the gene encoding the enhanced green fluorescent protein (EGFP).

Day 1	Prepare HEK293T cells (~3.0×10^6 cells/10 cm cell culture dish × 18 dishes) 24 h before transfection.	
Day 2	Mix four kinds of plasmids as follows:	
	pCAGkGP4.1R	108 μg
	pCAG RTR2	36 μg
	pCAGGS-FuGs*	36 μg
	pCL20C-MSCV-EGFP	180 μg

(continued)

	Add 900 μL of 2.5 M $CaCl_2$ to this plasmid mixture and dilute with distilled water up to 9 mL. Add 9 mL of 2× HBSP while vortexing and incubate at room temperature for about 5–10 min (DNA solution). Drop the DNA solution into a culture medium (1 mL of DNA solution/dish) and incubate.
	*The pCAGGS-FuG-B or -B2 plasmid is used to produce HiRet vector [21, 29, 34]. The pCAGGS-FuG-C, -FuG-E, or -FuG-E (P440E) plasmid is used to produce NeuRet vector [22–24, 35].
Day 3	After 18 h, remove the medium, wash with phosphate-buffered saline (PBS), add 6 mL of fresh medium, and incubate.
Day 4	After 24 h, collect the medium, centrifuge at 1500 × *g* for 5 min, and filtrate the supernatant with a 0.45 μm filter. Centrifuge the filtrated medium at 6000 × *g* for 16–20 h.
Day 5	Remove the supernatant and suspend the pellet, including viral particles, with 5 mL of PBS. Suspend gently to avoid foaming. Purify viral vectors by using the following ion-exchange chromatography method.

1. Apply 5 mL of the crude viral solution to a Sepharose Q FF ion-exchange column attached to the AKTA prime plus chromatography apparatus system, and elute with a linear 0.0–1.5 M NaCl gradient. Collect 2 mL of eluate per polypropylene round-bottom tube serially.
2. Monitor the UV curve shown by the absorbance at 280 nm, and collect the peak fractions containing viral particles (generally, from fraction numbers 24–34).
3. Concentrate the purified viral vector solution by ultrafiltration using a Vivaspin 10 K MWCO filter* to ~100 μL.

   *Add 3 mL of 1% bovine serum albumin in PBS to the Vivaspin filter and incubate for 1 h at room temperature before use. Discard the blocking solution and wash it with PBS when using the filter
4. Extract viral RNA from 5 μL of vector solution in accordance with the manual attached to the NucleoSpin RNA virus kit.
5. Measure the copy number of the RNA genome in accordance with the manual attached to the Lenti-X qRT-PCR titration kit. The condition of quantitative real-time reverse transcription-PCR is as follows: 42 °C for 20 min; 95 °C for 3 min; and 40 cycles each of 95 °C for 15 s and 60 °C for 30 s.
6. Dispense the vector solution into aliquots (~10 μL), and store at −80 °C.

3.2 Intracranial Injection of HiRet/NeuRet Vectors

We here describe the protocol for HiRet/NeuRet vector injection into the mouse brain. Male C57BL/6 J mice (8–12 weeks of age) are used for the vector injection. Animal care and handling

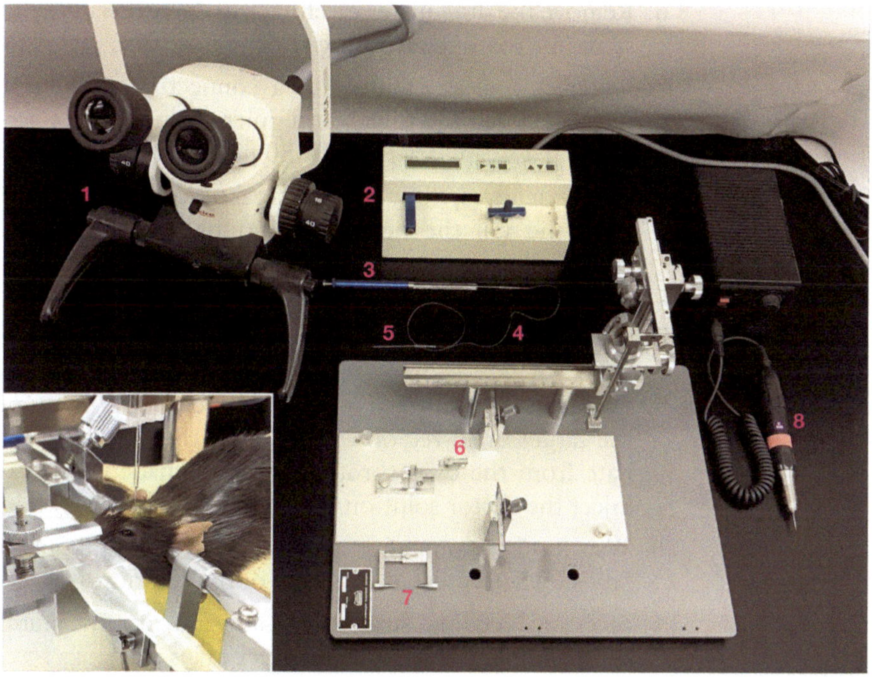

Fig. 3 A set of tools for injecting viral vectors into the brain. (1) Surgical stereomicroscope; (2) infusion pump; (3) microsyringe; (4) connecting tube; (5) glass capillary; (6) nose fixing clamp; (7) auxiliary ear bar; and (8) dental drill. Inset shows the intracranial injection of vector solution into the mouse brain under isoflurane anesthesia

procedures are conducted in accordance with the guidelines established by the Experimental Animal Center of Fukushima Medical University. We make all efforts to minimize both the number of animals used and their suffering. The injection apparatus and the state of intracranial injection are shown in Fig. 3.

1. Anesthetize a mouse with 1.5% isoflurane, and place it in a stereotaxic frame.
2. Fix the anesthetized mouse horizontally and firmly to a stereotaxic instrument with the auxiliary ear bar and nose fixing clamp.
3. Expose the skull by making an incision in the scalp, remove the periosteum with the swab, and dry well.
4. Mark the position on the skull surface, which is determined by the anteroposterior and mediolateral coordinates of a target brain region from the bregma, with a sterile surgical marker. The area of the marked position should be determined according to the size of the target region. Refer to the mouse brain atlas [36] for the coordinates.
5. Make a hole in the marked position using a dental drill, being careful to avoid hemorrhaging.

6. Remove the dura mater carefully with tweezers under a surgical stereomicroscope, exposing the brain surface.
7. Connect a glass capillary to a connecting tube, and subsequently attach the tube to the microsyringe. The tip diameter of the glass capillary is about 50 μm.
8. Fill the glass capillary with mineral oil or Fluorinert, withdraw from the vial the necessary amount of the HiRet/NeuRet vector solution into the capillary, and then connect the microsyringe to the infusion pump. The volume of the vector solution should be optimized according to the size of the target region.
9. Lower the glass capillary by using a stereotaxic instrument into the target brain region estimated by the dorsoventral coordinate from the dura through the hole made in the skull, and inject the vector solution at a rate of 0.05–0.1 μL/min.
10. Wait for 5–10 min after injection, and then pull out the capillary slowly to prevent backflow of the solution.
11. Stop any bleeding, suture the scalp, and keep the animal warm until complete anesthetic recovery takes place.
12. Three to 4 weeks after the vector injection, the transgene expression reaches the maximum level.

4 Notes

In this section, we discuss some points that, based on our experience, are important to ensure that experiments using HiRet/NeuRet vectors are conducted smoothly and efficiently.

When the lentiviral payload exceeds 8 kb, the packaging efficiency of viral vectors appears to be decreased by an extensive amount. Therefore, we recommend that transgenes should be designed to be no larger than 8 kb. The yield of lentiviral vectors is also dependent on transfection efficiency in HEK293T cells. In cases of low yield, we regularly change to a new batch of cells and prepare a fresh 2× HBSP solution. It is important to accurately adjust the pH of the 2× HBSP solution to within the range of 7.05–7.10. In addition, lentiviral vectors can be directly concentrated by ultracentrifugation of the filtrated culture medium [37]. Injection of viral vectors prepared by this simple technique frequently causes damage to the brain tissue because this vector solution is contaminated with cell debris. Thus, HiRet/NeuRet vectors should be purified using ion-exchange chromatography before the concentration process is started. Generally, lentiviral vectors are known to be stable for at least 6 months when stored at −80 °C (*see*: https://en.vectorbuilder.com/products-services/service/lentivirus-packaging.html). On the other hand, our preliminary storage experiments show that HiRet/NeuRet vectors are stable for 1 month at −80 °C, but the functional titer is decreased to approximately 70% in 2 months. Therefore, we recommend that these vectors are used at the latest 1 month after the storage. In addition, repeated freezing and thawing should be avoided.

In order to inject vectors into the correct site, it is necessary to carefully and accurately adjust the injection coordinates. Examining previous literature is important to verify the pathway connections prior to the viral injection. Subsequently, it is recommended to confirm the neural connection between two brain regions by using neural tracers such as biotinylated dextran amine (for anterograde labeling) and cholera toxin B subunit (for retrograde labeling). When HiRet/NeuRet vectors encoding the EGFP gene are injected into a brain region, the transgene is expressed in neurons projecting to the injection site through retrograde gene transfer. Transgene products are transported through axons in the anterograde direction into the nerve terminal regions. Therefore, we can verify the injection site by detecting EGFP fluorescence or immuno-positive signals localized in the nerve terminals. Retrograde gene expression is usually observed in projecting brain regions 3–4 weeks after local injection of HiRet/NeuRet vectors into the brain. Because lentiviral genomes are integrated into host genomes, the transgene expression remains stable as long as host cells exist. In cases where gene expression level is low, the use of specific gene promoters may improve the expression degree. In addition, to enhance the cell specificity of retrograde gene transfer, the use of the HiRet/NeuRet vector system in combination with the Cre-dependent conditional gene expression method is effective [29, 35].

Because the retrograde gene transfer efficiency of HiRet/NeuRet vectors depends on the neural pathway, it is necessary to assess whether these vectors work well in the neural pathway of interest before starting experiments. In rodents, HiRet/NeuRet vectors show the high efficiency of retrograde gene transfer in various neural pathways, such as the corticostriatal and thalamostriatal pathways, but display only modest efficiency, for example, in the nigrostriatal pathway [21, 22]. On the other hand, these vectors convey transgenes in a retrograde direction with high efficiency in the nigrostriatal pathway of non-human primates [21, 22, 24–26]. These findings indicate that the efficiency of retrograde gene transfer of HiRet/NeuRet vectors is also dependent on animal species.

Finally, HiRet/NeuRet vectors have been applied not only to optogenetics [27–31] and chemogenetics [29, 32, 33] but also to other experimental techniques; immunotoxin-mediated cell targeting [29, 34, 38], conditional suppression of signaling cascade [35], and reversible synaptic transmission blockade [39–44]. HiRet/

NeuRet vectors will provide a more powerful tool for the detailed analysis of neural functions in the future. We expect the HiRet/ NeuRet vector system to be more widely disseminated via this chapter and will significantly contribute to advancements in the field of neuroscience.

Acknowledgments

This work was supported by a grant-in-aid from Japan Agency for Medical Research and Development (JP17dm0207052 to Ka.K.). We thank Dr. R. Fukabori for providing photographs and Dr. Y. Iguchi for his critical reading of this manuscript.

References

1. Naldini L, Blömer U, Gage FH et al (1996) Efficient transfer, integration, and sustained long-term expression of the transgene in adult rat brains injected with a lentiviral vector. Proc Natl Acad Sci U S A 93:11382–11388
2. Mitrophanous K, Yoon S, Rohll J et al (1999) Stable gene transfer to the nervous system using a non-primate lentiviral vector. Gene Ther 6:1808–1818
3. Butler SL, Johnson EP, Bushman FD (2002) Human immunodeficiency virus cDNA metabolism: notable stability of two-long terminal repeat circles. J Virol 76:3739–3747
4. Thomas CE, Ehrhardt A, Kay MA (2003) Progress and problems with the use of viral vectors for gene therapy. Nat Rev Genet 4: 346–358
5. Kafri T (2004) Gene delivery by lentivirus vectors an overview. Methods Mol Biol 246:367–390
6. Wong LF, Goodhead L, Prat C et al (2006) Lentivirus-mediated gene transfer to the central nervous system: therapeutic and research applications. Hum Gene Ther 17:1–9
7. Cockrell AS, Kafri T (2007) Gene delivery by lentivirus vectors. Mol Biotechnol 36:184–204
8. Nectow AR, Nestler EJ (2020) Viral tools for neuroscience. Nat Rev Neurosci 21:669–681
9. Cronin J, Zhang XY, Reiser J (2005) Altering the tropism of lentiviral vectors through pseudotyping. Curr Gene Ther 5:387–398
10. Kobayashi K, Inoue KI, Tanabe S et al (2017) Pseudotyped lentiviral vectors for retrograde gene delivery into target brain regions. Front Neuroanat 11:65
11. Kobayashi K, Kato S, Kobayashi K (2018) Genetic manipulation of specific neural circuits by use of a viral vector system. J Neural Transm (Vienna) 125:67–75
12. Kato S, Kobayashi K (2020) Pseudotyped lentiviral vectors for tract-targeting and application for the functional control of selective neural circuits. J Neurosci Methods 344: 108854
13. Joshi S, Joshi RL (1996) Molecular biology of human immunodeficiency virus type-1. Transfus Sci 17:351–378
14. Nielsen MH, Pedersen FS, Kjems J (2005) Molecular strategies to inhibit HIV-1 replication. Retrovirology 2:10
15. Pluta K, Kacprzak MM (2009) Use of HIV as a gene transfer vector. Acta Biochim Pol 56:531–595
16. Haery L, Deverman BE, Matho KS et al (2019) Adeno-associated virus technologies and methods for targeted neuronal manipulation. Front Neuroanat 13:93
17. Callaway EM, Luo L (2015) Monosynaptic circuit tracing with glycoprotein-deleted rabies viruses. J Neurosci 35:8979–8985
18. Junyent F, Kremer EJ (2015) CAV-2--why a canine virus is a neurobiologist's best friend. Curr Opin Pharmacol 24:86–93
19. Pomeranz LE, Reynolds AE, Hengartner CJ (2005) Molecular biology of pseudorabies virus: impact on neurovirology and veterinary medicine. Microbiol Mol Biol Rev 69:462–500
20. Fenno LE, Mattis J, Ramakrishnan C et al (2014) Targeting cells with single vectors using multiple-feature Boolean logic. Nat Methods 11:763–772
21. Kato S, Kobayashi K, Inoue K et al (2011) A lentiviral strategy for highly efficient retrograde gene transfer by pseudotyping with fusion

envelope glycoprotein. Hum Gene Ther 22: 197–206

22. Kato S, Kuramochi M, Takasumi K et al (2011) Neuron-specific gene transfer through retrograde transport of lentiviral vector pseudotyped with a novel type of fusion envelope glycoprotein. Hum Gene Ther 22:1511–1523
23. Kato S, Kobayashi K, Kobayashi K (2014) Improved transduction efficiency of a lentiviral vector for neuron-specific retrograde gene transfer by optimizing the junction of fusion envelope glycoprotein. J Neurosci Methods 227:151–158
24. Kato S, Sugawara M, Kobayashi K et al (2019) Enhancement of the transduction efficiency of a lentiviral vector for neuron-specific retrograde gene delivery through the point mutation of fusion glycoprotein type E. J Neurosci Methods 311:147–155
25. Tanabe S, Inoue KI, Tsuge H et al (2017) The use of an optimized chimeric envelope glycoprotein enhances the efficiency of retrograde gene transfer of a pseudotyped lentiviral vector in the primate brain. Neurosci Res 120:45–52
26. Tanabe S, Uezono S, Tsuge H et al (2019) A note on retrograde gene transfer efficiency and inflammatory response of lentiviral vectors pseudotyped with FuG-E vs. FuG-B2 glycoproteins. Sci Rep 9:3567
27. Matsuda T, Hiyama TY, Niimura F et al (2017) Distinct neural mechanisms for the control of thirst and salt appetite in the subfornical organ. Nat Neurosci 20:230–241
28. Morishima M, Kobayashi K, Kato S et al (2017) Segregated excitatory-inhibitory recurrent subnetworks in layer 5 of the rat frontal cortex. Cereb Cortex 27:5846–5857
29. Kato S, Fukabori R, Nishizawa K et al (2018) Action selection and flexible switching controlled by the intralaminar thalamic neurons. Cell Rep 22:2370–2382
30. Nomura K, Hiyama TY, Sakuta H et al (2019) [Na+] increases in body fluids sensed by central Nax induce sympathetically mediated blood pressure elevations via H+-dependent activation of ASIC1a. Neuron 101:60–75
31. Isa K, Sooksawate T, Kobayashi K et al (2020) Dissecting the tectal output channels for orienting and defense responses. eNeuro 7: ENEURO.0271-20.2020
32. Ishida A, Kobayashi K, Ueda Y et al (2019) Dynamic interaction between cortico-brainstem pathways during training-induced recovery in stroke model rats. J Neurosci 39: 7306–7320
33. Hayashi T, Akikawa R, Kawasaki K et al (2020) Macaques exhibit implicit gaze bias anticipating others' false-belief-driven actions via medial prefrontal cortex. Cell Rep 30:4433–4444
34. Kato S, Kuramochi M, Kobayashi K et al (2011) Selective neural pathway targeting reveals key roles of thalamostriatal projection in the control of visual discrimination. J Neurosci 31:17169–17179
35. Kobayashi K, Sano H, Kato S et al (2016) Survival of corticostriatal neurons by Rho/Rho-kinase signaling pathway. Neurosci Lett 630:45–52
36. Paxinos G, Franklin KBJ (2008) The mouse brain in stereotaxic coordinates, 3rd edn. Academic Press, San Diego
37. Kato S, Inoue K, Kobayashi K et al (2007) Efficient gene transfer via retrograde transport in rodent and primate brains using a human immunodeficiency virus type 1-based vector pseudotyped with rabies virus glycoprotein. Hum Gene Ther 18:1141–1151
38. Inoue K, Koketsu D, Kato S et al (2012) Immunotoxin-mediated tract targeting in the primate brain: selective elimination of the cortico-subthalamic "hyperdirect" pathway. PLoS One 7:e39149
39. Kinoshita M, Matsui R, Kato S et al (2012) Genetic dissection of the circuit for hand dexterity in primates. Nature 487:235–238
40. Sooksawate T, Isa K, Matsui R et al (2013) Viral vector-mediated selective and reversible blockade of the pathway for visual orienting in mice. Front Neural Circuits 7:162
41. Ishida A, Isa K, Umeda T et al (2016) Causal link between the cortico-rubral pathway and functional recovery through forced impaired limb use in rats with stroke. J Neurosci 36: 455–467
42. Tohyama T, Kinoshita M, Kobayashi K et al (2017) Contribution of propriospinal neurons to recovery of hand dexterity after corticospinal tract lesions in monkeys. Proc Natl Acad Sci U S A 114:604–609
43. Kinoshita M, Kato R, Isa K et al (2019) Dissecting the circuit for blindsight to reveal the critical role of pulvinar and superior colliculus. Nat Commun 10:135
44. Vancraeyenest P, Arsenault JT, Li X et al (2020) Selective mesoaccumbal pathway inactivation affects motivation but not reinforcement-based learning in macaques. Neuron 108: 568–581

Chapter 3

Generation of High-Titer Defective HSV-1 Amplicon Vectors

Rachael Neve

Abstract

Strategies for gene delivery into neurons, either to study the molecular biology of brain function or for gene therapy, must utilize vectors that persist stably in postmitotic cells and that can be targeted both spatially and temporally in the nervous system in vivo. Herpes simplex virus type 1 (HSV-1) possesses multiple features that make it an ideal vector for genetic intervention in the nervous system. In particular, it accepts large molecules of exogenous DNA; it infects nondividing cells from a wide range of hosts with high efficiency; it enables the strong expression of foreign genes; it is episomal and thereby does not cause integration effects; its infection of postmitotic cells is stable; and its particles can be concentrated to relatively high titers. Here, we describe the considerations involved in the preparation and use of a herpes simplex virus type 1 (HSV-1) amplicon as a vector for gene transfer into neurons both in vitro and in vivo.

Key words HSV-1, Amplicon vector, In vivo, Virus vector, Packaging, Helper virus, Replication incompetent, Brain, Transduction, dl1.2, 2–2 cells, Sucrose gradient, High titer

1 Introduction

There are two types of replication-deficient herpes simplex virus type 1 (HSV-1) vectors: vectors in which the foreign DNA of interest is cloned into the viral genome itself and those that are comprised of a plasmid (amplicon) carrying minimal HSV-1 sequences that allow it to be packaged into virus particles with the aid of a helper virus [1]. A number of genes within the wild-type HSV genome are dispensable for its growth in cells in vitro. Roizmann and his colleagues capitalized on this finding to create recombinant HSV-1 viruses that could be used as vectors for gene transfer into cells [2]. This type of genetically engineered genomic vector, with modifications, has been used by multiple investigators. Its chief disadvantage is that it has not been possible to generate, using this methodology, the high titers needed for the study of behavioral changes following its in vivo injection into the brains of animals.

Mark A. G. Eldridge and Adriana Galvan (eds.), *Vectorology for Optogenetics and Chemogenetics*,
Neuromethods, vol. 195, https://doi.org/10.1007/978-1-0716-2918-5_3,

The idea of the amplicon vector originated during the characterization of defective HSV-1 particles that arose and interfered with HSV-1 stocks passaged at high multiplicities of infection (MOI) [1, 3]. Analysis of the genomes of these defective HSV-1 particles revealed that they retained only a minimal subset of DNA sequences from the wild-type genome. These sequences included an origin of DNA replication and a cleavage/packaging site [1, 3]. It was discovered that incorporation of these two sequences into a plasmid (the "amplicon") conferred on the plasmid the ability to be replicated and packaged into virus particles when it was transfected into a cell that was super-infected with a "helper virus," which supplied HSV replication and virion assembly functions in *trans*. The plasmid sequences that were packaged into virus particles consisted predominantly of 150 kb concatemers of the original plasmid [1, 3–5].

The chief advantages of this amplicon vector type are that cloning manipulations are relatively easy due to the small size of the plasmid (5–10 kb) and that titers up to 3 logs higher than those achieved with genomic vectors are routinely obtained. The essential HSV sequences on the amplicon include an origin of DNA replication (*oriS*) and the packaging "a" sequence. Regulation of the expression of the recombinant gene can be accomplished with HSV or cellular promoters. Because HSV-1-based amplicon vectors contain only a small proportion of the viral genome, their packaging into virus particles requires the participation of additional HSV-1 genes that are expressed in *trans*. These genes are expressed by replication-defective mutants of HSV-1 (helper viruses), as described in this unit. Members of the HSV-1 vector community continue to try to develop helper-free packaging protocols, under the impression that the helper virus present in the amplicon preps might interfere with subsequent manipulations or be toxic (e.g., *see* Ref. [6]). However, the publication of more than 250 papers over 25 years of using amplicon vectors in the brain with the appropriate controls have established that helper virus does not interfere with subsequent experiments in the brain, nor is it toxic. During the initial use of amplicon vectors in vivo, amplicon vectors sometimes caused limited toxicity due to the carryover of cell debris and accompanying toxins during the packaging process. However, improvements in the amplicon packaging procedure were quickly instituted, eliminating any lingering toxicity in the preparations of amplicon viruses. Moreover, continued research in this laboratory has resulted in the development of HSV vector backbones that can be used for infection at the site of injection (Fig. 1), for retrograde circuit-based studies and monosynaptic labeling (Fig. 2, e.g., *see* Refs. [7, 8]), and for expression in zebrafish (Fig. 3, e.g., *see* Ref. [9]). It can express large genes (up to about 15 kb) and is also an optimal vector for CRISPR manipulations [10].

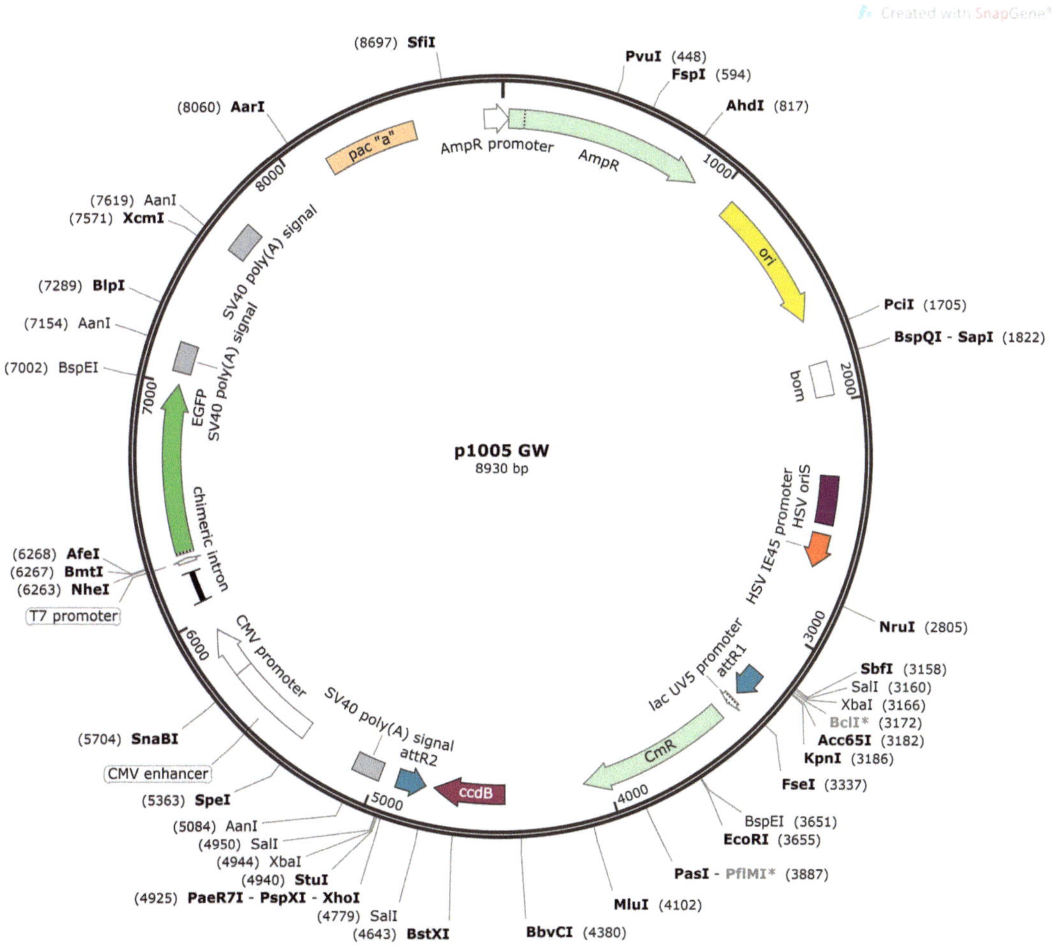

Fig. 1 Map of p1005-GW, the prototypical short-term HSV from which all other ST HSVs are derived

2 Basic Protocol 1

Basic Protocol 1 describes the generation of helper virus stocks. Preparation of recombinant amplicon vector particles by transfection of amplicon and superinfection of helper virus into cells, and harvesting of packaged particles, is delineated in Basic Protocol 2. Thorough characterization of each amplicon viral vector stock involves measuring (1) the helper virus plaque-forming units per milliliter (pfu/mL) on 2–2 cells and (2) the amplicon stock infectious units per ml (iu/mL) on PC12 cells. The Support Protocols detail methods for determining titers of helper virus by plaque assay and of amplicon stocks by vector assay.

NOTE: All incubations are performed in a humidified 37 °C, 10% CO_2 incubator unless otherwise specified. All other procedures are performed in a laminar flow hood to maintain sterility.

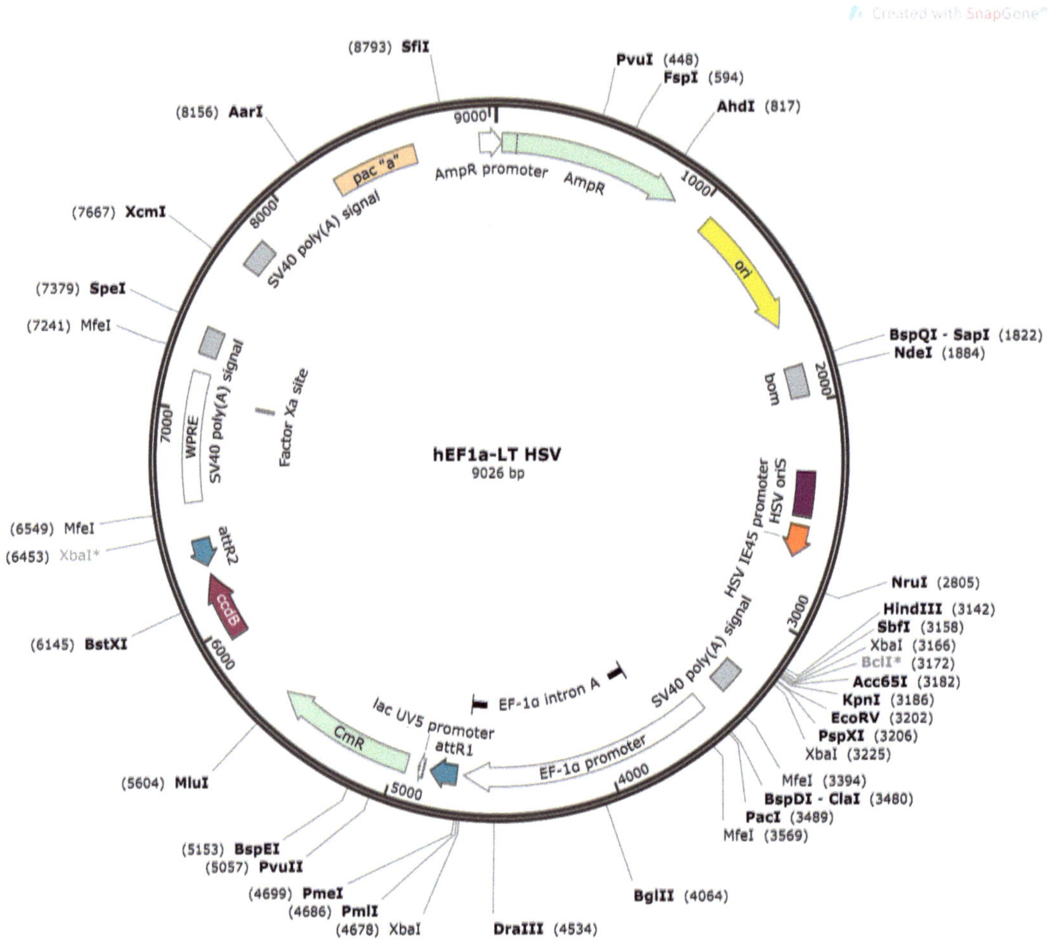

Fig. 2 Map of hEF1a LT HSV, a prototypical retrograde HSV from which all other retrograde HSVs and monosynaptic labeling HSVs are derived

2.1 Preparation of Helper Virus Stocks

Seed stocks of individual plaque-purified virus isolates are generated. These seed stocks are stored at −70 °C and are subsequently used for inoculation of large-scale helper virus preparations.

2.1.1 Materials

Adherent 2–2 cells (Vero cells containing IE2 (ICP) gene and promotor [10], maintained in supplemented DMEM/10% FBS.

D-PBS with Ca^{+2}/Mg^{+2}, hereafter referred to as D-PBS.

Ca^{+2}/Mg^{+2}-free (CMF) D-PBS.

TrypLE Express (Life Technologies).

Supplemented DMEM (Corning 10-017-CM) containing 2%, 5%, or 10% (v/v) FBS (Seradigm) and 1% penicillin + streptomysin).

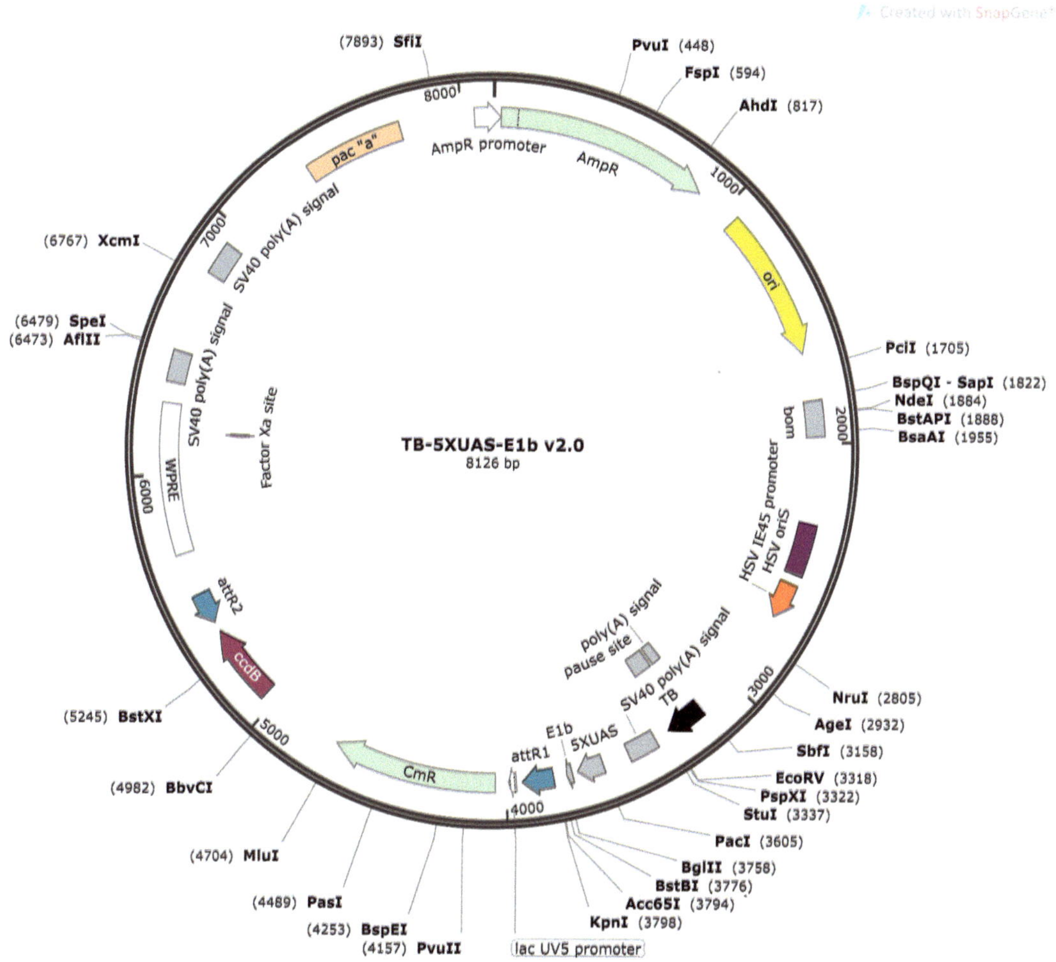

Fig. 3 Map of TB-5XUAS E1b v2.0, a prototypical zebra fish HSV from which all other zebra fish vectors are derived

5*dl*1.2 helper virus stock (HSV-1 (KOS) strain with a partial deletion in IE2 (ICP27) [11].

Plaque agarose (see recipe).

Laminar flow hood.

60 and 100 mm tissue culture plates.

Cell lifters.

15 mL polypropylene conical tube with a plug seal.

Dry ice/ethanol bath.

15 mL polystyrene conical tubes.

Cup-type sonicator, e.g., Misonix.

2.1.2 Make a Plaque Plate

1. Passage 2–2 cells by trypsinization by first removing the medium in which the cells are maintained (DMEM/10% FBS) and washing the plate(s) twice with CMF-DPBS.
2. Apply a thin layer of TrypLE Express and incubate for 10 min at 37 °C.

   2–2 cells are highly adherent and difficult to get off the dish, so patience is needed.

3. Count cells with a hemocytometer and seed a 60 mm tissue culture plate with 1×10^6 cells in 5 mL supplemented DMEM/10% FBS. Culture overnight in a humidified 37 °C incubator with 10% CO_2.
4. Remove 3 mL medium from the culture and add an amount of 5*dl*1.2 helper virus stock that will give well-separated plaques (~50 pfu/plate). Allow the virus to adsorb at 37 °C for ≥90 min, but <6 h.
5. Remove medium and overlay cells with 3 mL plaque agarose that is no warmer than 40 °C. Allow the agarose to solidify completely, return the plate to the 37 °C incubator, and incubate overnight.
6. The next day, add 2 mL supplemented DMEM/2% FBS on top of the agarose. Incubate overnight at 37 °C.
7. The following day, remove the medium and replace it with another 2 mL supplemented DMEM/2% FBS.

2.1.3 Make a Seed Stock

8. One to three days after infection (**step 4**), prepare several 60 mm plates of uninfected 2–2 cells as in **steps 1–3**.
9. The following day (2–4 days after infection, when plaques are visible), pick a single plaque from the agarose-overlaid plate using a pipette tip to stab through the agar and into the cells and eject the plug into the medium of one 6 mm plate of uninfected cells. Repeat with remaining uninfected plates. Culture plates at 37 °C until >95% of the cells show cytopathic effects (CPE).

   It is advisable to pick several different plaques to ensure that at least one isolate has the appropriate phenotype for the virus (i.e., that it has not undergone spontaneous mutation).

   Cells should show CPE after 1 or 2 days depending on the amount of virus obtained from the plaque. The cells should round up but remain adherent.

10. Harvest cells by scraping them into the medium with a cell lifter and transferring them, with the medium, to 15 mL polypropylene conical tubes with plug seals (one tube/plate).

    The preparation can be stored for up to several days at ≤−70 °C at this stage.

11. Freeze/thaw three times using a dry ice/ethanol bath and a 37 °C water bath, being careful to minimize the time the cells are thawed at 37 °C.

    As a rule, it takes about 10 min to freeze the stock and 10–15 min for each thaw.

12. Transfer thawed cells to 15 mL polystyrene conical tubes and sonicate for 2 min in a cup-type sonicator (power setting 6, 50% duty cycle, 1-s cycles) to disrupt the cells and release the virus.

13. Immediately centrifuge 5 min at low speed (1000 × *g* in a refrigerated or room temperature benchtop centrifuge) to pellet cell debris but not virus particles.

14. Transfer each supernatant to a fresh screw-cap tube for virus storage. Measure virus titer (*see* Subheading 2) and store at ≤−70 °C to use as seed stock for future viral preparations.

    Seed stocks prepared as outlined here usually have titers of 0.5–1 × 10^7 pfu/ml and can be stored for several years or more at ≤−70 °C.

2.1.4 Amplify the Seed Stock

15. Seed four 100 mm tissue culture plates with 2 × 10^6 2–2 cells per plate in 10 mL supplemented DMEM/10% FBS and incubated 1 day at 37 °C.

16. Change medium to 10 mL supplemented DMEM/2% FBS and added 50 μL of seed stock virus. Incubate at 37 °C.

    The 2–2 cells should reach a density of 0.5–2 × 10^7 cells/plate by the second day.

    The cells should be infected at a multiplicity of infection (MOI) ≤ 0.1. At high MOI, spontaneously arising defective viruses are able to grow and interfere with virus replication, resulting in lower pfu/mL.

17. Harvest the cells (*see* **steps 10–14**) when they have rounded up but still remain adherent. Measure virus titer (*see* Subheading 3) and store virus in aliquots at ≤−80 °C.

    Helper stocks prepared as outlined here usually have titers of ~1 × 10^8 pfu/mL and can be stored for several years or more.

3 Support Protocol 1

3.1 Titration of Helper Virus by Plaque Assay

Plaque assays should be performed on all amplicon virus stocks because the vector/helper ratio is an important indicator of the quality of the stock. A vector/helper ratio of 1:1 is desired. This procedure uses crystal violet to count plaques grown on 2–2 cells. The plaques will be clear against a stained background.

3.1.1 Additional Materials (also See *Subheading 2)*

Plaque-fixing solution: 5% (v/v) methanol/10% (v/v) acetic acid (store at room temperature)

Crystal violet stain (see recipe)

Dissecting microscope

1. Plate 2–2 cells in 60 mm tissue culture plates (1×10^6 per plate) in 5 mL supplemented DMEM/10% FBS (*see* Subheading 2, **steps 1–3**) and grow for 1 day in a humidified 37 °C, 10% CO_2 incubator. Prepare one plate for each dilution of virus stock to be titered using a series of tenfold dilutions up to 10^7.
2. Infect cells by removing 3 mL medium from each plate and adding up to 100 μL of the appropriate virus dilution. Allow the virus to adsorb at 37 °C for ≥90 min, but <6 h.
3. Remove medium and overlay cells with 3 mL plaque agarose that is no warmer than 40 °C. Allow the agar to solidify completely, return plates to the 37 °C incubator, and incubate overnight.
4. Add 2 mL supplemented DMEM/2% FBS on top of the agarose and incubate overnight.
5. The next day, replace the medium with another 2 mL supplemented DMEM/2% FBS and incubate overnight.
6. Three days after infection, examine the plates under a microscope for plaques. If they are too small to reveal clear holes in the cell monolayer, leave the plates for another day before staining.
7. Remove medium and fix cells with 3 mL plaque-fixing solution for ≥15 min.
8. Remove the plaque-fixing solution and discard agarose with a quick wrist flick (into a disposal container).
9. Add 1 mL crystal violet stain and leave for 5 min.
10. Remove stain, wash the plate with 1 mL water, decant, and allow it to air dry.

    The stain can be reused.
11. Count the number of plaques and use the dilution factor of the virus stock to calculate the virus plaque-forming titer in pfu/mL.

    Plaques should be macroscopically visible as holes, but it is advisable to use a microscope to avoid any plaques.

4 Basic Protocol 2

4.1 Packaging Amplicon into Virus Particles

Host cells are transfected with the recombinant HSV-1 amplicon vector. They are then superinfected with helper virus to provide the HSV-1 gene products in *trans* that are required for packaging the amplicon into virus particles. The resultant virus population is a mixture of helper virus and amplicon particles. This mixture is passaged three times on 2–2 cells. Because the amplicon is packaged as tandemly repeated concatamers, each amplicon virus particle has multiple replication origins. The amplicon viruses, therefore, have a selective growth advantage over the helper viruses, which have a single origin of replication, and the ratio of amplicon to helper virus increases with each passage.

After the third passage (P3), the virus is purified and concentrated, first by banding it on a sucrose step gradient to separate it from cellular debris and then by pelleting the virus from the band and resuspending it in a small volume.

4.1.1 Materials

Adherent 2–2 cells maintained in supplemented DMEM/10% FBS.

Supplemented DMEM (see recipe) containing 5% or 10% (v/v) FBS, room temperature and prewarmed to 37 origins C.

Seradigm FBS.

D-PBS, prewarmed to 37 °C.

CMF-D-PBS.

D-PBS with calcium and magnesium.

TrypLE Express (Life Technologies).

≥50 ng/μL DNA for transfection, purified with a Qiagen column or its equivalent and resuspended in TE (10 mM Tris–HCl, 1 mM EDTA).

Opti-MEM (Life Technologies), prewarmed to 37 °C.

Plus Reagent (Life Technologies).

LipofectAMINE 2000 (Life Technologies).

5*dl*1.2 helper virus amplified from seed stock (*see* Subheading 2).

Defined fetal bovine serum (Seradigm—no substitutes).

10%, 30%, and 60% (w/v) sucrose solutions (see recipe).

Laminar flow hood.

60 mm, 100 mm, and 150 mm tissue culture plates.

Cell lifters.

Samco 232-1S fine tipped transfer pipettes.

1 M solution HEPES, pH 7.3 (USB).

Dry ice/ethanol bath.

15 and 50 mL polypropylene conical tubes.

Cup-type sonicator.

Ultra-Clear 25 × 89 mm (SW28) and 14 × 89 mm (SW41) tubes.

SW28 and SW41 (or SW40) ultracentrifuge rotors, or their equivalent.

18-G needles.

3 mL syringes.

Additional reagents and equipment for counting cells with a hemocytometer and for measuring virus titer (*see* Subheading 4).

Important: use lubricant-free microfuge tubes for the entire procedure. I use Costar 3207 tubes.

4.1.2 Prepare 2–2 Cells

1. Maintain 2–2 cells at 37 °C in a humidified, 10% CO_2 incubator in supplemented DMEM/10% FBS. Passage by trypsinization with TrypLE (*see* Subheading 2, **steps 1–3**), being careful not to allow the cells to grow beyond confluence. When a fresh aliquot of cells is thawed, passage at least two times before the cells are plated for transfection.

   If the cells grow beyond confluence, they become suboptimal for transfection and infection by virus and should be discarded.

   Large numbers of cells are needed throughout the packaging procedure. To minimize the number of plates used, it is helpful to maintain the stock cells in 150 mm plates. If 0.5–1 × 10^6 cells are passaged per plate, it will be 3–4 days before they become confluent.

2. Two days before transfection, plate 2–2 cells at 3 × 10^5 per 60 mm tissue culture plate in 5 mL supplemented DMEM/10% FBS (one plate of cells per DNA construct).

   See the "packaging schedule" at the end of the chapter to help plan which steps to carry out each day.

   The goal is to have 80% confluent cells for transfection. It is helpful to plate three sets of plates, one with 2.5 × 10^5 cells per plate, one with 3 × 10^5 cells per plate, and one with 3.5 × 10^5 cells per plate, to ensure that the correct density of cells is achieved.

4.1.3 Transfect Cells

3. Dilute 2 μg of ≥50 ng/μL DNA in 250 μL of room temperature Opti-MEM in a sterile 1.5 mL microcentrifuge tube (one tube per plate—i.e., per DNA construct) at room temperature.
4. Add 8 μL of Plus Reagent to each tube, and mix by flicking with the fingers. Leave the mixture at room temperature for 15 min, no longer.
5. Dilute 12 μL LipofectAMINE 2000 in 250 μL room temperature Opti-MEM in a second tube (one per plate), let it sit for 5 min, and then add it to the 250 μL DNA/Plus Reagent mixture in the first tube. Leave the DNA/LipofectAMINE 2000 mixture for 20–45 min at room temperature to allow liposomes to form.
6. Remove medium from the plates, wash once with 2 mL prewarmed (37 °C) Opti-MEM, remove wash, and replace with another 2 mL Opti-MEM.

7. Add the ~500 μL DNA/LipofectAMINE mixture to the cells dropwise and evenly over the whole plate. Incubate for 5–7 h at 37 °C.
8. Wash cells three times with 2 mL prewarmed 37 °C D-PBS.

   NOTE: *D-PBS used to wash cells should contain Ca^{2+} and Mg^{2+}. Calcium- and magnesium-free DPBS (CMF-DPBS) is used only when cells are washed prior to trypsinization (see* ***step 1****). This wash step is necessary to remove all residual lipofectA-MINE, which will otherwise inhibit the replication of the virus the following day.*
9. Replace D-PBS with 5 mL prewarmed, supplemented DMEM/10% FBS and incubate cells 15–17 h at 37 °C.

4.1.4 Superinfect and Harvest Transfected Cells (P0) by Osmotic Lysis

10. Remove medium from plates and add 5 mL prewarmed, supplemented DMEM/5% FBS.
11. Add 1 × 10^6 pfu of 5*dl*1.2 helper virus seed stock and incubate at 37 °C until 95% of the cells show cytopathic effects (CPE) and have rounded up (~26–30 h).
12. Remove the medium from each plate to a 15 mL conical polypropylene tube. Add 1 mL 0.1× D-PBS to the cells on the plate and leave for 2 min for the cells to swell.
13. Use a Samco 232-1S fine-tipped transfer pipet with a built-in bulb to collect the lysate and transfer it to the tube containing the medium. Add 100 μL of 10× CMF DPBS to each tube to restore ionic strength.

    The optimum MOI for superinfection by helper virus following transfection is between 0.2 and 0.6.

    The harvested cells can be stored at −80 °C at this stage for at least several days.
14. Centrifuge cells 5 min at low speed (1350 × *g*) in a refrigerated or room temperature benchtop centrifuge to pellet cell debris but not virus particles. Collect supernatant.

    *The P0 supernatant can be used immediately for the P1 infection (****step 16****) or stored at −80 °C for at least several days.*

4.1.5 Amplify Virus Stock (P1)

15. Two days before P1 infection, plate fresh 2–2 cells at 5 × 10^5 cells per 60 mm tissue culture plate in 5 mL supplemented DMEM/10% FBS. Incubate at 37 °C.
16. Two days later, replace the medium with 4 mL DMEM/5% FBS and add ~4 mL of the P0 supernatant. Incubate at 37 °C until the cells show 95% CPE (~21–24 h).
17. Harvest and process cells (P1) as in **steps 12–14**.

The P1 virus stocks can be frozen at this point, but it is preferable to proceed immediately to the P2 infection so as not to disrupt the flow of the packaging procedure.

4.1.6 Amplify Virus Stock (P2)

18. Plate fresh 2–2 cells: 1.2×10^6 per 100 mm dish in 10 in 10 mL DMEM + 10% FBS (two dishes per sample). Incubate them at 37°C for 2 days.
19. Two days later, replace the medium on the 2–2 cells with 6.0 mL DMEM + 5% FBS per dish, and add 4 mL P1 supernatant per dish. Incubate the cells overnight at 37 °C.
20. Harvest the P2 viruses when they show 95% CPE (about 21–24 h). Remove the medium from each pair of plates to a 50 mL tube, trying to avoid transferring cells. Add 2 mL of 0.1× DPBS to the cells on each plate and leave it on the cells at room temperature for at least 2 min. Use a Samco 232-1S transfer pipette to collect the lysate and transfer it to the tube containing the medium. Add 400 μL CMF 10× DPBS to restore the ionic strength of the medium .

   CMF 10× DPBS is used because, unlike CMF 10× DPBS, 10× DPBS that contains Ca^{+2} and Mg^{+2} has an acid pH when diluted and will kill the pH-sensitive viruses.

   The P2 virus stocks can be frozen at this point, but it is preferable to proceed immediately to the P3 infection so as not to disrupt the flow of the packaging procedure.

4.1.7 Amplify Virus Stock (P3)

21. Plate fresh 2–2 cells: 1.2×10^6 per 100 mm dish in 10 mL DMEM + 10% FBS (6 dishes per sample). Incubate the cells at 37 °C for 2 days.
22. Two days later, replace the medium on the 2–2 cells with 6.0 mL DMEM + 5% FBS per dish, and add 4 mL P2 supernatant per dish. Incubate the cells overnight at 37 °C.
23. Harvest the P3 viruses when they show 95% CPE (about 21–24 h). Scrape up the cells with a cell lifter and transfer the cells and medium to two 50 mL polypropylene conical tubes per sample (30 ml/tube). Add 1.5 mL 1 M HEPES, pH 7.3, to each tube, screw on the cap, and mix by turning the tubes upside down once.

   The HEPES buffers the lysates from becoming acidic during the subsequent freeze-thaw cycles and thereby optimizes the retention of viral titer.

24. Freeze-thaw the cells three times using a dry ice/ethanol bath and a 37 °C water bath. Minimize the time that they are in the 37 °C water bath.

25. Following the third freeze of the cells, the crude cell lysates may be stored at ≤−70 °C until they are ready to be processed by sucrose gradient centrifugation.

4.1.8 Purify and Concentrate the Virus

26. Transfer the crude cell lysates to 50 mL *polystyrene* tubes (very important) and sonicate them for 2 min in a cup sonicator (power setting 6, 50% duty cycle, 1 s cycles).

    Optional: first, sonicate the lysates for 1 min in the 50 mL polypropylene tubes they are originally in. This doubles the final titer.

27. Transfer the sonicated lysates back to polypropylene tubes.

    This is necessary because polystyrene tubes will crack during the subsequent spin.

28. Centrifuge the lysates at 1800 × *g* for 10 min to make a very firm cell debris pellet.
29. Prepare sucrose step gradients in 40 mL (25 × 89 mm) ultra-clear SW28 centrifuge tubes at room temperature by layering the following sucrose solutions into the tube (starting at the bottom): 6 mL 60% sucrose in D-PBS, 7 mL 30% sucrose in DPBS, and 3 mL 10% sucrose in D-PBS. Three tubes are needed for each virus sample.
30. Load 20 mL crude virus onto the gradient and spin 1 h at 112,000 × *g* (SW28 rotor, 25 K rpm) at 15 °C.
31. Immobilize each tube in a clamp on a ring stand and place it in front of a black background. The virus will appear as a milky, sharp band at the 30%/60% interface, while debris will collect in a diffuse band close to the 10%/30% interface.
32. Using a 3 mL syringe with an 18-G needle attached, pierce the tube underneath the band, with the beveled edge of the needle pointed upward. Slowly pull the band into the syringe in a volume of 2 mL. Transfer this to an 11.5 mL (14 × 89 mm) *polypropylene* SW41 tube and discard the remainder.

    The viruses stick to ultra-clear tubes when they are pelleted in them. Up to 90% of the virus is lost if ultra-clear rather than propylene SW41 tubes are used to spin it down.

33. Dilute the virus in each tube with 9.5 mL D-PBS and gently mix the contents of each tube by pipetting it up and down.
34. Centrifuge at 125,000 × *g* (SW41 or SW40, 26 K rpm, 15 °C) for 40 min. Carefully aspirate the supernatant.
35. For ST (short-term viruses, add 60 μL D-PBS 10% sucrose + 25 mm HEPES to the opalescent pellet in each tube, and resuspend the pellet by shaking it in a rack on a platform shaker at 4 °C overnight. For retrograde viruses, add 60 μL to each tube, and for viruses expressing the rabies coat protein, add

20 μL to each tube. Very briefly triturate the virus the following day before dispensing it into aliquots in prelubricated microfuge tubes and storing it at −80 °C or lower.

The ST viruses will yield a titer of ~1.5×10^8 i.u/mL, the retrograde viruses a titer of ~1×10^9 i.u./mL, and the viruses expressing rabies a titer of ~5×10^9 i.u./mL.

The HEPES in the sucrose, with prelubricated microfuge tubes, prevents the viruses from sticking to the tubes when the virus stocks are thawed and then refrozen.

Stocks are stable indefinitely at −80 °C

It is very difficult to resuspend if the virus is triturated without an overnight incubation in the buffer. Moreover, the mechanical disruption necessary to resuspend the virus under these conditions will greatly decrease the viability of the virus. Therefore, slow resuspension overnight at 4 °C is recommended.

To determine virus titer, see Subheading 5.

5 Support Protocol 2

5.1 Titration of Amplicon Virus by AMPLICON Vector Assay

Infectious amplicon virus is titered in this procedure. An immunocytochemical assay can be used to detect transgenes, or the virus can be visualized by co-expression of a fluorescent protein. PC12 cells are used for vector assays because they are round and easy to distinguish as single cells when positive for expression. Expression is usually quite high in this cell line, though it will vary with different proteins. Use your favorite immunostaining protocol.

5.1.1 Materials

20 μg/mL poly-D-lysine solution.

PC12 cells.

DMEM/10% HS/5% FBS: Supplemented DMEM (see recipe) without G418, containing 10% (v/v) horse serum and 5% (v/v) FBS.

Amplicon stock to be measured (*see* Subheading 4).

PBS with and without 10 mM EDTA.

TBS (Tris-buffered saline).

4% (w/v) paraformaldehyde solution.

Laminar flow hood.

24-well tissue culture plates.

21-G needle.

Dissecting microscope.

Additional reagents and equipment for counting cells with a hemocytometer.

5.1.2 Infect and Fix Cells

1. Add 500 µL of 20 µg/mL poly-D-lysine solution to each well of a 24-well tissue culture plate and incubate ≥5 min at room temperature to coat. Aspirate completely before using.

   Poly-D-lysine is used to ensure that the PC12 cells adhere to the culture surface.

2. Harvest PC12 cells by trypsinization (*see* Subheading 2, **steps 1–3**) and pass through a 21-G needle to dissociate aggregates.
3. Count cells with a hemocytometer and seed them in the coated 24-well plate at 3×10^5/well in 500 µL DMEM/10% HS/5% FBS. Prepare four wells for each amplicon stock to be titered. Incubate overnight in a humidified 37 °C, 10% CO_2 incubator.
4. The next day, change the medium, and add 0, 0.5, 1, 2, and 5 µL amplicon stock to the four wells. Incubate at 37 °C overnight.

   The uninfected well acts as a negative control for background staining.

5. The following day, wash cells once with 500 µL PBS and fix with 500 µL of 4% paraformaldehyde solution for ≥15 min at room temperature.

   If the paraformaldehyde is old (e.g., *>7 days), or if the cells are not fixed for a sufficient period of time, the cells will collapse and shrink, causing background staining problems.*

6. Wash cells once with 500 µL TBS (for immunocytochemistry).

   Phosphate inhibits alkaline phosphatase. This is a critical issue if using the antibody–alkaline phosphatase (AP) staining procedure.

   For immunocytochemical staining, use your favorite immunostaining protocol.

5.1.3 Record and Analyze Data

1. Estimate the percentage of stained cells in one or more of the wells infected for each amplicon stock, and use this information to extrapolate the titer of the stock. For example, if 20% of the cells in a well infected with 2 µL of the virus are stained, and if it is assumed that each stained cell represents one infectious unit of the virus, then it can be inferred that the 2 µL of virus contained 6×10^4 (20% of 3×10^5 cells plated/well) infectious units. Thus, the titer of the stock would be 3×10^4 iu/µL or 3×10^7 iu/mL.

5.2 Reagents and Solutions

Use deionized, distilled water in all recipes and protocol steps.

5.2.1 Alkaline Phosphatase (AP) Buffer

100 mM Tris–HCl·Cl, pH 8.5.

150 mM NaCl.

5 mM $MgCl_2$.

Store up to 1 year at room temperature.

5.2.2 AP Substrate Solution

Prepare a 50 mg/mL 5-bromo-4-chloro-3-indoyl phosphate (BCIP) solution by dissolving one tablet (Sigma B-0274) in 0.5 mL 100% dimethylformamide. Prepare fresh.

Prepare a 10 mg/mL nitroblue tetrazoleum (NBT) solution by dissolving one tablet (Sigma N-5514) in 1 ml of water. Prepare fresh.

Make AP substrate solution by adding 330 μL NBT solution to 10 mL AP buffer (see recipe). Mix and then add 33 μL BCIP solution. Prepare fresh.

5.2.3 Crystal Violet Stain

Prepare a solution containing 0.5% (w/v) crystal violet and 0.2% (w/v) sodium acetate. Adjust the pH to 3.6 with acetic acid before bringing it to the final volume. Filter through 1 MM Whatman paper and store for up to several months at room temperature.

The stain can be reused several times.

5.2.4 Dulbecco's Modified Essential Medium, Supplemented

Dulbecco's modified essential medium (DMEM) (e.g., *Corning cat. no. 10-017-CM) contains:*

1% (w/v) penicillin/streptomycin.

4 mM glutamine.

500 μg/mL G418 (neomycin analog, optional).

Store up to 2 months at 4 °C.

It is not necessary to maintain the cells in the presence of G418 at all times. The authors have grown the cells without selective pressure for at least short periods of time with good results. Because G418 may affect cellular metabolism in ways other than conferring G418 resistance, and because it is expensive, the author often includes it in the medium only for the first two passages of the cells after they are thawed, after which the cells can be maintained without G418 for the remainder of the packaging procedure.

It is important to note that this packaging procedure has been optimized using Seradigm FBS. When another FBS is used, the timing of the procedure changes significantly (e.g., *following transfection and superinfection, the cells do not show CPE for up to 48 h). Therefore, it is essential to use Seradigm FBS throughout the packaging procedure.*

5.2.5 Dulbecco's Phosphate-Buffered Saline (D-PBS)

D-PBS and CMF-D-PBS are commercially available. For the packaging procedure, especially, commercial D-PBS is strongly recommended to ensure the utmost purity.

5.2.6 Paraformaldehyde Solution, 4% (W/V)

Add 20 g paraformaldehyde to 300 mL H_2O and heat to 55–60 °C. Slowly add 1 M NaOH dropwise over ~10 min until the solution becomes clear, then cool the solution to room temperature. Use pH paper to check that the pH is 7.0–7.5 (add more NaOH if necessary). Add 100 mL of 0.5 M sodium phosphate buffer, pH 7.0, and then add water to a final volume of 500 mL (final 0.1 M phosphate; final pH 7.0–7.5). Store up to 1 week at 4 °C.

CAUTION: Paraformaldehyde is toxic, and the preparation process involving heating results in considerable vaporization, which increases the hazard. It is essential to use appropriate safety procedures, such as working in a fume hood.

5.2.7 Plaque Agarose

Prepare supplemented DMEM that is double-strength (2×), and place it and a bottle of FBS into a 42 °C water bath. Prepare a solution of molten 2% (w/v) tissue culture-grade agarose (e.g., SeaPlaque, FMC Bioproducts) in water and place it into the 42 °C water bath. Allow temperatures to equilibrate. Mix reagents together at a ratio of 25:25:1 agarose/2× DMEM/FBS. Prepare fresh, remove just before required, and use when the temperature is <40 °C.

5.2.8 Poly-D-Lysine Solution, 20 μg/mL

Prepare a 1 mg/mL stock solution of poly-D-lysine (mol. wt. 70,000–150,000; Sigma) in water. Filter sterilize and store up to a year or more at −20 °C. Dilute to 20 μg/mL in water immediately prior to use.

5.2.9 Sucrose Solution, 60%, 30%, and 10% (W/V)

Place 100 mL of a 10× stock solution of CMF D-PBS into a 1 L beaker. Add water to 750 mL. Then add 600 g sucrose, mix, and add water to 1 L. Add 0.45 mL of 2 M $MgCl_2$ and 0.25 mL $CaCl_2$, for a final concentration 0f 0.1 g/L of each. Adjust pH to 7.3–7.4 with 1 M NaOH. Filter the sucrose through a 0.45 μm unit. Make 30% and 10% solutions by diluting the 60% sucrose with D-PBS with calcium and magnesium. Store up to 1 year at 4 °C. Warm the solutions to room temperature before use.

Filtration of the sucrose solution is very slow; be patient.

6 Commentary

6.1 Background Information

HSV-1 vector systems were historically limited by their cytotoxicity and relatively low titers. Despite the fact that replication-incompetent helper viruses cannot progress through the lytic cycle in normal cells, cytotoxicity from toxins in the raw cellular lysates could occur. An HSV-1 packaging protocol designed to minimize cytotoxicity and optimize the ratio of vector to helper virus is presented in this unit. Viruses produced using this protocol are not toxic [12–14], and a vector/helper virus ratio of 1:1 is obtained.

The key features of this protocol are the use of an improved helper virus/host cell combination, the utilization of LipofectAMINE 2000 and Plus Reagent (Life Technologies) rather than calcium phosphate for transfection, and the purification and concentration of virus particles away from cellular debris following generation of the crude viral lysate. *IE2 (ICP27)* deletion mutants, particularly the 5*dl*1.2 helper virus/2–2 host cell combination used in the present protocol, have been shown to be more effective in the packaging procedure than other helper viruses/host cell combinations [13] and virtually never generate wild-type HSV-1. (This author has done thousands of preparations using this helper/virus combination for over 27 years and has never detected wild type virus). Transfection of vector DNA was found to be more efficient and reproducible using LipofectAMINE rather than the calcium phosphate method [15]. Finally, the purification of the virus away from cellular debris on discontinuous sucrose gradients and concentration by ultracentrifugation (as described in this unit) eliminated cytotoxicity and increased the titers of the virus stocks.

6.2 Critical Parameters and Troubleshooting

6.2.1 Choosing the Right Helper Virus and Host Cell Combination

Different helper viruses and the appropriate complementary cell lines may be used. The authors have used two systems not described in this unit: the *IE 3* deletion mutant d120 grown on E5 cells and the *IE 3* deletion mutant D30EBA grown on RR1 cells [15]. However, packaging with either of these two systems usually yields titers ~tenfold lower than when the 5*dl*1.2/2–2 system is used. The vector/helper ratio is also consistently lower using the *IE 3* deletion mutants than when the *IE 2* mutant 5*dl*1.2 is used. These differences are likely related to differential efficiency of growth of the different mutants since packaging with wild-type HSV-1 yields very high vector/helper ratios.

6.2.2 Optimizing Transfection and Infection Efficiencies

Precise knowledge of the growth state of the cells is crucial for ensuring good efficiencies of transfection and infection. This fact cannot be overemphasized. As a guideline, do not split the cells 2 days a row. The cells should be washed at least twice with Ca^{+2}- and Mg^{+2}-free DPBS (CMF-DPBS) before they are treated with TrypLE for 10 min. Maintain the cells at 37 °C in a humidified, 10% CO_2 incubator in supplemented DMEM with 10% (v/v) FBS. When a fresh aliquot of cells is thawed, it should be passaged at least two times before the cells are plated for transfection. Be careful always to keep the cells from growing past confluence; if the cells grow past confluence, they become suboptimal for transfection and for infection by virus and should be discarded.

High transfection efficiency is important to obtain good vector titers. Cells should be 80% confluent when they are transfected. Transfection with LipofectAMINE 2000 and Plus Reagent results in higher packaging efficiencies than with other reagents such as Lipofectin or calcium phosphate. DNA quality is another important variable. The DNA used for the transfection must be prepared by Qiagen columns or their equivalent. The final titer of the vector virus depends directly on the efficiency of the initial transfection.

As a control for transfection efficiency, always package a sample of HSV expressing a fluorescent protein so that you can simply visualize fluorescence in live cells. At the time of the P0 harvest, following transfection, and following superinfection with a helper virus, the lysate may be titered for vector on PC12 cells. Do not continue with preparations that have titers $<10^4$ iu/mL.

6.2.3 Optimizing the Packaging Procedure

The freeze/thaw cycles of the viral lysates must be done in polypropylene tubes, as polystyrene tubes will crack, and the resultant direct contact of the lysate with CO_2 will lower the pH of the medium and make the virus inviable. However, the lysates must be transferred to polystyrene tubes for the sonication because cup sonicators are ineffective on samples in polypropylene vessels.

Maintain sterile technique throughout the packaging procedure.

6.2.4 Maintaining Virus Viability

Virus particles are thermolabile. Up to 50% loss in titer can be expected after storage of unpurified virus at −70 °C for 1 year, although only ~10–20% loss in titer is seen after 1 year for sucrose gradient-purified virus. Storage of the virus at −80 °C or lower will prevent complete loss of titer over time. Concentrated virus in high-protein solutions has higher stability. After one thawing and refreezing of the virus stored at 80 °C, less than 10% loss can be expected, depending on the size of the aliquot, ambient temperature, and other factors. Thaw the virus rapidly in a 37 °C water bath and refreeze in a dry ice/ethanol bath. As a rule, do not thaw and refreeze a given aliquot of the virus more than four times.

6.3 Anticipated Results

Thorough characterization of the viral vector stock involves measuring infectious vector units per milliliter (iu/mL) on PC12 cells (*see* Subheading 5), as well as measuring helper plaque-forming units per milliliter (pfu/mL) on 2–2 cells (*see* Subheading 3). The best viral stocks have a vector titer of 2–5 × 10^8 iu/mL and a vector/helper virus ratio of 1:1.

6.4 Time Considerations

After each infection, the time that must elapse to achieve cytopathic effects in 95% of the cells (95% CPE) is variable, usually ranging from 21 to 24 h. It is critical to monitor the cells at all times and to resist the temptation to harvest the virus prematurely. Vector titers of 5 × 10^8 iu/mL and a vector/helper ratio of 1:1 can be obtained only if care is taken throughout the procedure to infect and harvest the cells at the appropriate times. The cells should be just confluent when they are infected and should show 95% CPE when they are harvested.

If performed without interruption, the packaging procedure can be performed in as few as 9 days from plating for transfection to obtaining sucrose gradient-purified virus (*see* Subheading 6.5 below). Because of the necessity of maintaining large numbers of confluent cells for the successive passages of the virus during the packaging procedure, it is logistically easiest to move through the procedure without breaks. However, if the procedure must be performed in stages, the crude virus lysates may be stored at ≤−70 °C after any or all of the virus harvests without affecting the final yield of the virus.

6.5 Packaging Schedule

Two days before transfection:

- Plate packaging cells at 3×10^5 per 60 mm tissue culture plate (P0).

Day 1:

- Perform transfections.
- Plate packaging cells at 5×10^5 per 60 mm tissue culture plate for P1 infection.

Day 2:

- Perform superinfection with a helper virus.
- Plate packaging cells at 1.2×10^6 per 100 mm tissue culture plate (2 plates per sample) for P2 infection.

Day 3:

- Harvest P0.
- Infect P1.
- Plate packaging cells at 1.2×10^6 per 100 mm tissue culture plate (6 plates per sample) for P3 infection.

Day 4:

- Harvest P1 by osmotic lysis.
- Infect P2.

Day 5:

- Harvest P2 by osmotic lysis.
- Infect P3.

Day 6:

- Harvest P3 by scraping up cells.

References

1. Spaete R, Frenkel N (1982) The herpes simplex virus amplicon: a new eukaryotic defective-virus cloning-amplifying vector. Cell 30:305–310
2. Roizman B, Jenkins FJ (1985) Genetic engineering of novel genomes of large DNA viruses. Science 229:1208–1214
3. Stow N, McMonagle E (1982) Propagation of foreign DNA sequences linked to a herpes simplex virus origin of replication. In: Gluzman Y (ed) Eucaryotic viral vectors. Cold Spring Harbor Laboratory Press, Cold Spring Harbor, pp 199–204
4. Spaete R, Frenkel N (1985) The herpes simplex virus amplicon: analyses of cis-acting replication functions. Proc Natl Acad Sci U S A 82: 694–698
5. Vlazny D, Kwong A, Frenkel N (1982) Site-specific cleavage/packaging of herpes simplex virus DNA and the selective maturation of nucleocapsids containing full-length viral DNA. Proc Natl Acad Sci U S A 79:1423–1427
6. Fraefel C et al (1996) Helper virus-free transfer of herpes simplex virus type I plasmid vectors into neural cells. J Virol 68:7190–7187
7. Fenno L et al (2014) Targeting cells with single vectors using multiple-feature Boolean logic. Nat Methods 11:763–782
8. Yonehara K et al (2013) The first stage of cardinal direction selectivity is localized to the dendrites of retinal ganglion cells. Neuron 79: 1078–1085
9. Satou et al (2022) A viral toolbox for conditional and transneuronal gene expression in zebrafish. eLife 2022 Jul 22;e77153. https://doi.org/10.7554/eLife.77153
10. Smith IL, Hardwicke MA, Sandri-Goldin RM (1992) Evidence that the herpes simplex virus immediate early protein ICP27 acts post-transcriptionally during infection to regulate gene expression. Virology 186:74–86
11. McCarthy AM, McMahan L, Schaffer PA (1989) Herpes simplex virus type 1 ICP27 deletion mutants exhibit altered patterns of transcription and are DNA deficient. J Virol 63:18–27
12. Carlezon Jr WA et al (1997) Sensitization to morphine induced by viral-mediated gene transfer. Science 277:812–814
13. Neve RL (1997) Introduction of glutamate receptor subunit 1 into motor neurons in vivo using a recombinant herpes simplex virus alters the functional properties of AMPA receptors. Neuroscience 79:435–447
14. Bursztajn S et al (1998) Overexpression in neurons of human presenilin-1 or a presenilin-1 familial Alzheimer disease mutant does not enhance apoptosis. J Neurosci 18:9790–9799
15. Lim F et al (1996) Generation of high-titer defective HSV-1 vectors using an IE 2 deletion mutant and quantitative study of expression in cultured cortical cells. BioTechniques 20:460–470

Chapter 4

Generation and Application of Engineered Rabies Viral Vectors for Neural Circuit Research

Masahiro Yamaguchi, Moe Iwata, Riki Kamaguchi, and Fumitaka Osakada

Abstract

To understand how the brain functions, it is essential to reveal how neural circuits in the brain give rise to perception, cognition, and behavior. Neural circuits in the brain operate over a vast range of spatiotemporal and computational scales, from high-level circuits that integrate information across multiple regions of the brain to microcircuits that perform simple input/output transformations within a specialized brain structure. Because naturally occurring rabies viruses (RABV) have a unique property in their ability to spread trans-synaptically between neurons in the retrograde direction, they are extremely valuable for elucidating neural circuits. Genetic modifications of RABV restrict target cell types and control trans-synaptic viral spread, allowing cell-type-specific monosynaptic circuit tracing in the brain. The monosynaptic circuit tracing system is based on the conditional expression of EnvA/TVA-targeting systems for directing the initial infection of G-deleted RABV vectors (RABVΔG) to particular cell types and complementation of RABV glycoprotein (RABV-G) *in trans* for trans-synaptically labeling directly connected, presynaptic neurons with RABVΔG. Herein, we introduce a step-by-step protocol for generating RABVΔG to study the circuit structure and function of normal and diseased brains.

Key words G-deleted rabies viral vectors, Trans-synaptic tracing system, Neural circuit, Brain

1 Introduction

1.1 History of Neuroanatomy

One of the major goals of neuroscience is to understand how the brain works to generate perception, cognition, and behavior. Neural circuits comprise networks of specific cell types that interact via long-range projections and local connections in precise patterns to perform computations responsible for perception, cognition, and behavior [1–3]. Therefore, unveiling the structure and function of neural circuits is pivotal to understanding how a normal brain functions and how a diseased brain malfunctions. Historically,

Masahiro Yamaguchi, Moe Iwata and Riki Kamaguchi contributed equally.

Mark A. G. Eldridge and Adriana Galvan (eds.), *Vectorology for Optogenetics and Chemogenetics*,
Neuromethods, vol. 195, https://doi.org/10.1007/978-1-0716-2918-5_4,

neuroscientists have employed anatomical studies to map neuronal connections between areas using anterograde and retrograde chemical tracers, revealing macroscale-wiring diagrams of the brain [4]. Those anatomical studies combined with electrophysiological recordings and lesion studies have contributed to our understanding of a global overview of the brain circuit architecture and function, including area function and information processing pathways. However, such conventional tracers did not have sufficient resolution to reveal neuronal connectivity at the level of cell types.

1.2 Application of G-Deleted Rabies Virus (RABVΔ G) for Neural Circuit Tracing

Wild-type rabies viruses (RABV) are beneficial for studying neural circuits because the vectors spread trans-synaptically between neurons, exclusively in the retrograde direction, making it possible to map neuronal connections in the brain [4–6] (Figs. 1a and 2a). Rabies glycoprotein (RABV-G) is essential for viral packaging and trans-synaptic spread [7, 8]. Therefore, deleting the RABV-G gene from the rabies viral genome allows both trans-synaptic spread control and target-specific infection [9–11] (Figs. 1b, c and 2a). The RABVΔG cannot spread following the initial infection unless an alternative source of RABV-G is present (Fig. 1b). However, supplying the missing RABV-G *in trans* in cells infected with RABVΔG permits the generation of new viral particles and

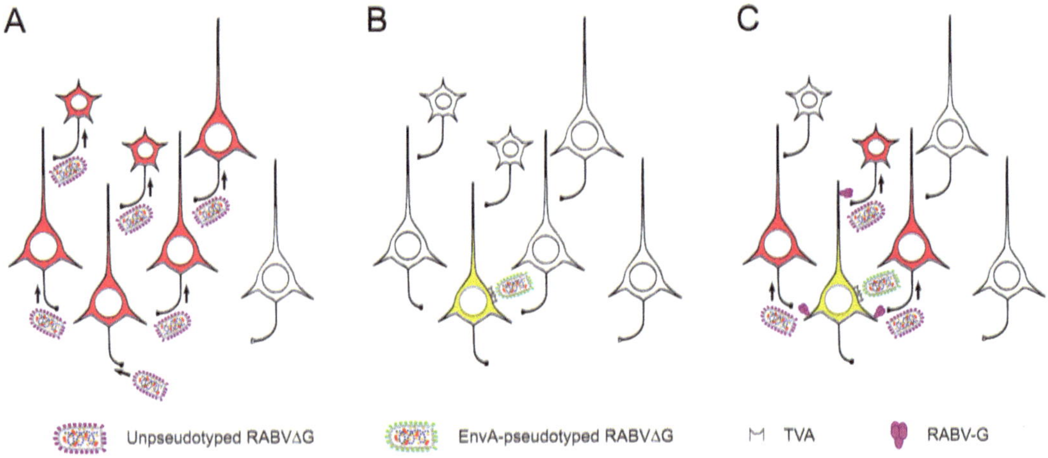

Fig. 1 Scheme of RABVΔG tracing. (**a**) Polysynaptic viral spread of wild-type rabies virus (RABV) in the nervous system. RABV particles display RABV-G on their surface, allowing them to infect presynaptic nerve terminals via receptors for RABV-G in the retrograde direction (red). RABV has broad tropism and can infect various neuronal types across mammalian species. (**b**) Deleting the RABV-G gene from the rabies viral genome eliminates trans-synaptic spread and endows the pseudotyping, allowing targeting of a particular cell subset using the EnvA/TVA system. EnvA-pseudotyped RABVΔG infects only TVA-expressing cells in mammals (yellow) because mammalian cells do not endogenously express TVA. (**c**) Monosynaptically restricted labeling of presynaptic networks of a particular cell subset with RABVΔG. When EnvA-pseudotyped RABVΔG infects neurons expressing TVA and RABV-G (yellow), transcomplementation with RABV-G allows RABVΔG to trans-synaptically spread to their presynaptic neurons (red)

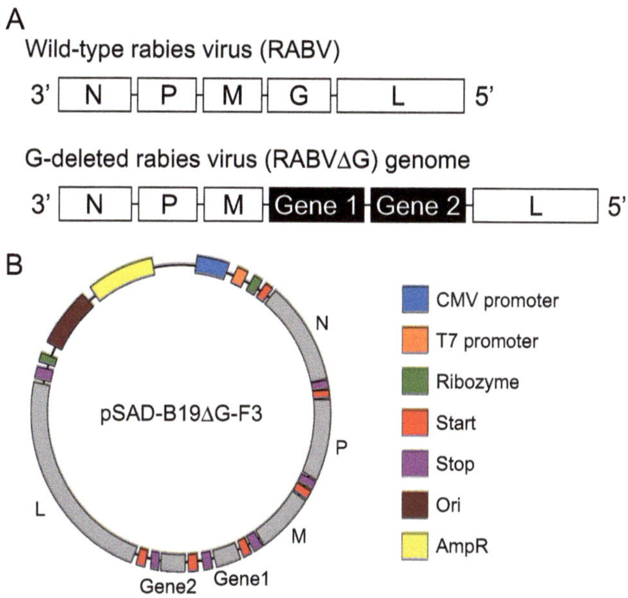

Fig. 2 Genomes of rabies viruses and RABVΔG vectors. (**a**) The genomes of wild-type and G-deleted rabies viruses. RABV is a non-segmented negative-stranded RNA virus belonging to the family Rhabdoviridae. The virion structure is a bullet-shaped particle, approximately 100–430 nm long and 45–100 nm in diameter. The rabies viral genome comprises single-stranded, linear, negative-sense RNA, 11–12 kb long. The five viral proteins N, P, M, G, and L are encoded in the genome in the order of 3′-N-P-M-G-L-5′. Additionally, two foreign genes can be inserted into the RABVΔG genomic vectors. N nucleoprotein, P phosphoprotein, M matrix protein, G glycoprotein, L large protein. (**b**) Map of the RABVΔG vector pSAD-B19ΔG-F3. One or two foreign genes can be inserted into the RABVΔG genomic vectors. Red and purple boxes indicate transcription start and stop sequences, respectively. Further information on pSAD-B19ΔG-F3 is provided in a previous publication by Osakada et al. [11]

subsequent infection of monosynaptic input cells directly connected to the initially infected neurons (Fig. 1c). RABVΔG can be pseudotyped with non-native envelope proteins to alter its tropism. For selectively targeting desired cell types in mammals, RABVΔG can be pseudotyped with an envelope protein from the avian sarcoma leukosis viruses, EnvA, which specifically binds to the avian-specific receptor TVA [9, 10]. Targeted expression of TVA in desired cell types allows specific infection with EnvA-pseudotyped RABVΔG (EnvA-RABVΔG) (Fig. 1b). Using a combination of RABV-G and TVA, complementation of RABV-G *in trans* in TVA-expressing neurons enables the trans-synaptic spread of RABVΔG to directly connect neurons presynaptic to the TVA- and RABV-G-expressing neurons (Fig. 1c). For multiplex

RABVΔG tracing, RABVΔG pseudotyped with oEnvA, oEnvB, or oEnvE, engineered envelope proteins derived from the avian sarcoma leukosis viruses, can be used for targeting discrete cell populations expressing oTVA, oTVB, or oTVE in the mammalian brain, respectively [12]. Using the RABVΔG tracing along with molecular tools, such as calcium/voltage sensors, neurotransmitter/neuromodulator sensors, optogenetic channels, or chemogenetic receptors, makes it possible to monitor and manipulate the activities of connectionally defined neuronal populations and relate neuronal connectivity to function [11, 13–16].

1.3 Overview and Rationale of Protocols

The development of methods for constructing genetically altered negative-strand RNA viruses de novo from cDNA has allowed the production of recombinant RABVΔG [7, 10–12, 17] (Fig. 2b). The SAD-B19 strain of RABV has been widely used as recombinant RABVΔG for neuroscience research [9–11]. For virus recovery from DNA plasmids (Subheading 3.1), the full-length positive-sense RNA of the viral genome is transcribed from a RABV genomic plasmid by T7 polymerase [10, 11] (Fig. 2b and **step 1** in Fig. 3). The 5′ and 3′ ends of the viral genome are processed by ribozymes. Deleting the G gene from the viral genome causes the inability to spread between neurons in the nervous system[6]. For virus production in vitro (Subheadings 3.1 and 3.2), RABV-G-expressing cells, namely B7GG cells, are used to propagate and generate RABVΔG particles coated with RABV-G by trans-complementation (**steps 2** and **3** in Fig. 3). RABVΔG can be pseudotyped to produce particles with non-native envelope proteins and to alter its tropism. RABVΔG pseudotyped with oEnvA, oEnvB, or oEnvE specifically infects cells expressing oTVA, oTVB, or oTVE, respectively [12]. For in vitro generation of pseudotyped RABVΔG (Subheading 3.3), BHK cells expressing EnvA are used to grow RABVΔG coated with EnvA by replacement of the envelope (**step 4** in Fig. 3). Herein we describe a protocol for generating EnvA-RABVΔG in BHK-EnvA cells. The protocol is applicable for pseudotyping with oEnvB or oEnvE [12]. For in vivo injection (Subheading 3.4), unpseudotyped RABVΔG and EnvA-RABVΔG are concentrated by ultracentrifugation (**step 5** in Fig. 3). Notably, the efficiency and purity of pseudotyped RABVΔG rely on the skills of experimenters. The viral titer and purity need to be determined for every viral preparation (Subheading 3.5 and **step 6** in Fig. 3). Measurement of infectious titers on cells, but not genomic titers by qPCR, is required because qPCR cannot distinguish pseudotyped RABVΔG from unpseudotyped RABVΔG that might be present in contaminated preparations. The preparation of pseudotyped RABVΔG contaminated with unpseudotyped RABVΔG will lead to a non-selective infection and should be discarded.

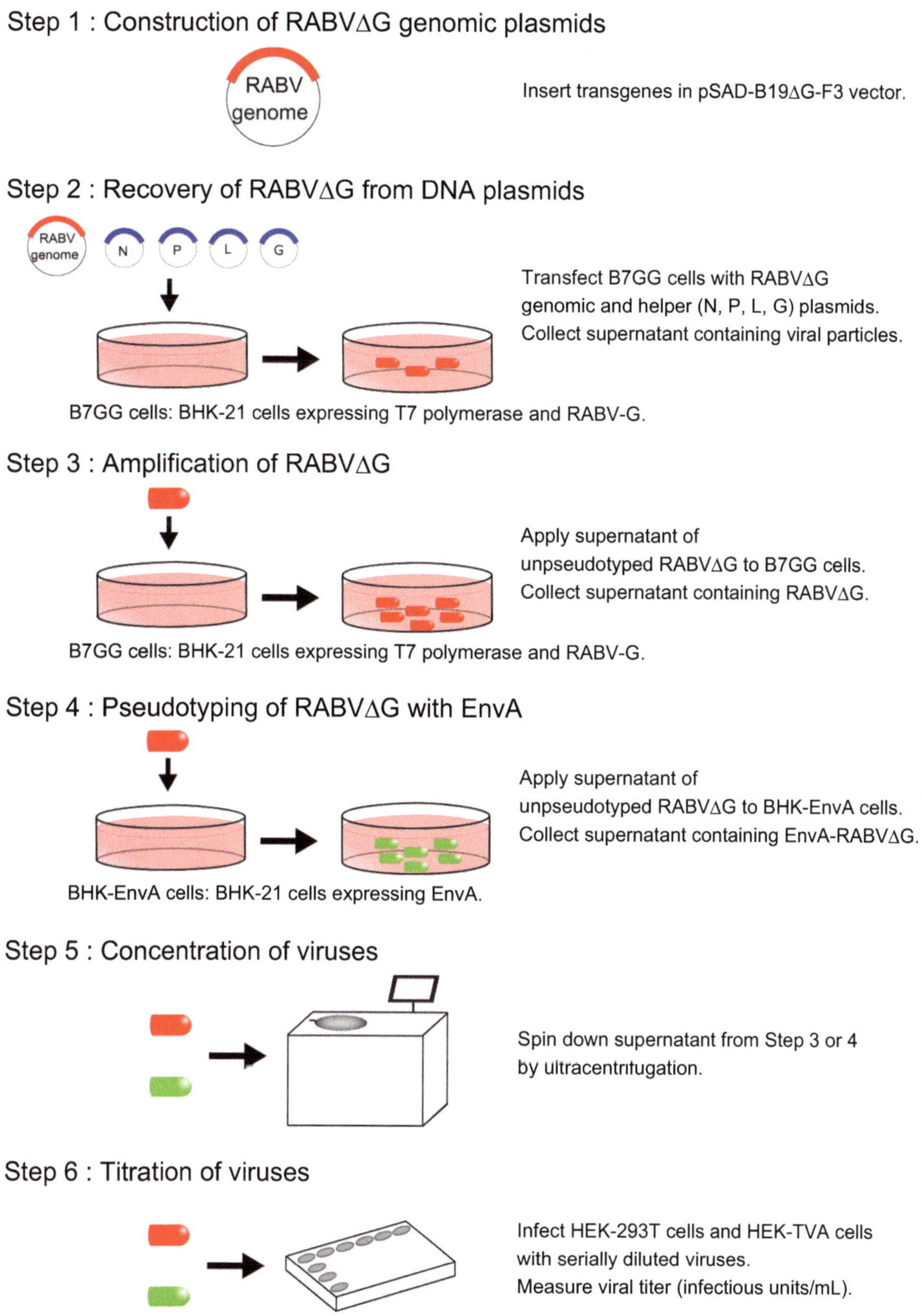

Fig. 3 Overview of production procedures. The generation of RABVΔG consists of six steps as follows: the construction of rabies viral genome, recovery of RABVΔG from plasmids (Subheading 3.1), amplification of RABVΔG in B7GG cells (Subheading 3.2), pseudotyping of RABVΔG in BHK-EnvA cells (Subheading 3.3), the concentration of RABVΔG (Subheading 3.4), and titration of viruses (Subheading 3.5)

2 Materials

Cells

B7GG cells: BHK-21 cells expressing T7 RNA polymerase, RABV-G, and histone-tagged GFP.

BHK-EnvA cells: BHK-21 cells expressing oEnvA, a chimeric envelope protein consisting of extracellular and transmembrane domains of EnvA fused to the cytoplasmic domain of RABV-G.

HEK293T cells: ATCC, cat. no. CRL-11268. These cells are used to titer unpseudotyped RABVΔG.

HEK293T-TVA800 cells: HEK293T-TVA800 cells are derived from HEK293T cells and express TVA. These cells are used to titer EnvA-RABVΔG.

Plasmids

pSADΔG-F3 (Addgene, ID. 32634).

pcDNA-SADB19N (Addgene, ID. 32630).

pcDNA-SADB19P (Addgene, ID. 32631).

pcDNA-SADB19L (Addgene, ID. 32632).

pcDNA-SADB19G (Addgene, ID. 32633).

Reagents

Fetal bovine serum (FBS; Sigma, cat. no. F7524): heat inactivation was carried out at 56 °C for 30 min, and FBS was divided into aliquots and stored at −20 °C for up to 1 year.

IMPORTANT: We recommend testing several lots beforehand and using the best lot. Generate viruses with several lots of FBS, compare their viral titers between lots, and select the best lot with the highest viral titer.

DMEM (Wako, cat. no. 043-30085).

Dulbecco's PBS, no calcium, no magnesium (PBS; Wako, cat. no. 045-29795).

0.25% Trypsin-EDTA (Wako, cat. no. 201-16945): divided into aliquots and stored at −20 °C for up to 1 year.

Hank's balanced salt solution, no calcium, no magnesium, without phenol red (HBSS, Wako, cat. no. 085-09355)

Lipofectamine 2000 (Thermo, cat. no. 11668-019)

20% Sucrose/HBSS: Mix 10 g of sucrose with 5 mL of 10× HBSS, adjust the total volume to 50 mL with distilled water and sterilize the solution by passing it through a 0.22 μm filter. The solution can be stored at 4 °C for several months.

Supplies

Plastic disposable pipettes (5, 10, 25, and 50 mL) (Thermo Scientific Nunc, cat. nos. 170355N, 170356N, 170357N, and 170358N, respectively).

Tubes (15 and 50 mL) (Thermo Scientific Nunc, cat. nos. 339650 and 339652, respectively).

Cryo vial (5 mL outer screw) (Simport, cat. no. T308-5A).

Culture dish (100 and 150 mm) (Thermo, cat. nos. 150466 and 150468).

24-well culture plate (Thermo Scientific, cat. no. 142475).

Vacuum filtration system (Steri filter unit with cap, Durapore low protein binding, 0.45 μm pore size, 250 mL) (Millipore, cat. no. S2HVU02RE).

Centrifuge tube, large (Beckman, cat. no. 326823).

Centrifuge tube, small (Beckman, cat. no. 326819).

Equipment

Biological safety cabinet (Panasonic, Thermo, or equivalent).

Incubators (37 °C and 5% CO_2 and 35 °C and 3% CO_2) for cell culture (Panasonic, Thermo, or equivalent).

Water bath.

Fluorescence microscope for cell culture (Zeiss, Nikon, or equivalent).

Centrifuge (Tomy, Kubota, or equivalent).

Ultracentrifuge (Beckman, cat. no. Optima XE-90).

Rotor with buckets (Beckman, cat. no. SW28).

Rotor with buckets (Beckman, cat. no. SW55).

Culture Medium

10% FBS-containing DMEM (DMEM/10% FBS): add 50 mL of FBS to 450 mL of DMEM and store at 4 °C for up to 3 weeks.

1% FBS-containing DMEM (DMEM/1% FBS): add 5 mL of FBS to 495 mL of DMEM and store at 4 °C for up to 3 weeks.

3 Step-by-Step Protocols for RABVΔG Production

3.1 Recovery of RABVΔG from cDNA

(1) Day 0 (seeding of B7GG cells)

1. Prepare a confluent culture of B7GG cells in a 100 mm dish (typically in the evening).

IMPORTANT: Use a sterile technique for cell culture. Replace pipettes every time to avoid cross-contamination. Do not allow the cells to remain without media for more than a few minutes. Carefully label all dishes, tubes, and bottles. To avoid cross-contamination between viruses, do not handle multiple viral preps at a time.

IMPORTANT: Create frozen stocks of B7GG cells at low passage.

2. Aspirate the medium from the dish using a Pasteur pipette and wash the dish with 10 mL of PBS prewarmed to 37 °C.
3. Add 5 mL of prewarmed 0.25% trypsin-EDTA and incubate the dish at 37 °C for 3 min.
4. Confirm that the cells are detached from the dish by tapping the dish; then, add 5 mL of DMEM, dissociate the cells by pipetting, and collect the cell suspension in a 50 mL conical tube.
5. To collect the remaining cells from the dish, apply an additional 5 mL of DMEM/10% FBS, and transfer the cell suspension into the 50 mL tube.
6. Centrifuge the 50 mL tube for 3 min at 174 × *g* (1000 rpm).
7. Aspirate the supernatant carefully, apply 10 mL of prewarmed DMEM/10% FBS, and dissociate the suspension into single cells by pipetting.
8. Count the cell number using a hemocytometer and calculate the cell density of the suspension. Then, prepare cell suspension at 3.3×10^5 cells/mL using DMEM/10% FBS.
9. Apply 10 mL of the cell suspension onto a new 100 mm dish. Then, incubate the dish at 5% CO_2 and 37 °C. This day is defined as day 0 of the Subheading 3.1.

   IMPORTANT: Low confluency at the time of transfection results in poor cell survival and low transfection efficiency. Higher confluency (approximately 80%) is desirable for viral recovery.

(2) Day 1 (transfection; preferably performed in the morning)

1. Aspirate the medium from the dish of B7GG cells seeded on day 0, add 6 mL of prewarmed FBS-free DMEM, and return the dish to the incubator. Prepare two 50 mL tubes (tubes A and B) and add 1.25 mL of prewarmed OPTI-MEM to each of them.
2. Prepare the transfection mixes as follows: add 30 μg of pSADΔG-DsRed, 15 μg of pcDNA-SADB19N, 7.5 μg of pcDNA-SAD-B19P, 7.5 μg of pcDNA-SADB19L, and

5 μg of pcDNA-SADB19G to 1.25 mL of OPTI-MEM of tube A. Add 112.5 μL of lipofectamine 2000 to 1.25 mL of OPTI-MEM of tube B.

3. Vortex each tube four times for 3 s and let both tubes stand at room temperature for 5 min.
4. Add the contents of tube A to tube B; then, vortex tube A + B five times for 3 s, and let tube A + B stand at room temperature for 15 min.
5. Add 2.5 mL of the transfection mix (tube A + B) evenly in a dropwise manner onto the 100 mm dish of B7GG cells prepared at **step 1**. Incubate the dish at 37 °C in 5% CO_2.
6. Six hours later, aspirate the medium, wash the dish with 10 mL of prewarmed PBS, and add 10 mL of DMEM/ 10% FBS. Then, incubate the dish at 37 °C in 5% CO_2.

   IMPORTANT: Examine cell conditions and fluorescent protein expression under a fluorescence microscope daily to assess infection efficiency and viral spread.

(3) Day 2

1. Remove the medium from the 100 mm dish of B7GG cells transfected on day 1 and add 5 mL of PBS prewarmed at 37 °C.

   IMPORTANT: Do not aspirate the medium containing viruses because there is a possibility of aerosol spread.
2. Remove the PBS and add 2 mL of prewarmed 0.25% trypsin-EDTA. Incubate the dish at 37 °C for 3 min.
3. Detach the cells from the dish by tapping, add 1 mL of DMEM/10% FBS prewarmed at 37 °C, dissociate the cells into single cells by pipetting them, and collect the cell suspension in a 50 mL conical tube.
4. Add 2 ml of DMEM/10%FBS to collect the remaining cells and transfer the media to the 50 mL conical tube.
5. Centrifuge the tube at 174 × g (1000 rpm) for 3 min.
6. Remove the supernatant, apply 1 mL of DMEM/10% FBS, and resuspend the pellet by pipetting.
7. Add 23 mL of DMEM/10% FBS to the tube and plate the 24 mL of cell suspension in a new 150 mm dish. Incubate the dish at 5% CO_2 and 37 °C after gently shaking the dish to make the cell distribution uniform.

(4) Day 3 (changing the medium)

1. Remove the supernatant from the 150 mm dish of B7GG cells, add 24 mL of new DMEM/1% FBS prewarmed at 37 °C, and incubate at 35 °C in 3% CO_2 (Fig. 4a).

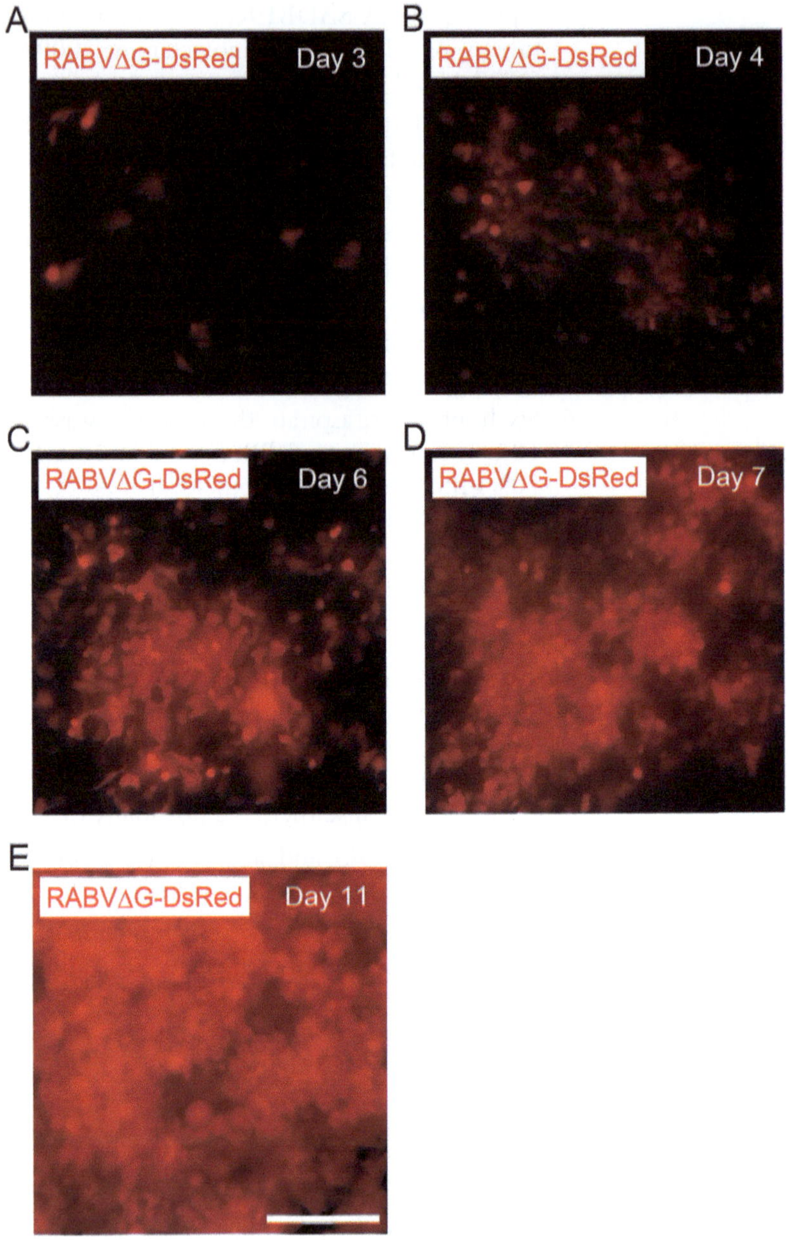

Fig. 4 Recovery of RABVΔG-DsRed from DNA plasmid. (**a**–**d**) Photomicrographs represent typical examples of recovery of RABVΔG-DsRed in B7GG cells on day 3 (**a**), day 4 (**b**), day 6 (**c**), day 7 (**d**), and day 11 (**e**). Scale bar: 200 μm

(5) Day 4

1. Examine cell conditions and fluorescent protein expression under a fluorescence microscope (Fig. 4b).

(6) Day 5 (addition of medium)

1. Add 4 mL of prewarmed DMEM/1% FBS into the dish (totally 28 mL of media in the dish) and incubate at 35 °C in 3% CO_2 for 2 days.

(7) Day 6

1. Examine cell conditions and fluorescent protein expression under a fluorescence microscope (Fig. 4c).

(8) Day 7 (the first collection of virus supernatant and change medium)

1. Collect the supernatant from the 150 mm dish in a 50 mL tube and add 24 mL of fresh prewarmed DMEM/1% FBS to the 150 mm dish. Incubate the dish at 35 °C in 3% CO_2 (Fig. 4d).
2. Centrifuge the collected virus supernatant at 694 × *g* (2000 rpm) for 10 min.
3. After centrifugation, collect about 28 mL of the supernatant carefully into a new 50 mL tube. Store the harvested virus supernatant at 4 °C.

(9) Day 9 (addition of medium)

1. Add 4 mL of fresh DMEM/1% FBS to the 150 mm dish and incubate at 35 °C in 3% CO_2.

(10) Day 11 (the second collection of virus supernatant)

1. Collect the second supernatant from the 150 mm dish in a 50 mL tube and centrifuge at 694 × *g* (2000 rpm) for 10 min (Fig. 4e).
2. After centrifugation, carefully collect the supernatant into a new 50 mL tube. Store the harvested supernatant at 4 °C.

3.2 Amplification of RABVΔG

<Amplification-1>

(1) Day 0 (seeding of B7GG cells)

1. Seed B7GG cells onto three 150 mm dishes at 8.0×10^2 cells/mm^2. This day is defined as day 0 of 3.2. Amplification-1 section.

(2) Day 1 (viral infection of B7GG cells)

1. Prepare three 150 mm dishes of 60–70% confluent B7GG cells (typically in the morning) on day 1. Aspirate the medium from the dish and add 3 mL of fresh DMEM/10% FBS prewarmed at 37 °C.

2. Mix the first and second supernatant (collected at Subheading 3.1, Recovery steps) and split the mixed supernatant into three parts; apply ~16 mL of the mixture to each 150 mm dish.
3. Maintain the dishes under a fully humidified atmosphere at 35 °C in 3% CO_2 for 6 h.
4. Six hours later (typically in the evening), remove culture media from the three dishes using a pipette and add 24 mL of prewarmed DMEM/1% FBS to each dish. Incubate the dishes at 35 °C in 3% CO_2.

(3) Day 2

1. Examine cell conditions and fluorescent protein expression under a fluorescence microscope to assess the initial infection efficiency (Fig. 5a).

(4) Day 3 (addition of medium)

1. Add 4 mL of fresh DMEM/1% FBS to each of the three 150 mm dishes of B7GG cells infected on day 1 (Fig. 5b). Incubate the dishes at 35 °C in 3% CO_2.

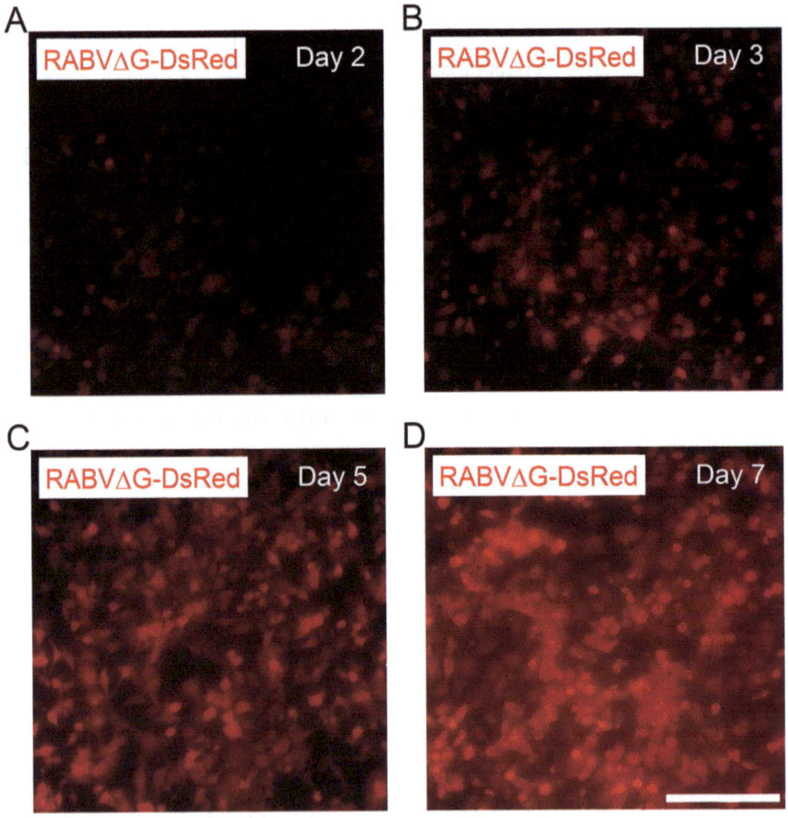

Fig. 5 Amplification of RABVΔG-DsRed. (**a**–**d**) Photomicrographs represent typical examples of amplification of RABVΔG-DsRed in B7GG cells on day 2 (**a**), day 3 (**b**), day 5 (**c**), and day 7 (**d**). Scale bar: 200 μm

(5) Day 5 (the first collection of virus supernatant and addition of medium)

1. Collect the supernatant from the three 150 mm dishes into two 50 mL tubes.
2. Add 24 mL of fresh DMEM/1% FBS prewarmed at 37 °C to each of the three 150 mm dishes (Fig. 5c). Incubate the three dishes at 35 °C in 3% CO_2.
3. Centrifuge the collected virus supernatant at 694 × *g* (2000 rpm) for 10 min.
4. After centrifugation, carefully collect supernatant into two new 50 mL tubes and discard pellets. Store the harvested supernatant at 4 °C.

(6) Day 7 (the second collection of virus supernatant)

1. Collect supernatant from the three 150 mm dishes of B7GG cells in two 50 mL tubes (Fig. 5d).
2. Centrifuge the collected supernatant at 694 × *g* (2000 rpm) for 10 min.

   IMPORTANT: Almost all cells must express DsRed for reaching an adequate titer.
3. After centrifugation, carefully collect the supernatant into two new 50 mL tubes and discard pellets. Store the harvested supernatant at 4 °C.

<Amplification-2>

(1) Day 0 (seeding of B7GG cells)

1. Seed B7GG cells at 8.0×10^2 cells/mm^2 onto ten 150 mm dishes. This day is defined as day 0 of 3.2. Amplification-2 section.

(2) Day 1 (viral infection of B7GG cells)

1. Prepare ten 150 mm dishes of 60–70% confluent B7GG cells (typically in the morning) on day 0. Aspirate the medium from the dishes and add 3 mL of fresh DMEM/10% FBS prewarmed at 37 °C to each dish.
2. Mix the first and second supernatant (collected at Subheading 3.2, Amplification-1 step), split the mixed supernatant into three parts, and apply ~16 mL of the mixture to each 150 mm dish. Maintain the dishes under a humidified atmosphere at 35 °C in 3% CO_2 for 6 h.
3. Six hours later, remove the medium from the ten dishes and add 24 mL of prewarmed DMEM/1% FBS to each dish. Incubate the dishes at 35 °C in 3% CO_2 for 3 days.

(3) Day 2

1. Examine cell conditions and fluorescent protein expression under a fluorescence microscope to assess the initial infection efficiency.

(4) Day 4 (addition of medium)

1. Add 4 mL of fresh DMEM/1% FBS to the ten 150 mm dishes of B7GG cells infected on day 1. Incubate the dishes at 35 °C in 3% CO_2.

(5) Day 6 (collection of the virus supernatant)

1. Collect the ten 150 mm dishes of B7GG cells, which were added to medium on day 4, and split them into five 50 mL tubes.

   IMPORTANT: Almost all cells must express DsRed to reach an adequate high titer.

2. Centrifuge the collected virus supernatant at 694 × g (2000 rpm) for 10 min.

3. After centrifugation, collect the supernatant carefully and filter the supernatant using a 0.45 μm filter.

4. Move to Subheading 3.3 for pseudotyping RABVΔG with EnvA or to Subheading 3.4 for concentrating supernatant containing unpseudotyped RABVΔG.

   IMPORTANT: Make frozen stocks of the viral supernatant (4.5 mL/cryotube) and record their amplification number; also, store at −80 °C. Amplify RABVΔG using the frozen stocks of viral supernatant by applying 4.5 mL of the frozen stock to a 150 mm dish of B7GG cells. RABVΔG is amplified in B7GG cells so that viral recovery from the plasmid DNA every time is not required. It is ideal for creating low-passage frozen stocks to avoid mutation accumulation in the rabies viral genome.

3.3 Pseudotyping of RABVΔG with EnvA

(1) Day 0 (seeding of BHK-EnvA cells)

1. Prepare three confluent 150 mm dishes of BHK-EnvA cells to prepare ten 150 mm dishes of BHK-EnvA cells at 8.0×10^2 cells/mm^2 that will be at a 60% confluence the next day (typically in the morning). This day is defined as day 0 of Subheading 3.3.

(2) Day 1 (viral infection of BHK-EnvA cells)

1. Prepare ten 150 mm dishes of 60–70% confluent BHK-EnvA cells (typically in the morning) on day 1 and aspirate the medium from the ten dishes of BHK-EnvA cells with a Pasteur pipette. Then, add 3 mL of fresh DMEM/10% FBS prewarmed at 37 °C.

2. Add 20 mL of the unpseudotyped RABVΔG prewarmed at 37 °C (collected at Subheading 3.2, Amplification-2 steps) into each dish. Maintain the dishes under a humidified atmosphere at 35 °C in 3% CO_2 for 6 h.

   IMPORTANT: The recommended amount for initial infection efficiency is an MOI of 0.9–1.0.
3. Six hours later, remove the medium from the five dishes.

   IMPORTANT: To protect the cells from unwanted stress, divide eight dishes into two parts and change the media from five dishes at a time. Repeat the procedure for the remaining five dishes later.
4. Add 20 mL of PBS prewarmed at 37 °C into each of the five dishes and wash the dishes to remove unpseudotyped virus added to the dishes.

   IMPORTANT: Washing with PBS, trypsinization, and subsequent centrifugation to remove unnecessary unpseudotyped RABVΔG viral particles are critical steps to obtain high-purity EnvA-RABVΔG preparations.
5. Add 20 mL of PBS prewarmed at 37 °C into each dish and wash the dishes to remove unpseudotyped virus added to the dishes.
6. Remove PBS, add 4.5 mL of 0.25% trypsin-EDTA prewarmed at 37 °C, and incubate the dishes at 37 °C for 3 min.
7. Confirm that the cells are detached from the dish by tapping the dish; then, add 4 mL of DMEM/10% FBS, dissociate the cells by pipetting, and collect the cell suspension in a 50 mL conical tube (about 40 mL).
8. To collect the remaining cells on the dishes, apply 7 mL of the medium to the first dish; transfer the cell suspension to the second, third, fourth, and fifth dish. Then, transfer the cell suspension to the 50 mL tube (about 50 mL of cell suspension in total).
9. Centrifuge the 50 mL tube for 3 min at 174 × g (1000 rpm).
10. Remove the supernatant of the 50 mL tube, resuspend with 10 mL of DMEM/10% FBS, and centrifuge the 50 mL tube for 3 min at 174 × g (1000 rpm).
11. During centrifugation, apply 18 mL of fresh DMEM/10% FBS to five new 150 mm dishes.
12. Remove the supernatant of the 50 mL tube, resuspend with 10 mL of DMEM/10% FBS, and split the cell suspension into five parts to add 62 mL to the five 150 mm dishes (20 mL of media per dish in total). Maintain the five dishes under a humidified atmosphere at 37 °C in 5% CO_2.

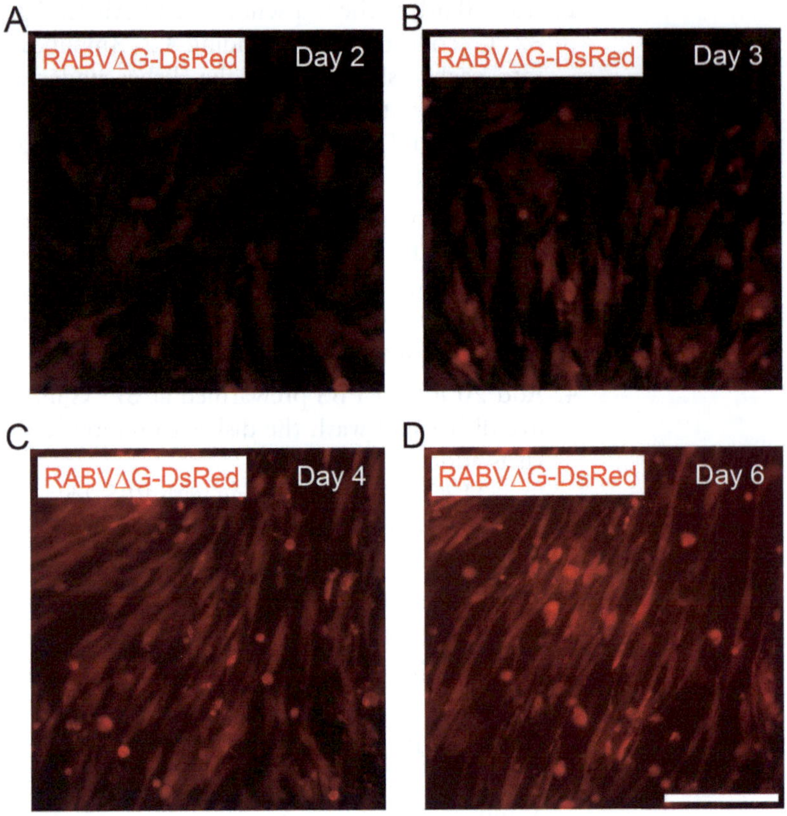

Fig. 6 Pseudotyping of RABVΔG-DsRed. (**a**–**d**) Photomicrographs represent typical examples of pseudotyping of RABVΔG-DsRed in BHK-EnvA cells on day 2 (**a**), day 3 (**b**), day 4 (**c**), and day 6 (**d**). Scale bar: 200 μm

13. Perform Procedures 3–12 in Subheading 3.3 (2) for the remaining five dishes to finally obtain ten 150 mm dishes of virus-infected BHK-EnvA cells in total.

(3) Day 2 (changing the medium)

1. Remove the medium from five dishes of virus-infected BHK-EnvA cells. Add 24 mL of fresh DMEM/10% FBS prewarmed at 37 °C (Fig. 6a). Maintain the dishes under a humidified atmosphere at 35 °C in 3% CO_2.

   IMPORTANT: Do not change the media from all 10 dishes at a time to avoid drying of cultured cells on dishes. Instead, change the media from approximately five dishes at a time.

2. Repeat Procedure 1 for the remaining five dishes of virus-infected BHK-EnvA cells.

(4) Day 3

1. Examine cell conditions and fluorescent protein expression under a fluorescence microscope to assess the initial infection efficiency (Fig. 6b).

(5) Day 4 (addition of medium)

1. Add 4 mL of fresh DMEM/10% FBS to the ten 150 mm dishes of BHK-EnvA cells (Fig. 6c). Incubate at 35 °C in 3% CO_2.

(6) Day 6 (collection of the virus supernatant)

1. Collect the supernatant from the ten 150 mm dishes of BHK-EnvA cells and split them into five 50 mL tubes (Fig. 6d).
2. Centrifuge the collected virus supernatant at 694 × *g* (2000 rpm) for 10 min.
3. After centrifugation, collect the supernatant carefully and filter the supernatant using a 0.45 μm filter.
4. Move to Subheading 3.4.

3.4 Concentration of the Viral Supernatant

1. For the first centrifugation, use the Beckman SW28 rotor.
2. Place six centrifuge tubes in six buckets of the SW28 rotor, apply 37 mL of the supernatant to each centrifuge tube, and cover the six buckets loosely with caps.
3. Weigh each bucket with the tube and cap on and balance them to approximately 0.01 g by adding the viral supernatant with a P200 micropipette in a biosafety cabinet.

   IMPORTANT: Be sure to balance all buckets; failure to balance the buckets will lead to equipment damage.

   IMPORTANT: Make frozen stocks of the remaining viral supernatant (4.5 mL/cryotube) stored at −80 °C and record their amplification number. Next, amplify RABVΔG using the frozen stocks of the viral supernatant. RABVΔG is amplified in B7GG cells to avoid viral recovery from the plasmid DNA every time. It is ideal for creating low-passage frozen stocks to avoid mutation accumulation in the rabies viral genome.
4. Place the buckets on the rotor and load the rotor on the ultracentrifuge; then, spin at 70,000 × *g* (19,400 rpm) for 2 h at 4 °C.
5. After centrifugation, move the buckets back to the biosafety cabinet, open the caps, and pull out the tubes. Remove the supernatant gently by decantation.
6. Resuspend each pellet in 300 μL of HBSS and collect all the virus suspensions into a tube.
7. For the second centrifugation, use the Beckman SW55 rotor.
8. Insert two round-bottomed tubes into the tube buckets.
9. Place 2.5 mL of 20% sucrose/HBSS in the round-bottomed tube of the bucket and apply the virus suspension gently onto the sucrose cushion. Rinse the tubes with 200 μL of HBSS to recover any remaining virus and transfer the suspension to the sucrose cushion again.

10. Weigh each bucket with the tube and cap on. Prepare a proper counterbalance containing HBSS and balance them to approximately 0.01 g by adjusting with HBSS using a P200 micropipette in a biosafety cabinet.

    IMPORTANT: Be sure to balance the buckets; failure to balance the buckets will lead to equipment damage.

11. Place the buckets on the rotor and load the rotor on the ultracentrifuge; then, spin the suspension at 50,000 × *g* (21,000 rpm) for 2 h at 4 °C.
12. After centrifugation, move the centrifuged tubes back to the hood and gently remove the supernatant by decantation.
13. Resuspend the viral pellets in 100 μL of HBSS, wrap the top of the tubes in parafilm, and store for 30–60 min at 4 °C.
14. Gently pipette the viral suspension, aliquot the virus (i.e., 4 μL/tube), and freeze them at −80 °C for future use.

3.5 Titration of Virus

(1) Day 0 (seeding of cells)

1. Prepare cell suspension of HEK293T and HEK293T-TVA cells at 3.0×10^5 cells/mL. Apply 500 μL of the cell suspension to each well of a 24-well plate. Incubate the plate at 37 °C in 5% CO_2.

   In the case of EnvA-pseudotyped RABVΔG, prepare both HEK293T and HEK293T-TVA cells to check their viral titer and purity.

(2) Day 1 (infecting HEK293T or HEK293T-TVA cells with serially diluted viral concentrates)

1. Make a tenfold serial dilution of the viral concentration from a dilution of 10^{-3}–10^{-9} in duplicates or triplicates.

   IMPORTANT: Change the tips before every dilution.

2. Add 250 μL of each viral dilution to wells gently.
3. Incubate the plate at 35 °C in 3% CO_2 for 2 days.

(3) Day 3 (counting the number of infected cells)

1. Observe fluorescent protein-expressing cells using a fluorescence microscope.

   IMPORTANT: If EnvA-pseudotyped RABVΔG infects HEK293T cells, the EnvA-pseudotyped RABVΔG prep is suspected to be contaminated with unpseudotyped RABVΔG and should not be used for circuit mapping.

2. Find wells containing the lowest number of fluorescent protein-expressing cells in the series of wells and count the number of fluorescent protein-expressing cells in the well. Calculate the average from duplicates or triplicates.

3. Calculate the infectious titer using the following formula.

Infectious units/mL = (no.of cells) × 4 × (dilution factor).

For example, if the number of DsRed$^+$ cells is 12, 27, and 30 in the 10^{-8} wells, the average of DsRed$^+$ cells is 23 cells per well, and its dilution factor is 10^8. The titer of this RABVΔG-DsRed prep becomes $23 \times 4 \times 10^8 = 9.2 \times 10^9$ (infectious units/mL).

4 Conclusion

The RABVΔG production protocol that we introduced here typically yields RABVΔG-DsRed at an infectious titer of 1.0×10^9–1.5×10^{10} and EnvA-RABVΔG-DxRed at a titer of 5.0×10^7–2.0×10^9 [10–12, 18]. EnvA-pseudotyped RABVΔG can target particular neuronal subsets, such as genetically defined cell types, projection-based cell types, or even a single neuron for circuit tracing [9, 12, 13, 19–21]. Trans-complementation with RABV-G allows trans-synaptic labeling of presynaptic neurons of the target cells with RABVΔG in a brain-wide manner. Notably, genetically identified cell types can be reproducibly targeted by different research groups using cell-type-specific Cre/Flp/tTA mouse lines to identify their direct monosynaptic inputs [20]. Such RABVΔG-based methods for elucidating neuronal connectivity across the whole brain at the resolution of cell types or single neurons will advance our understanding of neural circuit organization. The conjunction of such methods with imaging and electrophysiology in living animals can reveal direct correlations between circuit structure and function and allow perturbation of connectionally defined neural populations by optogenetics and chemogenetics [11, 13, 14].

Neurons are diverse and can be classified into many subtypes, each of which plays a distinct role in brain function. Although cell morphology and electrophysiological properties have been used for cell type classification criteria, recent advances in single-cell RNA sequencing technology have proposed transcriptome-based taxonomy of cell types [22, 23]. For cell-type-specific circuit mapping with RABVΔG, Cre-expressing mouse lines are the most powerful way to reproducibly access genetically defined cell types to restrict RABV-G and TVA expression [20]. Many Cre driver lines are available for targeting neurotransmitter-defined neurons [24–27] (ChAT-Cre, DR1-Cre, DR2-Cre, Sert-Cre, DAT-Cre, VGAT-Cre, and VGLUT2-Cre), inhibitory neuron subtypes [28] (parvalbumin-Cre, somatostatin-Cre, and vasoactive intestinal peptide-Cre), cortical layer-specific neurons [29] (Sepw1-Cre, Nr5a1-Cre, Tlx3-Cre, and Ntsr1-Cre), and hippocampal subdomain-specific neurons [30] (Dox10-Cre and Map3k15-

Cre). There are many strategies to deliver the RABV-G and TVA genes in Cre-expressing cells of these transgenic mouse lines. When designing experiments, we should consider the amount of RABV-G and the leak expression of TVA in starter cells to guarantee the efficiency and specificity of rabies viral tracing. Higher expression levels of RABV-G in starter cells can label more presynaptic cells. Even undetectable, low levels of TVA can lead to viral infection and a reduction in the specificity of viral tracing because of the high sensitivity of TVA. Using helper viral vectors, particularly AAVs, is the most convenient way to deliver the RABV-G and TVA genes for trans-synaptic labeling and a marker gene for labeling starter cells. However, AAVs have limitations in expressing transgenes in number and size. The helper AAVs provide many strategies in the trans-synaptic labeling of RABVΔG to detect starter cells and their presynaptic cells. One needs to choose an optimal strategy depending on the goal of the experiments. Here we introduced five major strategies focusing on their pros and cons (Table 1).

Table 1
Advantages and limitations of helper AAVs for transsynaptic RABVG tracing. Each strategy has its advantages and limitations in detecting and quantifying starter cells and labeling the efficiency of presynaptic cells

Strategy	Helper AAVs	Quantitativity of starter cells	Labeling efficiency of presynaptic cells	References
1	AAV-FLEX-XFP-2A-TVA-2A-RABV-G or AAV-FLEX-TVA-XFP-2A-RABV-G	High	Low	Wall et al. [20, 25]; Suzuki et al. [12]
2	AAV-FLEX-TVA-XFP + AAV-FLEX-RABV-G	Low	High	Miyamichi et al. [32]; Ogawa et al. [26]
3	AAV-FLEX-TVA-XFP + AAV-FLEX-H2B-XFP-2A-RABV-G	Medium	Medium	Faget et al. [27]
4	AAV-FLEX-TVA-2A-XFP-2A-RABV-G + AAV-FLEX-RABV-G or AAV-FLEX-TVA-XFP-2A-RABV-G + AAV-FLEX-RABV-G	High	High	Szőnyi et al. [33]
5	Use mouse lines expressing TVA, RABV-G, and XFP from a gene locus of the mouse genome	High	Depend on a marker gene locus	

1. Introduce all RABV-G, TVA, and marker genes (e.g., GFP) from a single helper viral vector [12, 19, 24]. This approach is the most straightforward. It allows a clear detection of GFP-positive starter cells expressing RABV-G and TVA with low efficiency in labeling their presynaptic cells. Furthermore, this clear labeling of starter cells as RABVΔG-infected GFP+ cells is suitable for quantification through trans-synaptic labeling.
2. Use two helper viral vectors; one helper viral vector introduces RABV-G alone in starter cells under strong promoters, such as the CAG promoter, making it possible to maximize trans-synaptic labeling efficiency. The other introduces both TVA and a marker gene (e.g., mCherry) [31, 32]. This approach using two helper viral vectors provides the highest efficiency in the trans-synaptic labeling but not 100% co-expression of RABV-G and TVA. A fraction of mCherry-positive cells become RABVΔG-infected RABV-G-expressing cells which are starter cells. Because RABV-G-expressing cells do not have any reporter, detection and quantification of starter cells by this approach are ambiguous.
3. Use two helper viral vectors. Introducing RABV-G and reporter fluorescence (e.g., GFP) under a strong promoter, such as the CAG promoter, in a helper viral vector to starter cells makes it possible to reliably label starter cells with reporter fluorescence and express a moderate level of RABV-G for trans-synaptic labeling [27]. Both TVA and a marker gene (e.g., mCherry) are introduced using another helper viral vector. This approach using two helper viral vectors does not give 100% co-infection. Starter cells can be defined as RABVΔG-infected GFP- and mCherry-expressing cells, making the detection of starter cells unambiguous, although trans-synaptic labeling efficiency by this approach is not as high as in 2.
4. Use two helper viral vectors (Fig. 7a). Injecting a helper viral vector expressing all RABV-G, TVA, and a marker gene (e.g., GFP) (the same vector in 1), along with another vector expressing RABV-G alone for boost expression of RABV-G (the same vector in 2), is versatile, allowing efficient trans-synaptic labeling and clear detection of starter cells expressing RABV-G, TVA, and the marker gene [33]. The expression level of RABV-G varies between starter cells depending on the co-infection efficiency with two helper viral vectors. Therefore, it is challenging to confirm which starter cells are co-infected with two viral vectors. However, starter cells are unambiguously labeled as RABVΔG-infected GFP-expressing cells (Fig. 7b). Furthermore, a higher number of presynaptic cells is labeled due to the high expression level of RABV-G (Fig. 7c, d).

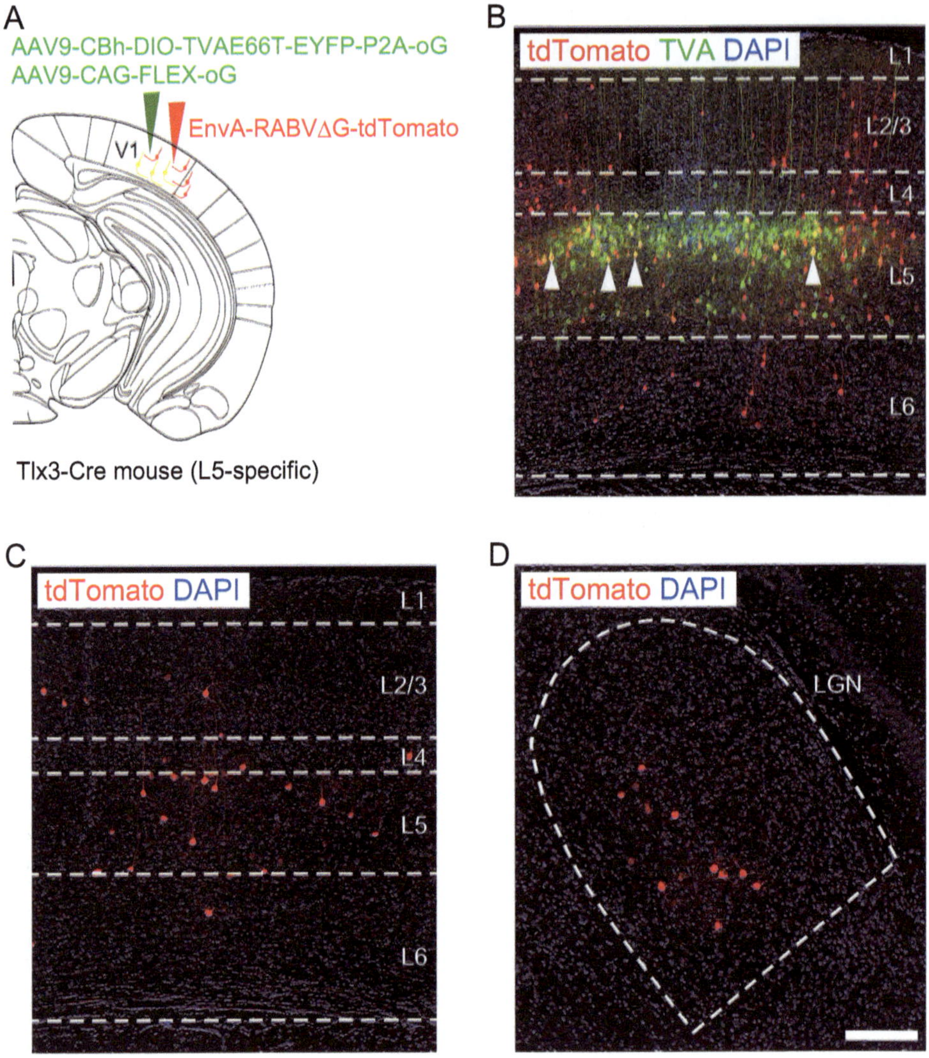

Fig. 7 In vivo tracing of EnvA-RABVΔG-DsRed. (**a**) Schematic illustration of the virus injection for monosynaptic RABVΔG tracing. Two helper AAVs were injected into the V1 of Tlx3-Cre mice in which the cortico-cortical layer 5 neurons specifically express Cre. Three weeks post-injection of the AAVs, EnvA-RABVΔG-DsRed was injected into the same location as the AAV injection. (**b**) YFP-positive cells in the V1 express TVA and oG. "Starter cells" express YFP and DsRed, as indicated by arrowheads. (**c**, **d**) Presynaptic cells express DsRed in the cortical (**c**) and subcortical areas, such as LGN (**d**). Scale bar: 100 μm

5. Use transgenic mouse lines that conditionally express RABV-G, TVA, and a marker gene in target cell populations Cre-dependently. However, the level of RABV-G expression and specificity of TVA expression vary depending on how the RABV-G and TVA are expressed, such as in a gene locus. Furthermore, the level of RABV-G expression from the mouse genome is not as high as the viral vectors. In contrast, the specificity of TVA expression from the mouse genome is higher than the viral vectors.

AAV expression after viral delivery takes 2–3 weeks to reach maximum expression of transgenes. Therefore, EnvA-RABVΔG should be injected 2–3 weeks after the injection of helper AAVs. The EnvA-RABVΔG infects TVA-expressing starter cells and then replicates and spreads to their presynaptic cells in the presence of RABV-G (oG: a RABV-G variant that allows efficient labeling of presynaptic neurons [34]) 7–10 days after the EnvA-RABVΔG injection. Notably, each strategy has its pros and cons, which is due to a trade-off between the ability to label starter neurons unambiguously and to spread across synapses efficiently.

Herein, we introduced a typical example of RABVΔG circuit tracing using helper AAVs (Fig. 7a–d). To map presynaptic networks of cortico-cortical projection neurons of layer 5 in V1, we used a Tlx3-Cre mouse line in which the cortico-cortical layer 5 neurons are labeled with Cre. We injected a mixture of AAV-CBh-DIO-oTVA-EYFP-P2A-oG and AAV-CAG-FLEX-oG (*see* Table 1, Strategy #4) into the primary visual cortex (V1) of the Tlx3-Cre mice. After sufficient time for viral transduction and protein expression from the AAVs (typically 3 weeks), we injected EnvA-RABVΔG-DsRed in the exact location of the AAV injection of the V1. Ten days later, we perfused the mice and prepared brain slices for anatomical analysis. Postsynaptic "starter" neurons were labeled as cells positive for both YFP and DsRed, while their presynaptic neurons were cells positive for DsRed only (Fig. 7b). The longer we waited after the EnvA-RABVΔG-DsRed injection, the more presynaptic neurons were labeled. However, RABVΔG-infected starter cells die 12–14 days after infection.

In summary, RABVΔG trans-complementation allows efficient labeling of presynaptic neurons that are directly connected to starter cells (e.g., a specific cell type and single neuron) expressing oG and TVA in a brain-wide manner. Injection of unpseudotyped RABVΔG will retrogradely label long-range projection neurons from the injection site. Furthermore, RABVΔG tracing can be combined with functional assays to monitor or manipulate neural activity, such as optical imaging, electrophysiology, optogenetics, and chemogenetics [11, 13–16]. These approaches outlined here are adaptable to even non-transgenic mammalian species, including ferrets and monkeys.

Acknowledgments

We thank members of the Osakada laboratory for stimulating discussions. This work was supported by Grants-in-Aid from the Japan Society for the Promotion of Science (F.O.), Brain/MINDS from the Japan Agency for Medical Research and Development (F.O.), and PRESTO and CREST from the Japan Science and Technology Agency (F.O.).

References

1. Bastos AM, Usrey WM, Adams RA, Mangun GR, Fries P, Friston KJ (2012) Canonical microcircuits for predictive coding. Neuron 76(4):695–711
2. Luo L, Callaway EM, Svoboda K (2018) Genetic dissection of neural circuits: a decade of progress. Neuron 98(2):256–281
3. Harris KD, Mrsic-Flogel TD (2013) Cortical connectivity and sensory coding. Nature 503(7474):51–58
4. Ugolini G (2010) Advances in viral transneuronal tracing. J Neurosci Methods 194(1):2–20
5. Ugolini G (1995) Specificity of rabies virus as a transneuronal tracer of motor networks: transfer from hypoglossal motoneurons to connected second-order and higher order central nervous system cell groups. J Comp Neurol 356(3):457–480
6. Callaway EM (2008) Transneuronal circuit tracing with neurotropic viruses. Curr Opin Neurobiol 18(6):617–623
7. Mebatsion T, Konig M, Conzelmann KK (1996) Budding of rabies virus particles in the absence of the spike glycoprotein. Cell 84(6): 941–951
8. Etessami R, Conzelmann KK, Fadai-Ghotbi B, Natelson B, Tsiang H, Ceccaldi PE (2000) Spread and pathogenic characteristics of a G-deficient rabies virus recombinant: an in vitro and in vivo study. J Gen Virol 81 (Pt 9):2147–2153
9. Wickersham IR, Lyon DC, Barnard RJ, Mori T, Finke S, Conzelmann KK, Young JA, Callaway EM (2007) Monosynaptic restriction of transsynaptic tracing from single, genetically targeted neurons. Neuron 53(5):639–647
10. Osakada F, Callaway EM (2013) Design and generation of recombinant rabies virus vectors. Nat Protoc 8(8):1583–1601
11. Osakada F, Mori T, Cetin AH, Marshel JH, Virgen B, Callaway EM (2011) New rabies virus variants for monitoring and manipulating activity and gene expression in defined neural circuits. Neuron 71(4):617–631
12. Suzuki T, Morimoto N, Akaike A, Osakada F (2019) Multiplex neural circuit tracing with G-deleted rabies viral vectors. Front Neural Circuits 13:77
13. Wertz A, Trenholm S, Yonehara K, Hillier D, Raics Z, Leinweber M, Szalay G, Ghanem A, Keller G, Rozsa B, Conzelmann KK, Roska B (2015) PRESYNAPTIC NETWORKS. Single-cell-initiated monosynaptic tracing reveals layer-specific cortical network modules. Science 349(6243):70–74
14. Tian J, Huang R, Cohen JY, Osakada F, Kobak D, Machens CK, Callaway EM, Uchida N, Watabe-Uchida M (2016) Distributed and mixed information in monosynaptic inputs to dopamine neurons. Neuron 91(6):1374–1389
15. Kim CK, Adhikari A, Deisseroth K (2017) Integration of optogenetics with complementary methodologies in systems neuroscience. Nat Rev Neurosci 18(4):222–235
16. Sabatini BL, Tian L (2020) Imaging neurotransmitter and neuromodulator dynamics in vivo with genetically encoded indicators. Neuron 108(1):17–32
17. Schnell MJ, Mebatsion T, Conzelmann KK (1994) Infectious rabies viruses from cloned cDNA. EMBO J 13(18):4195–4203
18. Okigawa S, Yamaguchi M, Ito KN, Takeuchi RF, Morimoto N, Osakada F (2021) Cell type- and layer-specific convergence in core and shell neurons of the dorsal lateral geniculate nucleus. J Comp Neurol 529(8):2099–2124
19. Masaki Y, Yamaguchi M, Takeuchi RF, Osakada F (2022) Monosynaptic rabies virus tracing from projection-targeted single neurons. Neurosci Res 178:20–32
20. Wall NR, Wickersham IR, Cetin A, De La Parra M, Callaway EM (2010) Monosynaptic circuit tracing in vivo through Cre-dependent targeting and complementation of modified rabies virus. Proc Natl Acad Sci U S A 107(50):21848–21853
21. Marshel JH, Mori T, Nielsen KJ, Callaway EM (2010) Targeting single neuronal networks for gene expression and cell labeling in vivo. Neuron 67(4):562–574
22. Hodge RD, Bakken TE, Miller JA, Smith KA, Barkan ER, Graybuck LT, Close JL, Long B, Johansen N, Penn O, Yao Z, Eggermont J, Hollt T, Levi BP, Shehata SI, Aevermann B, Beller A, Bertagnolli D, Brouner K, Casper T, Cobbs C, Dalley R, Dee N, Ding SL, Ellenbogen RG, Fong O, Garren E, Goldy J, Gwinn RP, Hirschstein D, Keene CD, Keshk M, Ko AL, Lathia K, Mahfouz A, Maltzer Z, McGraw M, Nguyen TN, Nyhus J, Ojemann JG, Oldre A, Parry S, Reynolds S, Rimorin C, Shapovalova NV, Somasundaram S, Szafer A, Thomsen ER, Tieu M, Quon G, Scheuermann RH, Yuste R, Sunkin SM, Lelieveldt B, Feng D, Ng L, Bernard A, Hawrylycz M, Phillips JW, Tasic B, Zeng H, Jones AR, Koch C, Lein ES (2019) Conserved cell types with divergent features in human versus mouse cortex. Nature 573(7772):61–68

23. Yuste R, Hawrylycz M, Aalling N, Aguilar-Valles A, Arendt D, Armananzas R, Ascoli GA, Bielza C, Bokharaie V, Bergmann TB, Bystron I, Capogna M, Chang Y, Clemens A, de Kock CPJ, DeFelipe J, Dos Santos SE, Dunville K, Feldmeyer D, Fiath R, Fishell GJ, Foggetti A, Gao X, Ghaderi P, Goriounova NA, Gunturkun O, Hagihara K, Hall VJ, Helmstaedter M, Herculano-Houzel S, Hilscher MM, Hirase H, Hjerling-Leffler J, Hodge R, Huang J, Huda R, Khodosevich K, Kiehn O, Koch H, Kuebler ES, Kuhnemund M, Larranaga P, Lelieveldt B, Louth EL, Lui JH, Mansvelder HD, Marin O, Martinez-Trujillo J, Chameh HM, Mohapatra AN, Munguba H, Nedergaard M, Nemec P, Ofer N, Pfisterer UG, Pontes S, Redmond W, Rossier J, Sanes JR, Scheuermann RH, Serrano-Saiz E, Staiger JF, Somogyi P, Tamas G, Tolias AS, Tosches MA, Garcia MT, Wozny C, Wuttke TV, Liu Y, Yuan J, Zeng H, Lein E (2020) A community-based transcriptomics classification and nomenclature of neocortical cell types. Nat Neurosci 23(12): 1456–1468
24. Sreenivasan V, Karmakar K, Rijli FM, Petersen CC (2015) Parallel pathways from motor and somatosensory cortex for controlling whisker movements in mice. Eur J Neurosci 41(3): 354–367
25. Wall NR, De La Parra M, Callaway EM, Kreitzer AC (2013) Differential innervation of direct- and indirect-pathway striatal projection neurons. Neuron 79(2):347–360
26. Ogawa SK, Cohen JY, Hwang D, Uchida N, Watabe-Uchida M (2014) Organization of monosynaptic inputs to the serotonin and dopamine neuromodulatory systems. Cell Rep 8(4):1105–1118
27. Faget L, Osakada F, Duan J, Ressler R, Johnson AB, Proudfoot JA, Yoo JH, Callaway EM, Hnasko TS (2016) Afferent inputs to neurotransmitter-defined cell types in the ventral tegmental area. Cell Rep 15(12): 2796–2808
28. Taniguchi H, He M, Wu P, Kim S, Paik R, Sugino K, Kvitsani D, Fu Y, Lu J, Lin Y, Miyoshi G, Shima Y, Fishell G, Nelson SB, Huang ZJ (2011) A resource of cre driver lines for genetic targeting of GABAergic neurons in cerebral cortex. Neuron 71(6): 995–1013
29. Gerfen CR, Paletzki R, Heintz N (2013) GENSAT BAC Cre-recombinase driver lines to study the functional organization of cerebral cortical and basal ganglia circuits. Neuron 80(6):1368–1383
30. Kohara K, Pignatelli M, Rivest AJ, Jung HY, Kitamura T, Suh J, Frank D, Kajikawa K, Mise N, Obata Y, Wickersham IR, Tonegawa S (2014) Cell type-specific genetic and optogenetic tools reveal hippocampal CA2 circuits. Nat Neurosci 17(2):269–279
31. Watabe-Uchida M, Zhu L, Ogawa SK, Vamanrao A, Uchida N (2012) Whole-brain mapping of direct inputs to midbrain dopamine neurons. Neuron 74(5):858–873
32. Miyamichi K, Shlomai-Fuchs Y, Shu M, Weissbourd BC, Luo L, Mizrahi A (2013) Dissecting local circuits: parvalbumin interneurons underlie broad feedback control of olfactory bulb output. Neuron 80(5):1232–1245
33. Szonyi A, Sos KE, Nyilas R, Schlingloff D, Domonkos A, Takacs VT, Posfai B, Hegedus P, Priestley JB, Gundlach AL, Gulyas AI, Varga V, Losonczy A, Freund TF, Nyiri G (2019) Brainstem nucleus incertus controls contextual memory formation. Science 364(6442):eaaw0445
34. Kim EJ, Jacobs MW, Ito-Cole T, Callaway EM (2016) Improved monosynaptic neural circuit tracing using engineered rabies virus glycoproteins. Cell Rep 15(4):692–699

Chapter 5

Quality Control for Adeno-Associated Viral Vector Production

Ina Ersing, Meghan Rego, Chen Wang, Yijun Zhang, Kate Harten DeMaio, Michelle Petrozzi, Alanna Fava, Gabrielle Clouse, Marcella Patrick, Karen Guerin, and Melina Fan

Abstract

Adeno-associated viral vectors (AAV) are frequently used by neuroscientists to deliver tools, such as biosensors and optogenetic and chemogenetic actuators, in vivo. Despite its widespread use, AAV vector characterization and quality control can vary between labs and viral vector cores, leading to variable results and irreproducibility. This protocol describes some of the characterization and quality control assays necessary to confirm an AAV vector's titer, genomic identity, serotype, and purity.

Key words Adeno-associated viral vectors, Quality control, rAAV, AAV, Viral vectors

1 Introduction

Recombinant adeno-associated virus (rAAV) vectors are increasingly popular gene delivery tools for use in research and therapeutically. They are relatively easily produced in a well-equipped lab or production facility, but before in vivo use, rAAV vectors should be extensively characterized to ensure proper dosing, vector genome integrity, and purity. The concentration of an rAAV vector is typically reported as a physical titer of the total viral particles present and does not take into account the presence of defective particles. The physical titer is frequently measured via quantitative polymerase chain reaction (qPCR) and reported as vector genomes per milliliter (VG/mL) or genome copies per milliliter (GC/mL) with values relative to a plasmid standard. Viral production facilities frequently design qPCR titration assays with primers and/or probes that target commonly occurring regions of the transfer

Ina Ersing and Meghan Rego contributed equally with all other contributors.

Mark A. G. Eldridge and Adriana Galvan (eds.), *Vectorology for Optogenetics and Chemogenetics*,
Neuromethods, vol. 195, https://doi.org/10.1007/978-1-0716-2918-5_5,

plasmid such as promoters, polyadenylation signals, the Woodchuck Hepatitis Virus Posttranscriptional Regulatory Element (WPRE), or the inverted terminal repeats (ITRs) [1]. By targeting common elements, once validated, the same assay can be used to titer multiple rAAV vectors in parallel. Production facilities using this method should validate the assay with an rAAV vector of known titer before use. Once validated, this rAAV reference virus should be included as a control in all titrations to confirm consistency across runs.

Recently, a droplet digital PCR-based titration (ddPCR) method was developed for rAAV vectors [2]. Unlike qPCR, ddPCR does not rely on a standard curve and provides an absolute measurement. In contrast, qPCR results vary depending on the specific plasmid used and the overall quality of the standard preparation leading to variation across run, users, and labs. In an analysis of replicate samples, ddPCR was determined to have a lower coefficient of variation and, therefore, a higher precision as compared to qPCR [3]. ddPCR is also less prone to PCR inhibitions than qPCR. This resistance to PCR inhibition is thought to be due to the partitioning reaction that occurs in ddPCR [4]. In ddPCR, a single droplet with minor to moderate PCR inhibition will still be read as positive, whereas in a minor to moderately inhibited qPCR more cycles are required to reach a signal above the threshold, causing underreported titers. Moreover, ddPCR is the superior titration method for self-complementary rAAV (scAAV) vector templates as it does not have the same self-annealing issues observed with qPCR. Previous studies have shown that the kinetics of ITR hairpin structure formation in the scAAV genome outcompetes those of primer and probe binding, impairing PCR efficiency and resulting in underreported titers [5, 6]. Though ddPCR is gaining popularity, especially among viral vector cores, qPCR is still the most commonly used titration assay.

In addition to the physical titer, some users may wish to determine the specific infectivity of the vector. The specific infectivity is the ratio of physical-to-infectious particles and assesses the presence of defective viral particles. The physical titer is measured by qPCR or ddPCR as described above, while a functional or infectious titer is determined using cell-based transduction assays. Stable cell lines are co-transduced with the rAAV vector and wild-type adenovirus, and transduced cells can be detected by a variety of methods such as flow cytometry or microscopy to detect fluorescent reporters, cellular staining to detect reporter enzyme expressions such as β-galactosidase, or PCR to detect of the gene of interest in target cells [7, 8]. Of note, transduction efficiencies often vary between cell lines and tissues; therefore, the functional titer determined in a cell-based assay may not transfer precisely to an in vivo system. Specific infectivity is not routinely measured for rAAV vectors used in basic research but is critical in determining dosing regimens for clinical use in order to minimize adverse immune responses [8].

Production facilities or laboratories that produce a number of different rAAV vectors should confirm the identity and serotype of the vector. The identity of the vector can be easily confirmed via PCR of the genomic DNA with primers targeting critical regions such as promoter elements or the genes of interest. PCR products can be analyzed on a gel to confirm size or sent for Sanger sequencing for a more thorough analysis. For labs producing recombinase-dependent vectors, it is recommended that they include additional reactions with primers inside and outside of the recombination sites to detect the presence of recombinase-independent recombination events. For a deeper analysis of vector genome integrity, labs may consider next-generation sequencing (NGS). In addition to sequence confirmation of the vector's genome, NGS has been used to detect contaminants, confirm serotypes, and measure recombination rates in rAAV vectors [9–12].

The serotype of an rAAV vector will determine its tropism or tissue specificity. It is therefore critical that production facilities routinely working with a number of different capsids confirm that the correct capsid was used for each vector. Serotype confirmation can be done in a number of ways. Historically, western blotting and mass spectrophotometry were used, but these methods have some limitations. Western blotting relies on antibodies that tend to cross-react due to the high homology between some serotypes, while mass spectrophotometry is capsid-specific but costly and labor-intensive.

One of the simplest methods for serotype confirmation is AAV-ID, a thermostability-based approach that uses melting temperature to differentiate between serotypes [13]. AAV-ID is fast and can be easily incorporated into most laboratories as it does not require specialized reagents or equipment. Of note, some serotypes naturally have similar melting temperatures and cannot be easily distinguished using this method. Moreover, the formulation buffer and purity of the vector will also affect the melting temperature. Recently, open-access Python software was developed to determine serotypes from NGS data [9]. The software takes advantage of the small amount of capsid plasmid DNA naturally packaged into the viral vector and probes NGS reads for serotype-specific, user-defined seed sequences. This software is open-access and available at GitHub in Addgene's Open Toolkit https://addgene.github.io/openbio/ .

Much like specific infectivity, rAAV vector purity affects transduction efficiency but in a serotype and tissue-specific manner [14]. Common impurities include packaging cell proteins and residual plasmid or host cell DNA and can be routinely detected by most labs. Overall DNA impurities can be estimated using spectrophotometry by subtracting the known OD260 and

OD280 contributed by the vector from the overall OD260 and OD280 values [14]. To measure specific DNA contamination, qPCR with primers and probes targeting specific genes or genetic elements of interest from the plasmid or host cells can be used.

Protein impurities are easily observed following SDS-PAGE and protein staining of the rAAV vector. In a pure rAAV vector, only three bands corresponding to the VP1, VP2, and VP3 capsid proteins should be visible. Additional bands in the stain suggest packaging cell protein contamination during vector purification.

The presence of endotoxins in viral vector preparations can also impact downstream applications. Endotoxins arise from the lipid A portion of bacterial lipopolysaccharides and can be introduced to the viral vector through low-quality plasmid DNA or bacterial contamination. The *Limulus polyphemus* amebocyte lysate turbidimetric assay detects endotoxin using a reagent derived from horseshoe crab blood that coagulates upon exposure to endotoxins. Even low activities of endotoxin can elicit an immune response underscoring the need for endotoxin screening before rAAV vector use in vivo [15].

rAAV vector contamination by microorganisms such as bacteria, yeast, or fungi should also be ruled out. To confirm sterility, production facilities can inoculate microbial growth media such as Fluid Thioglycollate Medium or Soybean-Casein Digest Medium with the rAAV vector and check for microbial growth after 2 weeks of cultivation. Production facilities may also consider checking vectors for mycoplasma contamination. Mycoplasma is a common tissue culture contaminant and can be difficult to detect during routine cell culture work. There are several methods to detect mycoplasma, including PCR-based assays targeting the 16S rRNA, luciferase-based assays that detect mycoplasma-specific enzymes and mycoplasma-specific indicator cell lines [16–18]. Mycoplasma contamination of producer cell lines can be difficult to treat, and it is recommended that viral production facilities routinely screen their cell lines before rAAV vector production.

The following protocol outlines some of the more widely used methods of characterizing rAAV vectors and assessing their purity and sterility (Fig. 1). Many of the methods described can be modified for other types of vectors, such as lentivirus or adenovirus [19]. The specific methods chosen do not require specialized lab equipment and can be easily applied to most laboratories and core facilities. However, they do not encompass an exhaustive list, and production facilities may prefer additional or alternative tests to meet their specific needs. It is worth noting that the efficiency of rAAV mediated in vivo transduction depends on numerous factors, including animal species, strain, age, viral delivery route, and targeting cell types. Differences in construct design, the transgene of interest, and the vector serotype make it difficult to directly predict an rAAV's performance in vivo. In addition to the assays described

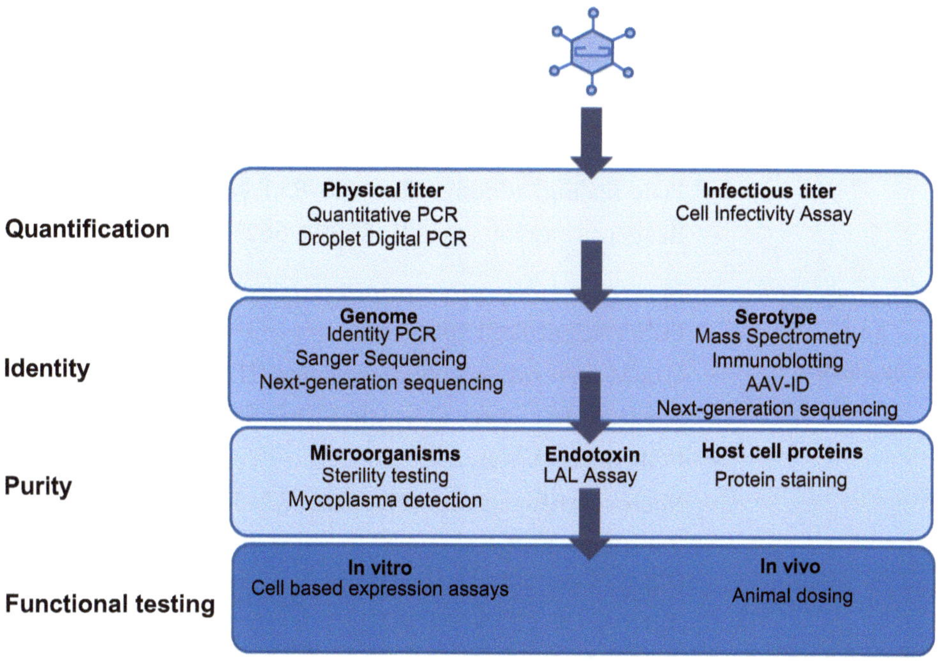

Fig. 1 Overview of the rAAV quality control workflow and assays. rAAV quality control consists of four general categories: quantification, identity confirmation, purity, and functional testing. A number of assays are possible for each category, and the choice of the assay will depend on the specific needs of the lab

in this chapter, for a specific experimental paradigm, we recommend that rAAV users design preliminary in vivo dosing experiments to determine the optimal dose before they begin any large-scale animal experiments.

2 Materials

2.1 Titration by Quantitative PCR

- PowerUp SYBR Green Master Mix or alternative universal SYBR master mix containing a high-quality DNA polymerase and a blend of dTTP/dUTP, protect from light (Thermo Fisher Scientific, A25776).
- Plasmid for the standard curve (containing ITRs or gene of interest).
- Yeast tRNA (Thermo Fisher Scientific, AM7119).
- 100 μM forward and reverse primer (targeting ITRs or gene of interest).
- AAV reference vector preparation, avoid multiple freeze-thaws, prepare individual aliquots for one-time use.
- RNase free DNase, keep on ice (Agilent Technologies, ST600031).
- DNase buffer (10×) (Thermo Fisher Scientific, B43).

- Nuclease-free water (Thermo Fisher Scientific, R0582).
- Microcentrifuge tubes (Neptune, 3745.X).
- 96 well optical plate (Bio-Rad, HSP9601).
- 96 well plate for dilutions (Bio-Rad, MLL9601).
- Plate sealing, adhesive film (Bio-Rad, MSB1001).
- Reagent reservoirs (VWR, 89094-662).

2.2 Confirmation of Packaged AAV Genome by Identity PCR

- Platinum Hot Start PCR 2× Master Mix, thaw on ice (Thermo Fisher Scientific, 13000013).
- 10 μM forward and reverse primers (IDT).
- Nuclease-free water (Thermo Fisher Scientific, R0582).
- PCR tubes (VWR, 89096-722).
- Microcentrifuge tubes (Neptune, 3745.X).
- Phosphate-buffered saline, 1× without calcium and magnesium (PBS) (Corning, 21-040-CV).
- Agarose (Bio-Rad, 1613101).
- Gel loading dye (New England Biolabs, J62157).

2.3 Serotype Determination via Melting Temperature (AAV-ID)

- 96-well plate (Bio-Rad, HSP9601).
- Microseal (Bio-Rad, MSB1001).
- SYPRO Orange 5000×, protect from light (Thermo Fisher Scientific, S6651).
- Phosphate-buffered saline, 1× with calcium and magnesium (Corning, 21-030-CV).
- Lysozyme solution, 0.25 mg/mL (Sigma Aldrich, L6876).

2.4 Vector Purity by SDS-PAGE and Silver Staining

- 4–12% NuPage Novex bis-tris mini gel, 1 mm thick, 10-well (Thermo Fisher Scientific, NP0321BOX).
- 20× MOPS SDS running buffer (Thermo Fisher Scientific, NP0001).
- 4× NuPage sample buffer (Thermo Fisher Scientific, NP0007).
- 10× NuPage sample reducing agent (Thermo Fisher Scientific, NP0009).
- PageRuler Plus Prestained protein ladder, optional (Thermo Fisher Scientific, 26619).
- Glacial acetic acid, corrosive to the skin and eyes; handle with extreme care (VWR, BDH3092-500 mL).
- Methanol, highly flammable and toxic, handle in a fume hood (VWR, BDH1135-1LP).
- Microcentrifuge tubes (Neptune, 3745.X).
- Gel loading tips (Neptune, 2016).

- 250 mL sterile bottles (Corning, 430281).
- SilverXpress silver staining kit, some contents are corrosive, handle with extreme care (Thermo Fisher Scientific, LC6100).
  - Solutions for silver staining (prepared freshly, according to kit instructions).

    Fixing solution: 90 mL deionized water + 100 mL methanol + 20 mL glacial acetic acid.

    Sensitizing solution: 105 mL deionized water + 100 mL methanol + 5 mL sensitizer (component of SilverXPress kit).

    Staining solution: 5 mL stainer A (component of SilverXPress kit) + 5 mL stainer B (component of SilverXPress kit) + 90 mL deionized water.

    Developing solution: 5 mL developer (component of SilverXPress kit) + 95 mL deionized water.

    Stopper solution: 5 mL stopper solution (a component of SilverXPress kit) directly added to staining solution when protein bands show sufficient intensity.
- ImageJ or alternative imaging software.

2.5 Limulus Amebocyte Lysate Chromogenic Endotoxin Test

- Pierce Chromogenic Endotoxin Quant Kit (Thermo Fisher Scientific, A39552).
- Stop reagent.
  - Glacial acetic acid, corrosive to the skin and eyes; handle with extreme care (VWR, BDH3092-500 mL): 25% v/v glacial acetic acid in water.
- Disposable sterile microplates (Corning, 9018).
- Reagent reservoirs (VWR, 89094-662).
- Disposable, endotoxin-free glass dilution tubes (Lonza, N207).

2.6 In vitro Sterility and Expression Assay

- Media, e.g., DMEM high glucose (Corning, 10-013-CV).
- HI-FBS (Seradigm, 89510-196).
- GlutaGRO, liquid 200 mM solution (with 8.5 g/L NaCl) (Corning, 25-015-CI).
- Flat-bottom 96-well tissue culture treated plate (Corning, 3596).
- 15 mL conical tubes (VWR, 21008-216).
- Microcentrifuge tubes (Neptune, 3745.X).
- 0.4% Trypan blue Solution (Thermo Fisher Scientific, T10282).
- Countess cell counting chamber slides (Thermo Fisher Scientific, C10228) (alternatively can use a hemocytometer).

3 Methods

3.1 Titration by Quantitative PCR

A qPCR titration assay with primers and/or probes targeting regions of the transfer plasmid or the ITRs and SYBR green detection is commonly used to determine the physical titer of rAAV preparations. All values are calculated relative to a plasmid standard.

1. Prepare plasmid standards for quantitative PCR titration.
   (a) Use purified plasmid DNA that contains the ITR sequence or your sequence of interest that you can target.
   (b) Linearize your selected plasmid by restriction digest. Verify your digest by running a sample on a 1% agarose gel.
   (c) Calculate the plasmid concentration in molecules/μL. Concentration (molecules/μL) = (DNA conc (μg/μL) × 10E-06 gram/MW (g/mol)) × 6.023E+23 molecules/mole.
   (d) Prepare plasmid DNA standard by diluting your linearized plasmid to 2E+09 molecules/μL in nuclease-free water with 4 μg/mL carrier RNA for stabilization. Store in small aliquots at −20 °C.
   (e) Use one aliquot to prepare tenfold serial dilutions that range from 2E+02 to 2E+08 molecules/μL for your qPCR standard curve for each qPCR assay.
2. Titration of viral vector genome titer by qPCR.
   (a) Prepare seven tenfold serial dilutions of plasmid standard stock (*see* Subheading 3.1, **step 1d**).
   (b) Thaw rAAV preparation of interest and rAAV reference vector preparation and keep on ice.
   (c) Digest all purified rAAV samples (including the rAAV reference sample) with DNase to eliminate any contaminating plasmid DNA carried over from the production process. Set up digests by mixing 5 μL vector sample with 39 μL deionized water, 5 μL 10× DNase buffer, and 1 μL DNase.
   (d) Gently mix and incubate for 30 min at 37 °C.
   (e) Heat inactivate for 15 min at 75 °C.
   (f) Dilute DNase-treated rAAV sample and rAAV reference sample according to the dilution scheme in Table 1.
      (i) If your sample is expected to have a titer <1E+12 GC/mL, use dilutions 3–6 to load onto your qPCR plate.
      (ii) If your sample is expected to have a titer >3E+13 GC/mL, use dilutions 4–7 to load onto your qPCR plate.

Table 1
Dilution series for AAV titration by qPCR

Dilution series	Sample volume (μL)	Volume of nuclease-free water (μL)	Dilution factor	Total dilution
Dilution 1	5 μL AAV stock	45 μL	10×	10×
Dilution 2	5 μL dilution 1	95 μL	20×	200×
Dilution 3	20 μL dilution 2	80 μL	5×	1000×
Dilution 4	20 μL dilution 3	80 μL	5×	5000×
Dilution 5	20 μL dilution 4	80 μL	5×	25,000×
Dilution 6	20 μL dilution 5	80 μL	5×	125,000×
Dilution 7	20 μL dilution 6	80 μL	5×	625,000×

Table 2
Example of a qPCR plate setup

	1	2	3	4	5	6	7	8
A B	1.00E+09	1.00E+08	1.00E+07	1.00E+06	1.00E+05	1.00E+04	1.00E+03	NTC
C D	AAV reference				Sample 3			
E F	Sample 1				Sample 4			
G H	Sample 2				Sample 5			

Plasmid standards are loaded in duplicate. The tenfold diluted plasmid standard ranges from 2E+02 to 2E+08 molecules/μL. For each standard, 5 μL are loaded for a range of 1E+03 to 1E+09 total molecules. For the AAV reference sample and all other AAV samples, four dilutions each are loaded in duplicate. A no template control (NTC) is included on the plate

(g) Prepare a master mix of 10 μL 2× Universal SYBR Master Mix, 0.15 μL 100 μM forward primer, 0.15 μL 100uM reverse primer, and 4.7 μL of nuclease-free water per reaction. Mix well.

(h) Set up and load the 96-well plate (Table 2).

  (i) Load 5 μL of each standard in duplicate.

  (ii) Load 5 μL of each sample in duplicate.

  (iii) Include a no template control (NTC = master mix + water).

  (iv) Add 15 μL of master mix per well and mix well by pipetting back and forth.

(v) Seal your 96-well plate with adhesive film.

(vi) Centrifuge plate at 3000 rpm for 2 min.

(vii) Run your qPCR instrument with the following protocol using SYBR detection: 98 °C, 3 min; 98 °C, 15 s; 58 °C, 30 s; read plate/repeat 39× from **step 3**/melt curve.

3. qPCR data analysis.

(a) Perform data analysis using your instrument's software. Determine the physical titer of your sample based on your standard curve and your sample dilutions.

(i) Set the program settings to an appropriate baseline to ensure that small amounts of background signal detected in initial PCR cycles will be removed.

(ii) Check your standard curve (Fig. 2a) and ensure that your linear correlation coefficient, R^2, is ~1, and your amplification efficiency is ~100%. Confirm that there is an appropriate difference in cycle threshold (Ct) values between each dilution of your standard curve (Fig. 2b).

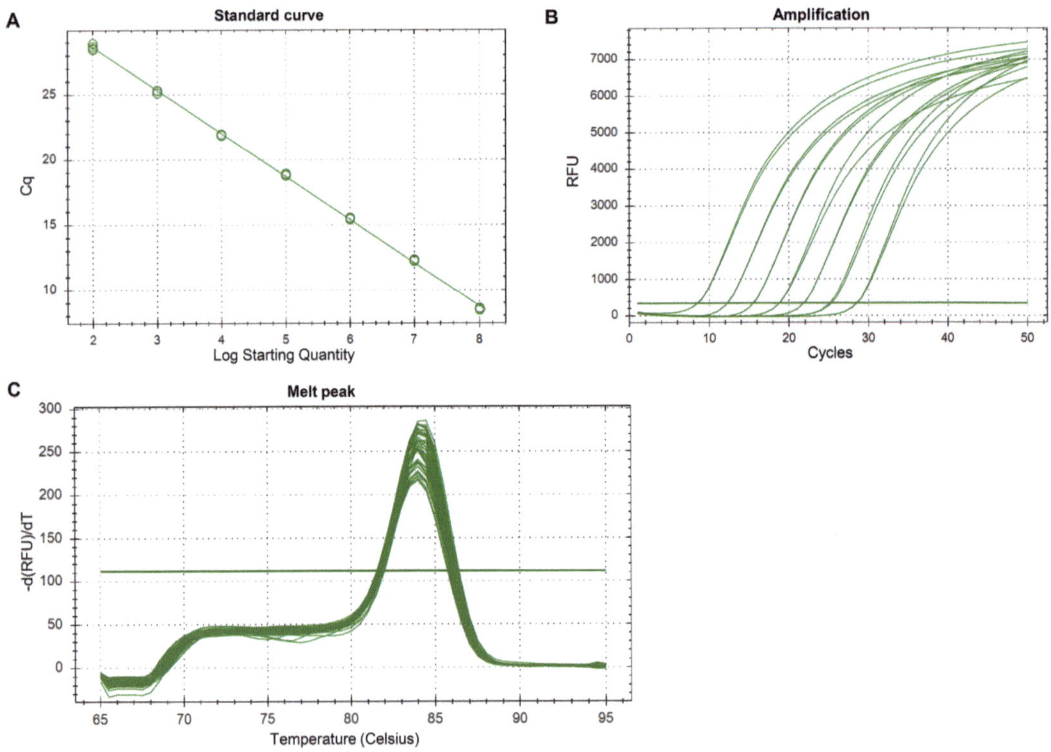

Fig. 2 (**a**) Example of plasmid standard curve with $R^2 = 0.999$, $E = 100.3\%$, slope $= -3.316$ and y-int $= 35.280$. (**b**) Example of amplification plots obtained from a plasmid standard curve. Each curve represents a standard dilution. (**c**) Example of melt curve analysis with a single peak

(iii) Make sure that a single peak is seen in your melt curve analysis (Fig. 2c). The existence of a second peak could indicate the presence of primer dimers and impact the Ct values of your samples.

(iv) Exclude any duplicates with a difference >0.5 in Ct value.

(v) Ensure that the Ct values for your NTCs are higher than any plasmid standard or sample Ct values on the plate.

(vi) You should observe differences in Ct values that make sense for your dilutions (~3.3 difference Ct for a tenfold dilution is appropriate).

3.2 Confirmation of Packaged AAV Genome by Identity PCR

Amplification of the packaged genomic DNA with primers targeting the promoter region or gene of interest can be used to confirm the identity of the vector.

1. Start by selecting a suitable primer pair for your construct of interest.

   (a) For non-FLEx/DIO constructs, target the area between the ITRs, including the fluorescent tag, with a forward and reverse primer pair of your choice (fwd1 + rev1) (Fig. 3a). Select at least one primer pair for confirmation by PCR.

   (b) For FLEx/DIO constructs, target the area between the *loxP/lox2272* sites with one primer pair of your choice (fwd2 + rev2) (Fig. 3b). In addition, select a primer pair that spans across the *loxP/lox2272* sites, with forward and

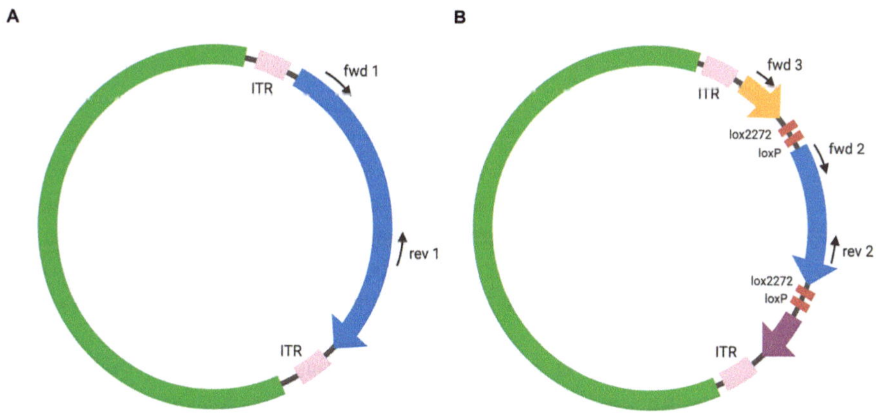

Fig. 3 Plasmid with selected primer pairs. (**a**) For an identity PCR for a non-FLEx/DIO construct, choose forward (fwd) and reverse (rev) primers that target your insert between the ITRs. (**b**) For an identity PCR for a FLEx/DIO construct, choose one primer pair that targets your insert area (fwd2 + rev2), one primer pair with one primer being outside of, and one in between the lox-sites (fwd3 + rev2) and one control primer pair (fwd3 + fwd 2)

reverse primers pointing toward each other (fwd3 + rev2). As a control, select a third primer pair that again spans across the *loxP/lox2272* sites, but with both primers pointing in the same direction (fwd3 + fwd2).

2. Dilute a small sample of your rAAV preparation in PBS.
   (a) For rAAV preparations with a titer between 1E+12 and 1E+13 GC/mL, prepare a tenfold dilution by diluting 5 μL of your rAAV in 45 μL of PBS.
   (b) For rAAV preparations with a titer higher than 1E+13 GC/mL, prepare a 25-fold dilution by diluting 4 μL of your rAAV in 96 μL of PBS.
3. Thaw PCR reagents on ice.
4. Prepare your master mix. Calculate the number of reactions needed and prepare a master mix for each primer pair according to polymerase manufacturer instructions. Make sure to include one NTC for each primer pair.
5. Set up PCR reactions.
   (a) Load 24 μL of master mix in appropriate wells.
   (b) Load 1 μL of diluted rAAV sample in one of the reaction tubes with your master mix.
   (c) For your NTC, load 1 μL of nuclease-free water in the other reaction tube with your master mix.
6. Run PCR protocol according to manufacturer instructions.
7. To analyze the result of your PCR, mix 10 μL of each PCR reaction with loading dye and run on a 1.2% agarose gel with an appropriate ladder.
   (a) For non-FLEx/DIO constructs, ensure that your product runs at the expected size.
   (b) For FLEx/DIO constructs, ensure that your product from the primer pair targeting the region within the *loxP/lox2272* sites and the primer pair spanning across the *loxP/lox2272* sites run at the expected size. Confirm that the control reaction, with the primers pointing in the same direction, does not show any bands on the gel. If unexpected bands show up in this reaction, you could be having issues with a PCR product due to recombination.

3.3 Serotype Determination via Melting Temperature (AAV-ID)—Protocol Adapted from Pacouret et al. [13]

AAV-ID is a thermostability-based approach that allows distinguishing between different AAV serotypes based on melting temperature.

1. Prepare a 50× solution of SYPRO Orange solution as follows:
   (a) Dilute 5 μL of SYPRO Orange 5000× into 495 μL phosphate-buffered saline, 1× with calcium and magnesium.

2. Aliquot 45 μL of purified rAAV into a 96-well plate.
3. Aliquot 45 μL of 0.25 mg/mL lysozyme solution into the 96-well plate as a positive control.
4. Add 5 μL of the 50× solution of SYPRO Orange to each well.
5. Pipet up and down 10–20× to mix.
6. Seal the plate with micro seal film.
7. Centrifuge the plate for 2 min at 2000 rpm.
8. Run the following PCR parameters:
   (a) Ramp: 25–99°C for 2 min.
   (b) Step and hold mode with 0.4°C increment (equivalent to ~0.2°C/min).
   (c) Reporter: Rox.
   (d) Quencher: None.
9. Analyze the data as follows:
   (a) Plot fluorescence as a function of temperature.
   (b) If necessary, normalize the signal between 0 and 100% as follows:
      (i) **Step 1**: $S = S - \min(S)$
      (ii) **Step 2**: $S = S/\max(S)/100$
   (c) Calculate and plot the derivative fluorescence signal to get the melting temperature.
      (i) $dF/dT(T) = (F(T + \Delta T) - F(T))/\Delta T$

3.4 Vector Purity by SDS-PAGE and Silver Staining

Potential residual protein impurities in rAAV preparations, e.g., from cellular proteins, can be detected by subjecting the sample to SDS-PAGE followed by a silver staining procedure. The presence of AAV capsid proteins VP1, VP2, and VP3 in correct stoichiometry and the absence of unwanted protein contaminants can be confirmed.

1. rAAV sample preparation:
   (a) Prepare a tenfold dilution of your rAAV vectors if your titer is >5E+12 GC/mL. Dilute 2 μL of rAAV in 18 μL of PBS.
   (b) Calculate the amount needed for 2E+10 vector particles and aliquot that amount into a fresh microcentrifuge tube. Add PBS to 13 μL.
   (c) Add 5 μL of 4× sample buffer to each sample.
   (d) Add 2 μL of 10× reducing agent to each sample.
   (e) Spin the sample for 1 min at 10,000 × *g* in a microcentrifuge.
   (f) Heat the sample for 5 min at 95 °C in a heat block.

(g) Spin the sample for 1 min at 10,000 × *g* in a microcentrifuge and place it on ice while preparing your gel.

2. Preparation of SDS-PAGE and gel loading:

   (a) Prepare 1× MOPS buffer by diluting 25 mL of 20× MOPS buffer in 475 mL of deionized water. Gently invert to mix.

   (b) Prepare the NuPage Novex bis-tris mini gel, place the gel in the electrophoresis chamber, and secure it. Check the manufacturer instructions on the correct orientation of the gel.

   (c) Fill the electrophoresis chamber with 1× MOPS running buffer.

   (d) Rinse each well with 200 μL 1× MOPS running buffer.

   (e) Optional: Load 5 μL of prestained protein ladder in the appropriate well.

   (f) Load 20 μL of each prepared rAAV sample in the appropriate wells.

   (g) Cover the electrophoresis chamber and attach it to a power supply.

   (h) Run the gel at a constant voltage (~150 V) until the dye from the loading buffer reaches the bottom of your gel (~1 h).

   (i) Turn off the power supply and remove the gel from the electrophoresis chamber.

   (j) Optional: Use a razor blade and cut the wells on top and the bottom of the gel where the dye is still visible.

3. Silver staining procedure:

   (a) Place the gel in a container and rinse for 5 min with deionized water while gently shaking.

   (b) Follow the staining procedure according to manufacturer instructions.

      (i) Fix gel in 200 mL fixing solution and shake gently for 10 min.

      (ii) Decant fixing solution and incubate gel in 100 mL sensitizing solution. Shake gently for 30 min.

      (iii) Decant solution and repeat the sensitizing step with a fresh sensitizing solution.

      (iv) Decant sensitizing solution and rinse gel with 200 mL deionized water while gently shaking for 10 min. Repeat the wash step.

      (v) Decant deionized water and stain gel with 100 mL staining solution for 15 min while gently shaking.

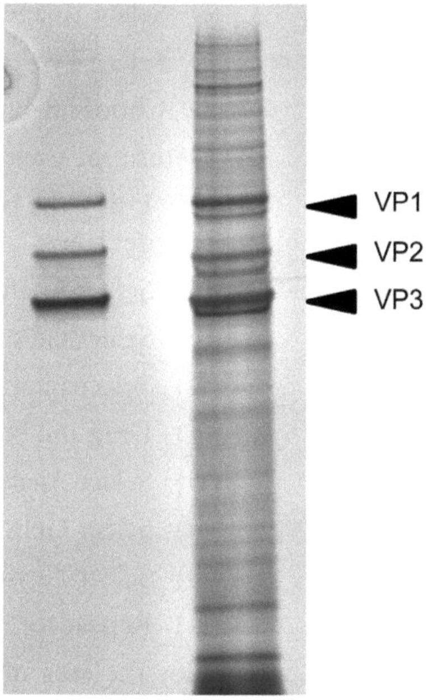

Fig. 4 Silver staining of purified (left) and non-purified rAAV (right)

(vi) Decant staining solution and rinse gel with 200 mL deionized water for 5 min while gently shaking. Repeat the wash step.

(vii) Develop gel with 100 mL developing solution while gently shaking. Observe the staining process and stop staining when reaching your desired band intensity for VP1, 2, and 3 by adding 5 mL stopper solution directly into the staining solution. Depending on the purity of the vector, protein impurities may or may not be visible. Incubate for 10 min.

(viii) Decant stopping solution and rinse the gel with 200 mL dH_2O for 10 min while gently shaking. Repeat the wash step two more times.

(ix) Dry gel or image directly with a gel imager under white light (Fig. 4).

4. Use ImageJ or a similar photo software to determine the relative intensity of the capsid bands to the overall lane.

   (a) Import the gel image into ImageJ.

      (i) Select File.

      (ii) Select Open.

      (iii) Choose the location of the file to open.

   (b) Change the image type to 8-bit.

(i) Select Image.
(ii) Select Type.
(iii) Choose 8-bit.

(c) Determine each lane of the gel as follows:
(i) Using the box tool, draw a box around the entire first gel lane.
(ii) Select Analyze.
(iii) Select Gels.
(iv) Select First Lane.
(v) Drag the box to the next lane.
(vi) Select Analyze.
(vii) Select Gels.
(viii) Select Next Lane.
(ix) Repeat for all lanes.

(d) Plot the area under the curves for the protein bands as follows:
(i) Select Analyze.
(ii) Select Gels.
(iii) Select Plot Lanes.

(e) One graph per lane will appear with peaks representing each protein band.

(f) Use the line tool to connect the bottom of each peak.

(g) Select the wand tool.

(h) Fill in each peak by selecting the center of the peak.

(i) Export results as a csv file.
(i) In the results table select File and Save As.

(j) Determine the relative abundance of the VP proteins as compared to the entire lane using the integrated density (ID) as follows:
(i) ((ID VP1 + ID VP2 + ID VP3)/(Total ID of Lane)) × 100

(k) Bands for VP1:VP2:VP3 should be visible in a ratio of 1: 1:10.

(l) VP protein bands should account for >90% of the protein abundance in the lane.

3.5 Limulus Amebocyte Lysate Chromogenic Endotoxin Test

A LAL chromogenic endotoxin test is used to detect bacterial endotoxins in rAAV preparations. Endotoxin is a lipopolysaccharide (LPS) molecule that can be found in the cell membrane of gram-negative bacteria. Since the presence of endotoxin in rAAV preparations can cause an immune response when injected into an animal model, it is crucial to confirm that no endotoxin contamination occurred during vector production.

1. Follow all steps of the procedure according to manufacturer instructions.
   (a) Allow all reagents to warm up to room temperature before use.
   (b) Reconstitute endotoxin standard in endotoxin-free water at 10 EU/mL.
      (i) Vortex the solution for at least 15 min.
   (c) Prepare desired set of standards from reconstituted endotoxin stock solution by preparing an initial dilution and at least four twofold serial dilutions in endotoxin-free glass vials, e.g., 0.01–0.1 EU/mL or 0.1–1 EU/mL.
      (i) Vortex each dilution for 1–2 min before proceeding.
   (d) Reconstitute lyophilized limulus amebocyte lysate (LAL) reagent in endotoxin-free water. Do not vortex. Avoid foaming.
   (e) Reconstitute lyophilized chromogenic substrate in endotoxin-free water.
      (i) Prewarm a sufficient amount of substrate to 37 °C for 5–10 min before use.
   (f) Pre-equilibrate a microplate in a heat block at 37 °C. Maintain the plate at 37 °C throughout the procedure.
   (g) In your assay, include the twofold serial dilutions of endotoxin as a standard, dilutions of the sample of interest and endotoxin-free water as a negative control.
      (i) Transfer 50 μL of each standard, rAAV sample, or endotoxin-free water into the appropriate wells while carefully avoiding contaminations.
      (ii) Add 50 μL of LAL reagent solution to each well.
   (h) Start your timer immediately once the LAL reagent solution is added to the first well. The appropriate time is indicated on the lysate vial.
   (i) Take the microplate out of the heat block and gently tap the side of the plate to mix.
   (j) Return the plate to the heat block and cover.
   (k) After the appropriate time indicated on the lysate vial, add 100 μL prewarmed chromogenic substrate solution per well.
   (l) Take the microplate out of the heat block and gently tap the side of the plate to mix.

(m) Return the plate to the heat block and cover.

(n) After 6 min, add 100 μL stop solution per well.

(o) Take the microplate out of the heat block adapter and gently tap the side of the plate to mix.

(p) Place the microplate in the plate reader and read the absorbance of each well at 405–410 nm.

(q) Assay analysis:

(i) Analyze data graphically according to kit manufacturer instructions by plotting the mean absorbance and correlating endotoxin concentrations of your standards into a graph and drawing a line for your standard curve. Determine the endotoxin concentration for each rAAV sample based on their absorbance using linear regression.

(ii) Alternatively, the data can be analyzed using a spreadsheet or calculator by calculating the mean absorbance for each of the four standards and determining the corresponding endotoxin concentration for each sample by linear regression.

3.6 In vitro Sterility and Expression Assay

An in vitro assay in which small amounts of rAAV preparation are used to transduce a target cell line. This assay can confirm in vitro expression of the expected fluorophore in fluorescently labeled rAAV and allow detection of any potential bacterial, fungal, or yeast contamination.

1. Determine the appropriate cell line for transgene expression based on the promoter in your construct (Table 3).
2. Warm-up media.
3. Prepare a tissue culture-treated 96-well plate in a biosafety cabinet.
4. Determine the number of wells required for testing your rAAV sample.

(a) If the vector is not recombinase-dependent:

(i) Plan to transduce duplicate wells with your vector.

(ii) Leave two wells untransduced as a negative control.

(b) If the vector is recombinase-dependent:

(i) Plan to transduce duplicate wells with the recombinase-dependent vector only.

(ii) Plan to transduce duplicate wells with recombinase-expressing rAAV only.

Table 3
Promoters along with their expression targets and recommended cell lines for testing

Promoter	Specificity	Suggested cell line for in vitro expression testing
CMV/CBA/CAG [20–22]	Ubiquitous	HEK293T
EF1a [23]	Ubiquitous	HEK293T
PGK [24]	Ubiquitous	HEK293T
Synapsin1 [22, 25–27]	Neuronal, broad	N2A
CamKIIa [22, 28, 29]	Neuronal, glutamatergic	N2A
Dlx enhancer [30]	GABAergic interneurons	N2A
GFAP [31, 32]	Astrocytes/Glia	N2A/U-87 MG
CD68 [33]	Microglia	C8-B4
TBG [34]	Hepatocytes	HepG2/Huh-7

(iii) Plan to transduce duplicate wells with the recombinase-dependent and recombinase-expressing vectors.

(iv) Leave two wells untransduced as a negative control.

(c) If the vector is expressing a recombinase:

(i) Plan to transduce duplicate wells with the recombinase-expressing vector only.

(ii) Plan to transduce duplicate wells with a recombinase-dependent rAAV only.

(iii) Plan to transduce duplicate wells with the recombinase-expressing and recombinase-dependent vectors.

(iv) Leave two wells untransduced as a negative control.

5. Prepare a cell stock solution to seed an appropriate amount of cells per well in 200 μL media each.

(a) For example, for HEK293T cells, seed 5000 cells per well in 200 μL DMEM. For N number of wells, add $(N + 1) \times 5000$ cells to $(N + 1) \times 200$ μL DMEM high glucose supplemented with 10% fetal bovine serum (D10) in a 15 mL conical tube. Mix gently.

6. Seed cell solution in appropriate wells on the 96-well plate.
7. Add 1 μL of rAAV sample to the appropriate wells.
8. Gently mix cells and rAAV by pipetting.
9. Return plate to incubator.

10. Sterility and expression testing analysis:
    (a) Look at each well under 20× magnification.
    (b) After 24–48 h, check for the presence or absence of bacterial or fungal contamination.
    (c) After 24–120 h, depending on the cell line and promoter, check the expression levels of your fluorescent tag.
    (d) Check all fluorescent filters to ensure only the expected fluorophores are being expressed.
        (i) For example, a GFP-expressing rAAV should not show any fluorescence in the RFP channel.
    (e) Take images of each well with all filters, using the 20× objective.
        (i) For example, brightfield, GFP, RFP, YFP.
        (ii) Ensure that recombinase-dependent vectors are only expressing the transgene and fluorophore in the presence of a recombinase-expressing rAAV.
    (f) Generate an overlay image.
11. Carefully examine the images and look for the following (Fig. 5):
    (a) Untransduced sample: There should be no observable bacterial, fungal, or yeast growth in the image. Bacterial growth is observed as dark specs in the background between cells; yeast is observed as round circles; and fungi typically grow as strands. No fluorescence should be seen in this well.
    (b) Recombinase-only sample: If the recombinase is not fluorescently tagged, there should be no fluorescence visible in the well, as in Fig. 5. If the recombinase has a fluorescent tag, such as RFP, you should observe RFP fluorescence but no fluorescence in the other filters. Ensure that your recombinase does not have the same fluorescent tag as your recombinase-dependent rAAV, or you will not be able to interpret the results.
    (c) Non-recombinase-dependent fluorescently tagged rAAV sample: You should observe fluorescence only with the appropriate filter. For example, if the rAAV has a GFP tag, fluorescence should be observed in the GFP filter only.
    (d) Recombinase-dependent fluorescently tagged rAAV only sample: Fluorescence should not be observed in any filter. If fluorescence is detected, it suggests that the rAAV has undergone recombinase-independent recombination. While low levels of recombinase-independent

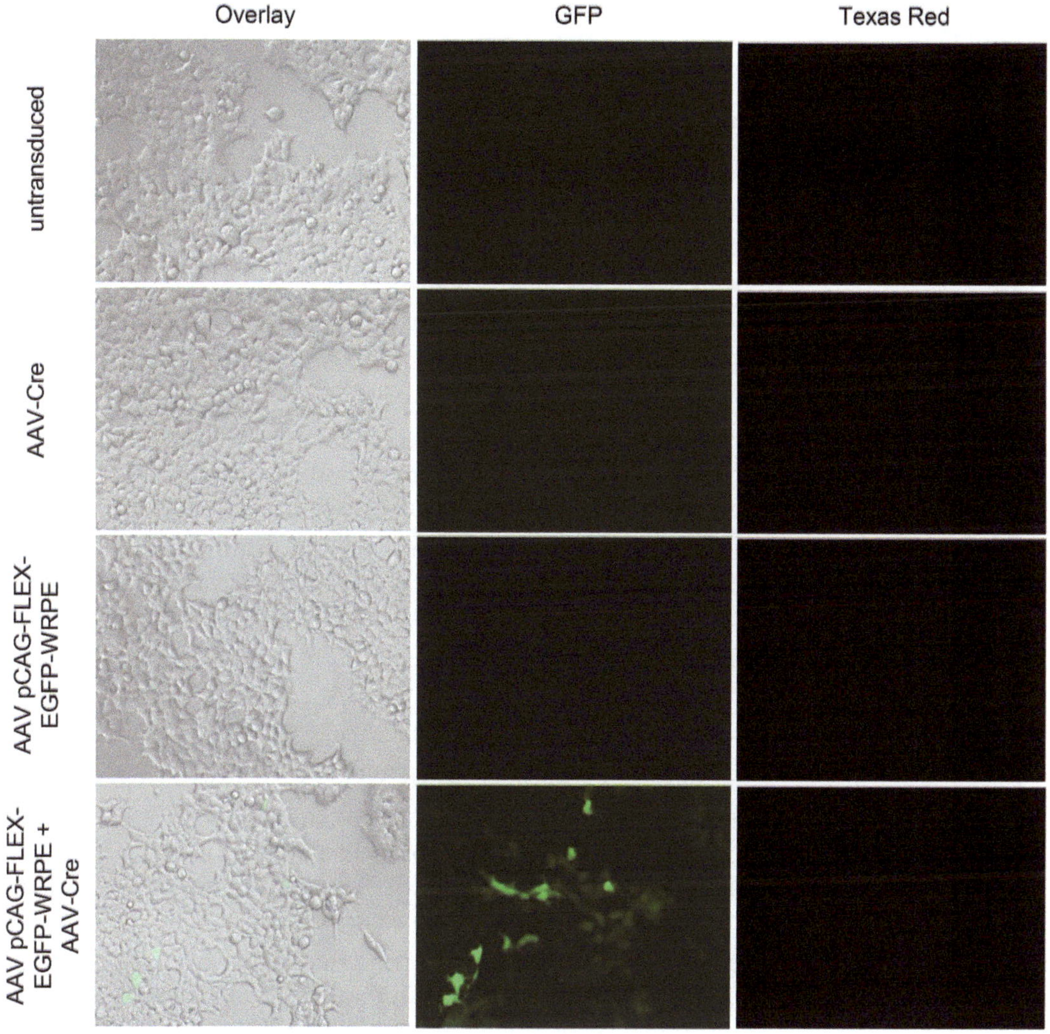

Fig. 5 In vitro expression images for AAV pCAG-FLEX-EGFP-WPRE with and without AAV-Cre in AAVpro 293T cells

recombination are expected, levels should be significantly lower than for the dual recombinase-dependent fluorescently tagged rAAV and recombinase sample.

(e) Recombinase-dependent fluorescently tagged rAAV and recombinase sample: Fluorescence should be observed for this sample in the appropriate channels. For example, if the recombinase-dependent fluorescently tagged rAAV has a GFP tag and the recombinase is untagged, then you should observe GFP fluorescence only. If the recombinase-dependent fluorescently tagged rAAV has a GFP tag and the recombinase has an RFP tag, then you should observe overlapping GFP and RFP fluorescence.

4 Notes

Titration by Quantitative PCR

- Do not treat your plasmid standards with DNase.
- The quality of the sample dilution series is critical. Make sure to pipette each dilution up and down at least 10 times and use at least half of the final volume to mix.
- Always include a no template control (NTC) in your run.
- Run plasmid standard and samples at least in duplicate.
- Plasmid standard: If only making one virus, use the transfer plasmid. If making many different ones, use a plasmid that contains a common element found in the different viruses. Store the standard in 4 μg/mL yeast tRNA to increase stability.
- Include a reference rAAV virus of a known titer in all assays to ensure run-to-run consistency. The reference virus should have a titer within one log of the expected titer of the sample.
- The following primer sequences can be used when detecting AAV2 ITRs [1]:
  - Forward ITR primer, 5′-GGAACCCCTAGTGATGGAGTT
  - Reverse ITR primer, 5′-CGGCCTCAGTGAGCGA
- Baseline removal: all samples will have some small amount of background signal that is most evident during initial PCR cycles. This background signal must be removed to determine the differences between samples accurately.
- Melt curve analysis: a single peak should be seen. The presence of a second peak at a temperature of ~70–75 °C usually indicates the presence of primer dimers which can increase the background signal and alter the Ct values of your samples.
- Pipetting less than 5 μL is not recommended. When preparing the master mix, prepare for enough reactions that the volume of each primer is >5 μL.

Confirmation of packaged AAV genome by identity PCR

- To reduce the level of plasmid recombination, use a stable, competent cell line for plasmid propagation and grow cultures at 30 °C.

Vector Purity by SDS-PAGE and Silver Staining

- Silver staining solution is considered hazardous and should not be poured down the sink.
- The gel should be completely submerged in the solution. A floating gel may not be stained properly.

- The VP proteins should make up at least 90% of the band intensity of the lane. Depending on the downstream applications, some laboratories may require higher levels of purity.

Limulus Amebocyte Lysate Chromogenic Endotoxin Test

- Reconstituted LAL reagent can be stored at −20 °C for up to 1 week. Freeze and thaw only once. Thaw immediately before use.
- Reconstituted endotoxin can be stored at 2–8 °C for up to 4 weeks. Vortex vigorously for 15 min before every use.
- The reconstituted chromogenic substrate can be stored at 2–8 ° C for up to 4 weeks.
- Protect chromogenic substrate from exposure to light.
- All materials used must be endotoxin-free.

In vitro Sterility and Expression Assay

- Before setting up this assay, determine the optimal cell density for your cell line in a 96-well plate.
- Recombinase activity can be driven by very low levels of protein. In some cases, when using a fluorescently tagged recombinase and fluorescently tagged recombinase-dependent rAAV, you may observe expression of the recombinase-dependent rAAV in cells that do not express observable levels of the tagged-recombinase. This does not necessarily represent recombinase-independent expression as the recombinase may be present in the sample but at levels too low to observe. In this case, refer to the recombinase-dependent fluorescently tagged rAAV only sample; if no fluorescence was detected in this sample, then the fluorescence observed in the dual recombinase-dependent fluorescently tagged rAAV and recombinase sample is likely being driven by low levels of the recombinase.

References

1. Aurnhammer C, Haase M, Muether N et al (2012) Universal real-time PCR for the detection and quantification of adeno-associated virus serotype 2-derived inverted terminal repeat sequences. Hum Gene Ther Methods 23:18–28
2. Lock M, Alvira MR, Chen S-J, Wilson JM (2014) Absolute determination of single-stranded and self-complementary adeno-associated viral vector genome titers by droplet digital PCR. Hum Gene Ther Methods 25: 115–125
3. Dobnik D, Kogovšek P, Jakomin T et al (2019) Accurate quantification and characterization of adeno-associated viral vectors. Front Microbiol 10:1570
4. Dingle T, Sedlak R, Cook L et al (2013) Tolerance of droplet-digital PCR versus real-time quantitative PCR to inhibitory substances. Clin Chem 59(11):1670–1672
5. Allay J, Sleep S, Long S et al (2011) Good manufacturing practice production of self-complementary serotype 8 adeno-associated viral vector for a hemophilia B clinical trial. Hum Gene Ther 22(5):595–604
6. Fagone P, Wright J, Nathwani A et al (2012) Systemic errors in quantitative polymerase chain reaction titration of self-complementary

adeno-associated viral vectors and improved alternative methods. Hum Gene Ther Methods 23(1):1–7
7. Zhen Z, Espinoza Y, Bleu T et al (2004) Infectious titer assay for adeno-associated virus vectors with sensitivity sufficient to detect single infectious events. Hum Gene Ther 15:709–715
8. Zeltner N, Kohlbrenner E, Clément N et al (2010) Near-perfect infectivity of wild-type AAV as benchmark for infectivity of recombinant AAV vectors. Gene Ther 17:872–879
9. Guerin K, Rego M, Bourges D et al (2020) A novel next-generation sequencing and analysis platform to assess the identity of recombinant adeno-associated viral preparations from viral DNA extracts. Hum Gene Ther 31:664–678
10. Maynard LH, Smith O, Tilmans NP et al (2019) Fast-Seq: a simple method for rapid and inexpensive validation of packaged single-stranded adeno-associated viral genomes in academic settings. Hum Gene Ther Methods 30:195–205
11. Tai PWL, Xie J, Fong K et al (2018) Adeno-associated virus genome population sequencing achieves full vector genome resolution and reveals human-vector chimeras. Mol Ther Methods Clin Dev 9:130–141
12. Lecomte E, Tournaire B, Cogné B et al (2015) Advanced characterization of DNA molecules in rAAV vector preparations by single-stranded virus next-generation sequencing. Mol Ther Nucleic Acids 4:e260
13. Pacouret S, Bouzelha M, Shelke R et al (2017) AAV-ID: a rapid and robust assay for batch-to-batch consistency evaluation of AAV preparations. Mol Ther 25:1375–1386
14. Ayuso E, Mingozzi F, Montane J et al (2009) High AAV vector purity results in serotype- and tissue-independent enhancement of transduction efficiency. Gene Ther 17:503–510
15. Sampath V (2018) Bacterial endotoxin-lipopolysaccharide; structure, function and its role in immunity in vertebrates and invertebrates. Agric Nat Resour 52:115–120
16. Uphoff CC, Drexler HG (2011) Detecting mycoplasma contamination in cell cultures by polymerase chain reaction. In: Methods in molecular biology. Humana Press, pp 93–103
17. Uphoff CC, Drexler HG (2012) Detection of mycoplasma contaminations. In: Basic cell culture protocols. Humana Press, pp 1–13
18. Spierenburg GT, Polak-Vogelzang AA, Bast BJEG (1988) Indicator cell lines for the detection of hidden mycoplasma contamination, using an adenosine phosphorylase screening test. J Immunol Methods 114:115–119
19. King D, Hitchcock T, Krishnan R (2021) Evolution of viral vector analytics for gene therapy manufacturing. In: Insights on successful gene therapy manufacturing and commercialization. The Cell Culture Dish, Inc, pp 34–31
20. Hitoshi N, Ken-ichi Y, Jun-ichi M (1991) Efficient selection for high-expression transfectants with a novel eukaryotic vector. Gene 108:193–199
21. Jun-ichi M, Satoshi T, Kimi A et al (1989) Expression vector system based on the chicken β-actin promoter directs efficient production of interleukin-5. Gene 79:269–277
22. Nathanson JL, Yanagawa Y, Obata K, Callaway EM (2009) Preferential labeling of inhibitory and excitatory cortical neurons by endogenous tropism of adeno-associated virus and lentivirus vectors. Neuroscience 161:441–450
23. Kim DW, Uetsuki T, Kaziro Y et al (1990) Use of the human elongation factor 1α promoter as a versatile and efficient expression system. Gene 91:217–223
24. Adra CN, Boer PH, McBurney MW (1987) Cloning and expression of the mouse pgk-1 gene and the nucleotide sequence of its promoter. Gene 60:65–74
25. Hoesche C, Sauerwald A, Veh RW et al (1993) The 5′-flanking region of the rat synapsin I gene directs neuron-specific and developmentally regulated reporter gene expression in transgenic mice. J Biol Chem 268:26494–26502
26. Glover CPJ, Bienemann AS, Heywood DJ et al (2002) Adenoviral-mediated, high-level, cell-specific transgene expression: a SYN1-WPRE cassette mediates increased transgene expression with no loss of neuron specificity. Mol Ther 5:509–516
27. Kügler S, Kilic E, Bähr M (2003) Human synapsin 1 gene promoter confers highly neuron-specific long-term transgene expression from an adenoviral vector in the adult rat brain depending on the transduced area. Gene Ther 10:337–347
28. Dittgen T, Nimmerjahn A, Komai S et al (2004) Lentivirus-based genetic manipulations of cortical neurons and their optical and electrophysiological monitoring in vivo. Proc Natl Acad Sci 101:18206–18211
29. Hioki H, Kameda H, Nakamura H et al (2007) Efficient gene transduction of neurons by lentivirus with enhanced neuron-specific promoters. Gene Ther 14:872–882
30. Dimidschstein J, Chen Q, Tremblay R et al (2016) A viral strategy for targeting and manipulating interneurons across vertebrate species. Nat Neurosci 19:1743–1749

31. Brenner M, Kisseberth W, Su Y et al (1994) GFAP promoter directs astrocyte-specific expression in transgenic mice. J Neurosci 14: 1030–1037
32. Lee Y, Messing A, Su M, Brenner M (2008) GFAPpromoter elements required for region-specific and astrocyte-specific expression. Glia 56:481–493
33. Greaves DR, Quinn CM, Seldin MF, Gordon S (1998) Functional comparison of the murine macrosialin and human CD68 promoters in macrophage and nonmacrophage cell lines. Genomics 54:165–168
34. Yan Z, Yan H, Ou H (2012) Human thyroxine binding globulin (TBG) promoter directs efficient and sustaining transgene expression in liver-specific pattern. Gene 506:289–294

Part II

Vector Modifications and Applications

Chapter 6

Vector Tropism

Shih-Heng Chen, Bo He, Sarwyn Singh, and Negin P. Martin

Abstract

Recombinant viruses expand the neurobiology toolbox and offer researchers an assortment of gene delivery options. Each type of virus varies in its capabilities and limitations, and therefore, there are important considerations in choosing the best viral vector system for each application. The suitability of a viral vector for gene delivery is dictated by factors such as genomic cargo capacity, duration and regulation of its transgene expression, immunogenicity, toxicity, and viral tropism. A myriad of permissive host factors is necessary to successfully express the viral genome and deliver transgenes. Viral tropism depends on the interaction of host and virion surface proteins for attachment and a concerted virus and host gene expression and assembly system. In this chapter, we hope to provide readers with a practical guide for selecting vectors for gene delivery in neuroscience applications.

Key words Virus, Tropism, AAV, Retrovirus, Lentivirus, Rabies delta-G, HSV, PRV, Sindbis

1 Introduction

The mammalian brain is a mosaic organ made up of neurons, glial cells, and stroma. Most neuroscience studies use engineered viruses to deliver light-sensitive effectors, chemogenetic sensors and receptors, tracers, and/or functional regulators to study the neural architecture or to investigate physiological functions in the nervous system. Neurons are typically very resistant to nonviral modes of gene delivery and are easily injured. The presence of the blood–brain barrier further complicates and limits widespread gene delivery throughout the brain. Therefore, viruses are often delivered to specific regions of the brain via stereotactic injections for localized gene delivery [1]. Very few recombinant viral vectors cross the blood–brain barrier. Methods for in vivo viral delivery are further discussed in Chaps. 7, 12, 13, and 16. In engineered viruses, genomic materials are replaced with desired transgenes and

Shih-Heng Chen and Bo He contributed equally.

Mark A. G. Eldridge and Adriana Galvan (eds.), *Vectorology for Optogenetics and Chemogenetics*, Neuromethods, vol. 195, https://doi.org/10.1007/978-1-0716-2918-5_6,

encapsulated in viral capsids and envelope proteins. Removal of genomic material often renders recombinant viruses infectious but nonpropagating. Following transduction, the viral genome is released into the host cell for expression. To select a suitable viral vector, researchers should consider the gene size capacity of a vector, duration and timing of gene expression, type of viral genome nucleic acid (RNA vs. DNA), transcriptional regulation, tissue and cell tropism, immunogenicity, and toxicity of the vector in the targeted in vivo model.

All viruses deliver a finite amount of genomic material. Adeno-associated viruses (AAVs) deliver a small genetic cargo of up to 4.2 kilobases (kb), whereas helper-dependent adenoviruses (HD-Ad) can accommodate 37 kb of genomic material. Separate AAV vectors may be needed to deliver multiple genes, and large genes or promoters often do not fit in a single AAV. Self-cleaving peptides (e.g., T2A and P2A) or internal ribosome entry sites (IRES) are helpful elements for expressing multiple genes from a single promoter to economize space in AAV vectors. Gene capacities of several popular recombinant viral vectors are summarized in Table 1 and are discussed in detail in Part 1 of this book.

Table 1
Viral vector genome capacity

Virus	Non-enveloped or enveloped	Genome	Transgene capacity
Adenovirus (AdV)—1st Gen	non-env	dsDNA	8 kb
Adenovirus (AdV)—2nd Gen	non-env	dsDNA	14 kb
Helper dependent adenovirus (HDAdV)	non-env	dsDNA	36 kb
HD canine adenovirus type 2 (CAV-2)	non-env	dsDNA	30 kb
Replication-defective herpes simplex virus type 1 (HSV-1)	env	dsDNA	30 kb
HSV-1-based amplicon	env	dsDNA	150 kb
Pseudorabies virus (PRV)	env	dsDNA	36 kb
Adeno-associated virus (AAV)	non-env	ssDNA	4.2 kb
Self-complementary AAV (SC-AAV)	non-env	ssDNA	2.1 kb
Lentivirus/retrovirus	env	dsRNA	8.5 kb
SADB19-Rabies dG, dGL, SiR, CVS-N2c	env	ssRNA	6 kb
Sindbis	env	ssRNA	6 kb

dsDNA double-stranded DNA, *ssDNA* single-stranded DNA, *dsRNA* double-stranded RNA, *ssRNA* single-stranded RNA

The duration and timing of gene expression should also be reviewed as part of vector selection. For example, recombinant Sindbis viruses reach robust expression after 24 h, so they are ideal for rapid visualization of dendrites and measuring changes in synapse formation in response to stimuli [2]. Nevertheless, the Sindbis virus is also very toxic and induces neuronal death within 3–5 days. So, it will not be an effective vector for long-term studies. In contrast, AAV cargo may take up to 2–3 weeks to detect, but its low toxicity allows for lasting gene expression.

Another important feature of a viral vector is the nature of its viral genome: RNA vs. DNA. The rapid expression of the Sindbis virus is due to its positive-strand RNA genome that mimics an mRNA and is transcribed in the cytoplasm shortly after viral entry. However, the limitation of RNA vectors such as Sindbis or Rabies virus is that transgene transcription cannot be regulated by promoters or recombinases such as Cre, Flp, or Dre. Lentiviruses and retroviruses are the only known class of RNA viruses that are reverse transcribed to DNA upon transduction and could carry transcriptional regulatory elements. Transcriptional regulators and recombinases are often used to select or exclude cells for gene expression. For example, the expression of a gene or promoter delivered by a lentivirus flanked by loxP could be regulated in cells that express Cre recombinases. Selective expression of promoters and Cre recombinases in specific cells or tissue of a transgenic animal could establish transcriptional tropism. Promoters for specific cell type targeting are discussed in Chap. 11.

The outer surface of a recombinant virus determines which species, tissue, and cell type it targets. Modifications to the proteins on the surface of viruses or overexpression of virus-binding proteins on cells are often used to create novel tropisms. For example, adenovirus Ad5 binds to coxsackievirus and adenovirus receptor (CAR) on host cells as its primary cellular receptor. Overexpression of CAR in skeletal muscles and lymphocytes of transgenic animals that are typically deficient in CAR creates a platform for Ad5 infection [3, 4]. Researchers typically introduce mutations in viral capsid proteins to change the viral tropism of non-enveloped viruses such as adenoviruses and adeno-associated viruses. This technique has been used extensively to generate a wealth of engineered AAV variants, as described in Chap. 9. Tropism of enveloped viruses is modified by exchanging envelope glycoproteins of one virus for another or pseudotyping. Researchers often pseudotype lentiviruses, retroviruses, and rabies-delta G viruses to alter their tropism.

Viruses bind and infect neurons through dendrites, axon termini, or soma for localized gene expression in the nervous system. However, the length of an axon is heavily protected from viral infection or stimuli by the myelin sheath. Specialized viral envelopes enable viruses or their cargo to be transported from axon termini

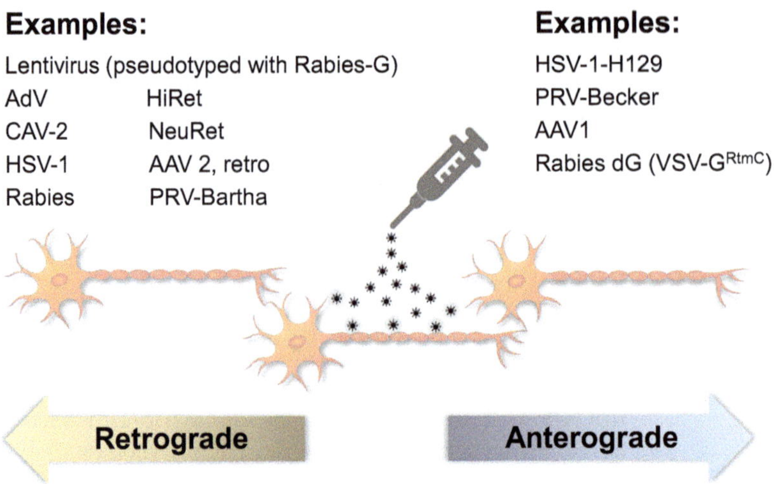

Fig. 1 Examples of vectors used for retrograde and anterograde neural tracing

toward the neuron body for retrograde transport. The movement of viruses or their cargo in the opposite direction toward the axonal termini is called anterograde transport. Examples of neurotropic recombinant viruses that shuttle along the axons are shown in Fig. 1. Viruses with retrograde/anterograde transport capabilities are often utilized for tracing neural tracks to map the nervous system architecture.

Several families of recombinant viruses are available for gene delivery to mammalian nervous systems: *Adenoviridae, Herpesviridae, Parvoviridae, Retroviridae, Rhabdoviridae*, and *Togaviridae.* In the following sections, we will describe the tropism and characteristics of these six broadly used recombinant vectors as a practical guide for the selection of gene delivery vectors.

2 *Adenoviridae* (e.g., Ad5, HD-Ad, and CAV-2)

2.1 Adenovirus

Adenovirus (AdV) is a non-enveloped double-stranded DNA virus. Adenoviruses, especially human adenovirus type 5 (HAdV5), are promising gene therapy vectors since they have broad tropism and offer robust gene expression in mammalian cells [5]. However, the high immunogenicity and cytotoxicity limit the applications of adenoviral vectors. Several generations of recombinant adenoviruses have been developed to reduce immunogenicity, extend gene expression, and increase cargo size. AdVs tropism allows for infection of neurons, astrocytes, microglia, oligodendrocytes, ependymal cells, and choroidal epithelial cells in mammalian brains [6–9]. HAdV5 not only expresses transgenes in infected neurons but also delivers transgenes via retrograde axonal transport to neural projections from the injected area (Table 2) [10]. The lytic

Table 2
Duration and direction of transgene expression in viral vectors

Virus	Initial expression	Direction	Duration
Adenovirus (AdV)	Weeks	Retrograde	Weeks to months
Canine adenovirus type 2 (CAV-2)	Weeks	Retrograde	Months to year
Herpes simplex virus type 1 (HSV-1)	Weeks	Retrograde (anterograde: HSV-1-H129)	Weeks to months
Pseudorabies virus (PRV)	Days	Bidirectional: PRV-Becker; retrograde: PRV-Bartha	Days to weeks
Adeno-associated virus (AAV or scAAV)	Days to weeks	Serotype dependent	Weeks to months
Lentivirus/retrovirus	Days	Pseudotype dependent	Weeks
SADB19-Rabies dG, dGL, SiR, CVS-N2c	Days to weeks	Retrograde	Days to weeks depending on toxicity
Sindbis	Hours to days	Injection site only	Days due to toxicity

infection of wild-type AdVs results in cell death after infection. Therefore, the essential E1 region of the genome was removed in the first generation of AdV vectors to reduce immunogenicity. In addition, the removal of the non-essential E3 region increased the cargo capacity to 8 kb. In the second generation, AdV, E2, and E4 regions were removed to further increase the cargo size to 14 kb [11]. However, most of the viral genes responsible for high immunogenicity and cytotoxicity are still present in the second-generation AdV vectors, and their toxicity limits the duration of transgene expression. To reduce the immune response against the Helper-Dependent Adenovirus (HDAd) vectors, all viral genes except the inverted terminal repeat (ITR) and packaging signal (ψ) were removed [12]. This deletion also dramatically increased the cargo size of the HDAd vector to 37 kb. The HDAd-infected cells do not present viral epitopes on their cell membranes reducing the immune response against AdV and prolonging transgene expression. The HDAd vectors provide superior nonintegrating and nonpropagating vectors that can deliver large genetic cargos to mammalian brain cells [13].

2.2 Canine Adenovirus Type 2 (CAV-2)

Preexisting immunity against human adenoviruses reduces the transduction efficiency of HAdVs and HDAd vectors for human gene delivery [14]. Recombinant nonhuman adenoviruses such as canine adenovirus type 2 (CAV-2) are promising vectors for clinical research and gene therapy due to the lack of adenovirus immunological memory in humans. CAV-2 is a safe and effective vector for laboratory use because it cannot replicate in human cells; however, it successfully delivers transgenes to rodents and nonhuman primates [15]. Studies show that recombinant CAV-2 vectors preferentially transduce neurons for local and retrograde transport of transgenes. They can be utilized as neural tracers for long-term gene expression in the rodent brain with very low immunogenicity. Replication-defective CAV-2 (E1/E3-deleted) vectors can deliver 7 kb of genetic cargo, and the helper-dependent CAV-2 (HD CAV-2) has a 30 kb cargo capacity [16]. The CAV-2 vectors enter the cells via coxsackievirus and adenovirus receptor (CAR). In the brain, CAR receptors are expressed on the surface of neurons. CAV-2 binds to the CAR at the neurons' presynaptic region and retrograde transport to deliver the transgene to the neuron body [17]. The expression of CAR varies among neurons of different brain regions. For example, CAV-2 only infects a subpopulation of neurons in the basal-lateral amygdala [18]. A recent study utilized AAV delivery of CAR in neurons to enable a high-efficiency, tropism-free CAV-2 gene delivery in various neuronal circuits [19]. Furthermore, due to the low immunogenicity of CAV-2 vectors, the transgene expression could persist for months in the infected neurons (Table 2) [20]. Moreover, a variety of tissues and the morphology of neurons were analyzed, and no major signs of virus-associated toxicity were noted [19, 21, 22].

3 *Herpesviridae* (e.g., HSV-1 and PRV)

3.1 Herpes Simplex Virus Type 1

Herpes simplex virus type 1 (HSV-1) is a human pathogen that infects over 90% of the population worldwide. HSV-1 has a broad cell tropism and infects almost every kind of cell, including neurons. Wild-type HSV-1 initially infects and replicates at the port of entry, such as oral or cornea mucosal tissues. Subsequently, the virus is retrogradely transported to the peripheral nervous systems (e.g., trigeminal ganglion) and the central nervous system (e.g., brain and spinal cord) and eventually establish latent infection in neurons for life [23]. Recombinant HSV vectors are directly injected into mammalian brains for in vivo gene delivery and neural tracing. HSV-1 infects neurons, astrocytes, oligodendrocytes, and neuronal progenitor cells at the injection site [23–25]. Most of the replication-defective recombinant HSV-1 vectors transport in the retrograde direction. The HSV-1 strain, H129, is the only strain reported to transport in an anterograde direction [26]. The preferential neural tropism and retrograde/anterograde transportability make HSV-1 an excellent vector for neural tracing and gene delivery to the nervous system (Table 2).

There are two HSV-1 vectors currently available to researchers: (1) replication-defective HSV-1 and (2) HSV-1-based amplicons [27]. The payload of replication-defective HSV-1 vector is 30 kb [28]. In subsequent generations of replication-defective HSV-1 vectors, one or several essential genes, such as immediate early genes (e.g., ICP4, ICP22, and ICP27) that prevent the virus from replicating and killing the infected cells, were removed [29]. However, replication-defective HSV-1 vectors still retain most viral genes and are toxic to cells.

The HSV-1-based amplicon vectors lack most of the viral genes, which dramatically reduces cytotoxicity. The HSV-1 amplicon vectors can accommodate up to 150 kb of foreign DNA, which could deliver one large gene or multiple genes to neurons [27, 28]. Moreover, the HSV-1 amplicon virus can package numerous amplicon plasmid copies as a head-to-tail concatemer in a single vector. The concatemers could enhance the expression of the transgene, and the low cytotoxicity of the amplicon vector could result in persistent transgene expression in neurons for months. Packaging amplicons into HSV-1 particles requires a helper virus or can be achieved by utilizing bacterial artificial chromosome (BAC) or cosmid containing the cloned HSV-1 genome in a helper virus-free system. The helper virus-dependent HSV-1 amplicon system produces high titers of HSV-1 particles but often contains helper virus contamination that could lead to elevated immune response and potential cytotoxicity. More recent generations of packaging systems use the Cre-lox*P* system to remove the helper virus's packaging signals. This strategy greatly reduces the helper virus contamination. However, there is often a very low level of helper virus present that limits HSV-1 amplicon use in gene therapy [30]. The current helper virus-free systems are limited to low-titer stocks [31]. For more details, please refer to Chap. 3.

3.2 Pseudorabies Virus(PRV)

Pseudorabies (PRV) is an alphaherpesvirus with a large range of host tropism, but it does not infect primates and humans [32]. PRV infects animals like swine, cattle, sheep, and rodents. Due to its preferential tropism for neurons and trans-synaptic transportation, PRV is an excellent tool for studying multisynaptic circuits (Table 2). Several recombinant PRV neuronal tracers were derived from PRV-Becker and PRV-Bartha strains [33, 34]. These recombinant PRVs are engineered to carry up to 30 kb of transgenes such as β-galactosidase, fluorophores, or recombinases that are used for neural tracing. PRV-Becker is highly virulent and is transported bidirectionally in the infected neurons. PRV-151, PRV-180, and PRV-181 are examples of retrograde and anterograde tracers derived from PRV-Becker [35, 36]. PRV-Bartha is less virulent than the Becker strain; the immune response to Bartha is reduced; infected animals live longer, allowing deeper neuronal circuit exploration. Unlike the Becker strain, and recombinants from Becker, which are bi-directionally transported, the recombinant PRVs derived from the Bartha strain are retrogradely transported in the infected neurons. Multiple PRV strains can infect the same neuron simultaneously and project to different regions of the brain or peripheral tissues [37, 38]. The major disadvantages of PRV strains are replication-competency, cytotoxicity, high immunogenicity, cellular fusion, and electrophysiological changes that are observed in infected cells [39].

4 *Parvoviridae* (e.g., AAV)

Recombinant adeno-associated viruses (AAVs) are small (20 nm in diameter), replication-defective, non-enveloped, and nonpathogenic viruses with a cargo capacity of 4.2 kb. AAVs permit efficient gene transfer as well as long-term and stable expression in many mammalian organs and tissues that could last for years [40]. Self-complementing AAVs (scAAVs) bypass the rate-limiting step of the second-strand synthesis in host cells and are, therefore, more rapidly triggering transgene expression. However, the scAAV design limits cargo capacity to 2.1 kb [41].

For AAV-mediated gene transfer in vivo, there are three layers of regulatory machinery to consider for viral tropism and transgene expression: AAV capsids and their receptors, tissue/cell-specific promotors, and post-transcriptional regulatory elements.

4.1 AAV Serotype Receptors and Tropisms

The binding of the AAV particles to the putative cell surface receptors is the initial step for many AAV infections. Some AAV particles utilize multiple receptors for cell entry, while others share the same receptors from different families. Although each AAV serotype can transduce almost all major cell types in the brain, including neurons, microglia, astrocytes, and oligodendrocytes, transduction efficiency varies significantly among serotypes for individual species, strains, and specific cell types [42, 43]. AAV serotypes 1, 2, 5, 8, 9, and rh10 are the most studied AAV serotypes in the CNS [44]. Table 3 summarizes the primary and co-receptors necessary for AAV serotype binding and entry into cells. Expression levels of these cell surface receptors dictate AAV tropism in the host. Thus, overexpression of these receptors can yield engineered tropism in the nervous system. Recent studies have identified the Adeno-associated virus receptor (AAVR) as a universal receptor that recognizes multiple AAV serotypes. AAVR-knockout mice are resistant to some AAV serotype infections [45, 46]. Several recent techniques for viral capsid modification have further enriched the AAV variant pool for neuroscientists to target specific tissue and brain regions for neurobiology applications [47, 48].

Table 3
AAV serotype tropism in the central nervous system

Serotype	Cellular receptors	Co-receptors	Tropism in mouse brain
1	Sialic acid		With CMV promoter, neurons and microglia in striatum, neurons and oligodendrocytes in hippocampus, and microglia and oligodendrocytes in cortex in the mouse brain [84]
2	HSPG, AAVR	FGFR1,αV-β5 integrin, LamR, HGFR, α5β1integrin, CD9	With CMV promoter, oligodendrocytes and neurons in hippocampus, neurons in cortex, and neurons in striatum [84]
3	HSPG, Laminin	FGFR1, HGFR, LamR	Minimal CNS tropism
4	Sialic acid		With RSV promoter, astrocytes in subventricular zone and glia overlying the rostral migratory stream neural tube [85]
5	Sialic acid, AAVR, clathrin, and caveolar	PDGFR	With CMV promoter, astrocytes and neurons in striatum and neurons and astrocytes in hippocampus, and oligodendrocytes and microglia in cortex [84]
6	Sialic acid and HSPG	EGFR	With CMV promoter, microglia and neurons in cortex, microglia and astrocytes in hippocampus and neurons in striatum [59, 84]
7	Unknown	Unknown	With GUSB promoter, neuronsin cortex, striatum, thalamus, and hippocampus [59]
8	Laminin	LamR	With CMV promoter, astrocytes, oligodendrocytes and neurons in striatum, oligodendrocytes, microglia, astrocytes and neurons in hippocampus, and astrocytes, oligodendrocytes, microglia and neurons in cortex [84]
9	Laminin, N-linked galactose	A putative integrin, LamR	With CMV promoter,neuronsin striatum, oligodendrocytes and neurons in hippocampus, and neurons, microglia and oligodendrocytes in cortex [84]

HSPG heparan sulfate proteoglycan, *FGFR1* fibroblast growth factor receptor 1, *LamR* laminin receptor, *HGFR* hepatocyte growth factor receptor, *PDGFR* platelet-derived growth factor receptor, *CMV promoter cytomegalovirus* immediate-early enhancer and *promoter, RSV promoter* the Rous sarcoma virus long terminal repeat (RSV) promoter, *GUSB* β-glucuronidase

4.2 Delivery Routes and AAV Tropisms

In cell culture, AAV 5 possesses a strong tropism for glial cells, whereas AAV 1, 6, and 7 show efficient transduction of neurons [49]. In vivo, AAV tropism is dependent on cell surface receptors and the administration routes.

4.2.1 AAV Intravenous Administration (IV)

AAV 9, rh.8, and rh.10 can cross the blood–brain barrier (BBB) and have been shown to drive significant transgene expression in neonatal neurons and adult astrocytic glial cells [50–52]. This led to preclinical studies using intravenous administration of AAV vectors as a novel therapeutic treatment for neuropathic lysosomal storage disorders (LSD) [44]. In contrast, AAVs 1, 2, 5, 6, and 8 show poor penetration and transgene expression in the brain after systemic administration [53, 54]. Several AAV variants such as PHP.B, PHP.eB, PHP.v1, PHP.s, CAP-B10, and CAP-B22 have been developed to cross the BBB for broad or cell-specific gene delivery to the mouse brain [55]. However, their performance is mouse strain-dependent [56].

4.2.2 Intraparenchymal Injection

Most AAV serotypes, including AAVs 1, 2, 5, 7, 8, 9, and rh.10, exhibit strong neuronal tropism following direct intraparenchymal brain injection except for AAV4, which preferentially infects glia [57–59]. Furthermore, AAVs 1, 5, and 9 demonstrate stronger neural tropism for spinal motor neurons than AAVs 2, 6, and 8 [60]. Therefore, AAVs 1, 9, and rh.10 were tested for use in gene delivery in a mouse model of neuropathic LSD [44]. AAV2 is often used for gene transfer in small brain regions due to its weak transduction and limited diffusion in brain tissue following intraparenchymal Injection.

4.2.3 Intra-cerebroventricular (ICV) Administration

Using this route of administration, AAV 4 transduces ependymal cells [61], AAV 6 mediates gene transfer in the spinal cord and motor neurons [60], and AAVs 7 and 9 mediate robust transduction in the cortical region and spinal cord [62].

4.3 Capsid Modification

Several approaches have been used to alter AAV tropisms for neuroscience research.

4.3.1 Hybrid Capsids

This technique swaps regions of capsid proteins to create novel chimeras with unique characteristics. For example, the substitution of heparin-binding residues from AAV 2 onto a similar region in AAV 5, which itself does not bind to heparan sulfate proteoglycan (HSPG), gives this chimeric AAV5/2 the ability to bind heparan sulfate [63]. Hybrid capsids of AAV 1 and AAV 2 (AAV1/2) have AAV 2 tropism and the AAV 1 ability to diffuse through brain tissue efficiently [64].

4.3.2 Peptide Insertion

An inserted peptide displayed on the AAV capsid can redirect AAV tropism. Since VP2 is not essential for AAV infectivity, the addition of a mitochondrial targeting sequence (MTS) in the VP2 region redirects AAV 2 virions to mitochondria [65]. This variant offers a promising gene therapy vector for treating neuropathic diseases related to dysfunctional mitochondria.

4.3.3 Capsid Shuffling and Directed Evolution

Capsid DNA from different AAV serotypes are mixed and randomly reassembled to produce chimeric capsid genomes with novel tropisms [66]. For example, a cell type-specific capsid selection method called Cre recombinase-based AAV targeted evolution (CREATE) was recently used to create AAV PHP.B, PHP.eB, and PHP.S in C57BL/6J mice. These AAV vectors demonstrated efficient and noninvasive gene delivery to the central and peripheral nervous systems after systemic delivery [55]. Please refer to AAV capsid evolution technology in Chap. 7 for a closer look at this technique and a list of generated variants.

4.4 Regulatory Elements to Refine Gene Targeting

AAV transduction efficiency and specificity could be further enhanced when a cell type-specific promoter is utilized in AAV vectors. For example, the glial fibrillary acidic protein (GFAP) promoter was found highly specific for expression in astrocytes following vector infusion into neonatal and adult brains [9]. Another example is the myelin basic protein (MBP) promoter that produces poor gene expression in oligodendrocytes in neonatal mice but has excellent expression if the vector is injected at postnatal day 10 [67]. AAV transduction efficiency can be further improved by microRNA (miRNA) target sequences at the post-transcriptional levels in the target tissues. Tissue-specific miRNAs target sequences in AAV vectors and are also highly efficient in reducing transgene off-target expression [68, 69]. For example, in a recent study, a target sequence for miR183 was introduced into the 3′ untranslated region of the cargo transgene mRNA of an AAV to restrict expression to dorsal root ganglion (DRG) neurons and to down-regulate transgene expression to lower toxicity [70]. In many AAV transfer vectors, Woodchuck hepatitis virus posttranscriptional regulatory elements (WPREs) have been introduced to increase transgene expression at the post-translational level [71–73].

5 *Retroviridae* (e.g., Lentiviruses and Gamma-retroviruses)

Recombinant lentiviruses and gamma-retroviruses (often referred to as retroviruses) are enveloped viruses with double-stranded RNA genomes. Following infection, their RNA genome is reverse transcribed by viral proteins to DNA, transported to the nucleus, and incorporated randomly into their host chromosomes. These random insertions could give rise to mutations that affect host function and are one of the drawbacks of using lenti/retroviral vectors. Another disadvantage of lentiviruses is that, unlike AAVs, their large size (100 nm) limits their spread through the extracellular space and restricts their use for gene delivery to mammalian brains [74]. Conversion of the viral RNA genome to DNA allows for the insertion of promoters in transfer vectors to control transgene expression. Similar to AAVs (described in Subheading 4.4), regulatory elements could be used to enhance and further refine gene expression in target cells. The maximum cargo size carried by lenti/retroviruses is 8.5 kb—almost twice that of AAV [75]. Lentiviruses are capable of infecting both dividing and nondividing cells. Retroviruses, however, cannot traverse through the nuclear pore to reach the nucleus and only infect dividing cells during mitosis. Therefore, retroviruses are typically used for gene delivery and identification of neural stem cells [76]. The applications of lentiviruses in neuroscience are discussed in Chap. 1 of this book.

During viral production, viral envelope glycoproteins are localized on the host plasma membrane for virion assembly and budding. Matured envelope proteins dictate lenti/retroviral tropism. It is possible to change lenti/retroviral tropism by modifying their envelope proteins or by substituting other viral envelope glycoproteins during production. This process is called pseudotyping. Not all envelope glycoproteins are compatible or a suitable match for lenti/retroviral pseudotyping. Examples of compatible glycoproteins for pseudotyping lentiviruses and their central nervous system tropisms are listed in Table 4. Glycoprotein G from the vesicular stomatitis virus (VSV-G) is frequently used to give broad tropism to pseudotyped lentiviruses. VSV-G binds to ubiquitously expressed LDL receptors on the surface of most mammalian cells. Envelope glycoproteins can also be spliced or modified to alter tropism or viral titer. For example, as described in Chap. 2, NeuRet/HiRet vector technologies use fusion envelope glycoproteins that combine the tropism of the rabies G protein with the VSV-G stability for the preparation of high titer lentiviruses [77].

6 *Rhabdoviridae* (e.g., SADB19-Rabies dG, dGL, SiR, and CVS-N2c)

Rabies viruses are neurotropic single-stranded RNA viruses that can be transported in the retrograde direction in axons (Fig. 1). They are highly toxic to neurons and cause neural death as they spread through the nervous system. The retrograde transport of the rabies virus makes it an ideal vector to trace neural pathways and study mammalian brain architecture. Recombinant SADB19 and CVS-N2c are two commonly used strains for gene delivery in neurobiology research [78]. CVS-N2c is more effective in

Table 4
Lentiviral pseudotypes with neural tropism

Glycoprotein	Binding target	Tropism
Chandipura virus and Piry virus G	Not known	Neural cells [86]
Lymphocytic choriomeningitis virus GP1 and GP2	a-dystroglycan	Dendritic cells, neural cells, glia, hepatocytes, anterograde transport [86–88]
Mokola virus G	Not known	Neural cells, oligodendrocyte precursors [86, 87]
Murine leukemia virus G	GLVR-2 or Ram-1/Pit-2	Neuroglial cells [87, 88]
Rabies virus G, NeuRet, HiRet	P75NTR, NCAM, nAchR	Neural cells and glia, retrograde transport [77, 86, 88, 89]
Ross river virus E1 and E2	Possible involvement of heparin sulfate or integrins	Neuroglial cells, fibroblast, hepatocytes Kupffer cells [86, 88]
Semliki forest virus G	Possible involvement of heparin sulfate or integrins	Neuroglial cells, fibroblast, hepatocytes, Kupffer cells [86, 90]
Sindbis virus E1 and E2	Possible involvement of heparin sulfate or integrins	Broad, neurons [91]
Vesicular stomatitis virus G	LDL receptor	Broad [86, 87, 89]

transducing neurons but it takes longer to produce these virions. To add safety, the glycoprotein (G) is often replaced with avian virus envelope proteins such as Env A or Env B, and target cells carry the corresponding avian protein receptors such as TVA or TVB. AAVs are often used to deliver TVA and TVB genes to specific neurons in the brain to target them with rabies vectors. The Env A or B pseudotyped rabies viruses can only bind neurons that express TVA or TVB on their surface for local gene expression. Target cells must also express G, in addition to TVA or TVB, to promote retrograde transport of the rabies viruses and their transgenes. Rabies glycoprotein G is often used to pseudotype other enveloped viruses to induce retrograde transport. Since rabies is an enveloped virus, it can be pseudotyped similar to lentiviruses, with a myriad of envelopes from other viruses to change its tropism or transport. Rabies dG (VSV-G^{RtmC}) is pseudotyped with an engineered surface protein that alters the direction of its axonal transport to produce an anterograde rabies vector [79].

Removal of G from recombinant rabies viruses provides space of up to 6 kb for a transgene. However, the RNA genome of rabies viruses does not allow for the addition of any transcriptional regulatory elements. Rabies-dG strains remain toxic and cause neuronal death within 1–2 weeks. Several strategies have been employed to decrease toxicity and prolong the gene expression of rabies vectors. In SADB19-dGL, the RNA-dependent RNA polymerase (L) gene has been removed, and self-inactivating rabies (SADB19-SiR) carries a PEST protein sequence that targets viral proteins for degradation [80, 81]. Reduction in expression of rabies proteins lowers toxicity in SiR vectors. Engineered rabies vectors are discussed in Chap. 6.

7 *Togaviridae* (e.g., Sindbis Virus)

Sindbis virus is a single-stranded positive-sense RNA virus with robust and rapid expression in a broad range of hosts [82]. Recombinant Sindbis vectors can carry a transgene cargo of up to 6 kb but are similar to other RNA viruses, and they do not allow for any transcriptional regulation. Nonpropagating recombinant Sindbis vectors have been developed that do not spread beyond the primary site of infection. They are effective vectors for in vitro and in vivo transduction of neurons and produce high levels of transgene expression within hours in infected cells. Although Sindbis viruses are highly toxic and immunogenic, they are often used in neurobiology research due to their rapid gene expression [2]. Their typical applications are short-term protein expression or delivery of factors to study the electrophysiological properties of neurons. Sindbis viruses have a significant effect on host gene expression and therefore are not ideal for gene regulation studies [83].

8 Summary

Virus transduction is initiated by the interaction of capsids or envelope glycoproteins on the exterior of virions with cell surface receptors. Successful transduction also depends on a permissive host environment for the expression of viral and delivered genes. Low immunogenicity and toxicity of a viral genome and its proteins assure adequate virus levels in host tissue for effective transduction and long-term expression of transgenes, respectively. The capabilities and limitations of several commonly used viral vectors are summarized in Table 5. The following chapters will explain and examine strategies for cell-specific targeting in the nervous system, altering vector tropism, and the latest advances in viral technology that have added valuable tools for neurobiology research.

Table 5
Capabilities and limitations of viral vectors

Virus	Tropism	Advantage	Limitation
Adenovirus (AdV)	Neuron, astrocytes, microglia, oligodendrocytes, ependymal cells, choroidal epithelium cells	Rapid and robust gene expression, HDAd has large capacity and long-term transgene expression	High immunogenicity of 1st and 2nd generation of AdVs, neutralized by host immune response, contamination of helper virus
Canine adenovirus type 2 (CAV-2)	Neurons	Broad host range, efficient axon terminal infection, long-term expression, low immunogenicity and cytotoxicity	Neuron populations that can be infected is limited by CAR expression
Herpes simplex virus type 1 (HSV-1)	Neurons and glial cells	Large cloning capacity, high transduction efficiency, low cytotoxicity with amplicon	Contamination of competent HSV or helper virus, high cytotoxicity in replication-defective HSV-1 vectors
Pseudorabies virus (PRV)	Neurons	Large capacity, transsynaptic transportation	Causes cellular fusion, electrophysiological changes of infected cells, high cytotoxicity
Adeno-associated virus (AAV)	Serotype/variant dependent	Low immunogenicity, diverse and engineered serotypes	Low genome capacity, antibodies against the serotype may limit transdcution
Lentivirus/ Retrovirus	Pseudotype dependent	Large capacity, diverse engineered pseudotypes	Immunogenic, not robust in vivo, retrovirus does not infect non-dividing cells
SADB19-Rabies dG, dGL, SiR, CVS-N2c	Neurons	Monosynaptic tracing	Highly immunogenic and toxic
Sindbis virus	Broad	Rapid and robust expression	Highly toxic, no transcriptional regulation

References

1. Keiser MS, Chen YH, Davidson BL (2018) Techniques for intracranial stereotaxic injections of adeno-associated viral vectors in adult mice. Curr Protoc Mouse Biol 8(4):e57
2. Bastrikova N et al (2008) Synapse elimination accompanies functional plasticity in hippocampal neurons. Proc Natl Acad Sci U S A 105(8):3123–3127
3. Nalbantoglu J et al (2001) Muscle-specific overexpression of the adenovirus primary receptor CAR overcomes low efficiency of

gene transfer to mature skeletal muscle. J Virol 75(9):4276–4282
4. Hurez V et al (2002) Efficient adenovirus-mediated gene transfer into primary T cells and thymocytes in a new coxsackie/adenovirus receptor transgenic model. BMC Immunol 3:4
5. Bett AJ, Prevec L, Graham FL (1993) Packaging capacity and stability of human adenovirus type 5 vectors. J Virol 67(10):5911–5921
6. O'Carroll SJ, Cook WH, Young D (2020) AAV targeting of glial cell types in the central and peripheral nervous system and relevance to human gene therapy. Front Mol Neurosci 13: 618020
7. Berns KI, Giraud C (1996) Biology of adeno-associated virus. Curr Top Microbiol Immunol 218:1–23
8. Rabinowitz JE, Samulski J (1998) Adeno-associated virus expression systems for gene transfer. Curr Opin Biotechnol 9(5):470–475
9. Xiang H et al (2018) Glial fibrillary acidic protein promoter determines transgene expression in satellite glial cells following intraganglionic adeno-associated virus delivery in adult rats. J Neurosci Res 96(3):436–448
10. Ridoux V et al (1994) Adenoviral vectors as functional retrograde neuronal tracers. Brain Res 648(1):171–175
11. Ricobaraza A et al (2020) High-capacity adenoviral vectors: expanding the scope of gene therapy. Int J Mol Sci **21**(10):3643
12. Palmer DJ, Ng P (2008) Methods for the production of first generation adenoviral vectors. Methods Mol Biol 433:55–78
13. Montesinos MS, Satterfield R, Young SM Jr (2016) Helper-dependent adenoviral vectors and their use for neuroscience applications. Methods Mol Biol 1474:73–90
14. Fausther-Bovendo H, Kobinger GP (2014) Pre-existing immunity against Ad vectors: humoral, cellular, and innate response, what's important? Hum Vaccin Immunother 10(10): 2875–2884
15. Martel AC et al (2020) Targeted transgene expression in cholinergic interneurons in the monkey striatum using canine adenovirus serotype 2 vectors. Front Mol Neurosci 13:76
16. Del Rio D et al (2019) CAV-2 vector development and gene transfer in the central and peripheral nervous systems. Front Mol Neurosci 12:71
17. Soudais C et al (2001) Preferential transduction of neurons by canine adenovirus vectors and their efficient retrograde transport in vivo. FASEB J 15(12):2283–2285
18. Senn V et al (2014) Long-range connectivity defines behavioral specificity of amygdala neurons. Neuron 81(2):428–437
19. Li SJ et al (2018) A viral receptor complementation strategy to overcome CAV-2 tropism for efficient retrograde targeting of neurons. Neuron 98(5):905–917 e5
20. Soudais C, Skander N, Kremer EJ (2004) Long-term in vivo transduction of neurons throughout the rat CNS using novel helper-dependent CAV-2 vectors. FASEB J 18(2): 391–393
21. Simao D et al (2016) Evaluation of helper-dependent canine adenovirus vectors in a 3D human CNS model. Gene Ther 23(1):86–94
22. Smith BF et al (2006) Administration of a conditionally replicative oncolytic canine adenovirus in normal dogs. Cancer Biother Radiopharm 21(6):601–606
23. Roizman B, Knipe DM, Whitley RJ (2007) Herpes simplex viruses. In: Knipe DM et al (eds) Fields virology. Lippincott Williams & Wilkins, New York, pp 2501–2601
24. Bello-Morales R et al (2014) The effect of cellular differentiation on HSV-1 infection of oligodendrocytic cells. PLoS One 9(2):e89141
25. Zheng W et al (2020) Patterns of herpes simplex virus 1 infection in neural progenitor cells. J Virol **94**(16):e00994–e00920
26. Zemanick MC, Strick PL, Dix RD (1991) Direction of transneuronal transport of herpes simplex virus 1 in the primate motor system is strain-dependent. Proc Natl Acad Sci U S A 88(18):8048–8051
27. Neve RL (2012) Overview of gene delivery into cells using HSV-1-based vectors. Curr Protoc Neurosci Chapter 4:Unit 4.12
28. Jacobs A, Breakefield XO, Fraefel C (1999) HSV-1-based vectors for gene therapy of neurological diseases and brain tumors: part II. Vector systems and applications. Neoplasia 1(5):402–416
29. Manservigi R, Argnani R, Marconi P (2010) HSV recombinant vectors for gene therapy. Open Virol J 4:123–156
30. Neve RL, Lim F (2013) Generation of high-titer defective HSV-1 vectors. Curr Protoc Neurosci Chapter 4:Unit 4.13
31. Laimbacher AS, Fraefel C (2012) Gene delivery using helper virus-free HSV-1 amplicon vectors. Curr Protoc Neurosci Chapter 4:Unit 4.14
32. Pomeranz LE, Reynolds AE, Hengartner CJ (2005) Molecular biology of pseudorabies virus: impact on neurovirology and veterinary

medicine. Microbiol Mol Biol Rev 69(3): 462–500
33. Ekstrand MI, Enquist LW, Pomeranz LE (2008) The alpha-herpesviruses: molecular pathfinders in nervous system circuits. Trends Mol Med 14(3):134–140
34. Enquist LW et al (1998) Infection and spread of alphaherpesviruses in the nervous system. Adv Virus Res 51:237–347
35. Demmin GL et al (2001) Insertions in the gG gene of pseudorabies virus reduce expression of the upstream Us3 protein and inhibit cell-to-cell spread of virus infection. J Virol 75(22): 10856–10869
36. del Rio T et al (2005) Heterogeneity of a fluorescent tegument component in single pseudorabies virus virions and enveloped axonal assemblies. J Virol 79(7):3903–3919
37. Card JP et al (2011) A dual infection pseudorabies virus conditional reporter approach to identify projections to collateralized neurons in complex neural circuits. PLoS One 6(6): e21141
38. Card JP, Enquist LW (2014) Transneuronal circuit analysis with pseudorabies viruses. Curr Protoc Neurosci 68:1.5.1–1.539
39. McCarthy KM, Tank DW, Enquist LW (2009) Pseudorabies virus infection alters neuronal activity and connectivity in vitro. PLoS Pathog 5(10):e1000640
40. Asokan A, Schaffer DV, Samulski RJ (2012) The AAV vector toolkit: poised at the clinical crossroads. Mol Ther 20(4):699–708
41. Choudhury SR et al (2017) Viral vectors for therapy of neurologic diseases. Neuropharmacology 120:63–80
42. Chen SH et al (2019) Recombinant viral vectors as neuroscience tools. Curr Protoc Neurosci 87(1):e67
43. Chen SH et al (2019) Production of viral constructs for neuroanatomy, calcium imaging, and optogenetics. Curr Protoc Neurosci 87(1):e66
44. Hocquemiller M et al (2016) Adeno-associated virus-based gene therapy for CNS diseases. Hum Gene Ther 27(7):478–496
45. Zhang R et al (2019) Divergent engagements between adeno-associated viruses with their cellular receptor AAVR. Nat Commun 10(1): 3760
46. Pillay S et al (2016) An essential receptor for adeno-associated virus infection. Nature 530(7588):108–112
47. Goertsen D et al (2022) AAV capsid variants with brain-wide transgene expression and decreased liver targeting after intravenous delivery in mouse and marmoset. Nat Neurosci 25(1):106–115
48. Buning H, Srivastava A (2019) Capsid modifications for targeting and improving the efficacy of AAV vectors. Mol Ther Methods Clin Dev 12:248–265
49. Royo NC et al (2008) Specific AAV serotypes stably transduce primary hippocampal and cortical cultures with high efficiency and low toxicity. Brain Res 1190:15–22
50. Foust KD et al (2009) Intravascular AAV9 preferentially targets neonatal neurons and adult astrocytes. Nat Biotechnol 27(1):59–65
51. Gray SJ et al (2011) Preclinical differences of intravascular AAV9 delivery to neurons and glia: a comparative study of adult mice and nonhuman primates. Mol Ther 19(6): 1058–1069
52. Yang B et al (2014) Global CNS transduction of adult mice by intravenously delivered rAAVrh.8 and rAAVrh.10 and nonhuman primates by rAAVrh.10. Mol Ther 22(7): 1299–1309
53. Towne C et al (2008) Systemic AAV6 delivery mediating RNA interference against SOD1: neuromuscular transduction does not alter disease progression in fALS mice. Mol Ther 16(6):1018–1025
54. Chen SJ et al (2013) Biodistribution of AAV8 vectors expressing human low-density lipoprotein receptor in a mouse model of homozygous familial hypercholesterolemia. Hum Gene Ther Clin Dev 24(4):154–160
55. Ravindra Kumar S et al (2020) Multiplexed Cre-dependent selection yields systemic AAVs for targeting distinct brain cell types. Nat Methods 17(5):541–550
56. Hordeaux J et al (2018) The neurotropic properties of AAV-PHP.B are limited to C57BL/6J mice. Mol Ther 26(3):664–668
57. Burger C et al (2004) Recombinant AAV viral vectors pseudotyped with viral capsids from serotypes 1, 2, and 5 display differential efficiency and cell tropism after delivery to different regions of the central nervous system. Mol Ther 10(2):302–317
58. Passini MA et al (2003) Intraventricular brain injection of adeno-associated virus type 1 (AAV1) in neonatal mice results in complementary patterns of neuronal transduction to AAV2 and total long-term correction of storage lesions in the brains of beta-glucuronidase-deficient mice. J Virol 77(12):7034–7040
59. Cearley CN, Wolfe JH (2006) Transduction characteristics of adeno-associated virus vectors expressing cap serotypes 7, 8, 9, and Rh10 in the mouse brain. Mol Ther 13(3):528–537

60. Snyder BR et al (2011) Comparison of adeno-associated viral vector serotypes for spinal cord and motor neuron gene delivery. Hum Gene Ther 22(9):1129–1135
61. Davidson BL et al (2000) Recombinant adeno-associated virus type 2, 4, and 5 vectors: transduction of variant cell types and regions in the mammalian central nervous system. Proc Natl Acad Sci U S A 97(7):3428–3432
62. Samaranch L et al (2013) Strong cortical and spinal cord transduction after AAV7 and AAV9 delivery into the cerebrospinal fluid of nonhuman primates. Hum Gene Ther 24(5): 526–532
63. Wu Z, Asokan A, Samulski RJ (2006) Adeno-associated virus serotypes: vector toolkit for human gene therapy. Mol Ther 14(3):316–327
64. Ip CW et al (2017) AAV1/2-induced overexpression of A53T-alpha-synuclein in the substantia nigra results in degeneration of the nigrostriatal system with Lewy-like pathology and motor impairment: a new mouse model for Parkinson's disease. Acta Neuropathol Commun 5(1):11
65. Yu H et al (2012) Gene delivery to mitochondria by targeting modified adenoassociated virus suppresses Leber's hereditary optic neuropathy in a mouse model. Proc Natl Acad Sci U S A 109(20):E1238–E1247
66. Gray SJ et al (2010) Directed evolution of a novel adeno-associated virus (AAV) vector that crosses the seizure-compromised blood-brain barrier (BBB). Mol Ther 18(3):570–578
67. von Jonquieres G et al (2013) Glial promoter selectivity following AAV-delivery to the immature brain. PLoS One 8(6):e65646
68. He B et al (2012) Enhancing muscle membrane repair by gene delivery of MG53 ameliorates muscular dystrophy and heart failure in delta-Sarcoglycan-deficient hamsters. Mol Ther 20(4):727–735
69. Qiao C et al (2011) Liver-specific microRNA-122 target sequences incorporated in AAV vectors efficiently inhibits transgene expression in the liver. Gene Ther 18(4):403–410
70. Hordeaux J et al (2020) MicroRNA-mediated inhibition of transgene expression reduces dorsal root ganglion toxicity by AAV vectors in primates. Sci Transl Med 12(569):eaba9188
71. Wang L et al (2016) Enhancing transgene expression from recombinant AAV8 vectors in different tissues using woodchuck hepatitis virus post-transcriptional regulatory element. Int J Med Sci 13(4):286–291
72. Xu R et al (2001) Quantitative comparison of expression with adeno-associated virus (AAV-2) brain-specific gene cassettes. Gene Ther 8(17):1323–1332
73. Galvan A et al (2021) Intracerebroventricular administration of AAV9-PHP.B SYN1-EmGFP induces widespread transgene expression in the mouse and monkey CNS. Hum Gene Ther. 32(11–12):599–615
74. Lerchner W et al (2014) Injection parameters and virus dependent choice of promoters to improve neuron targeting in the nonhuman primate brain. Gene Ther 21(3):233–241
75. Salmon P, Trono D (2007) Production and titration of lentiviral vectors. Curr Protoc Hum Genet Chapter 12:Unit 12.10
76. Gage FH, Temple S (2013) Neural stem cells: generating and regenerating the brain. Neuron 80(3):588–601
77. Kato S, Kobayashi K, Kobayashi K (2014) Improved transduction efficiency of a lentiviral vector for neuron-specific retrograde gene transfer by optimizing the junction of fusion envelope glycoprotein. J Neurosci Methods 227:151–158
78. Osakada F, Callaway EM (2013) Design and generation of recombinant rabies virus vectors. Nat Protoc 8(8):1583–1601
79. Haberl MG et al (2015) An anterograde rabies virus vector for high-resolution large-scale reconstruction of 3D neuron morphology. Brain Struct Funct 220(3):1369–1379
80. Chatterjee S et al (2018) Nontoxic, double-deletion-mutant rabies viral vectors for retrograde targeting of projection neurons. Nat Neurosci 21(4):638–646
81. Ciabatti E et al (2017) Life-long genetic and functional access to neural circuits using self-inactivating rabies virus. Cell 170(2): 382–392 e14
82. Xiong C et al (1989) Sindbis virus: an efficient, broad host range vector for gene expression in animal cells. Science 243(4895):1188–1191
83. Uyaniker S et al (2019) The effects of Sindbis viral vectors on neuronal function. Front Cell Neurosci 13:362
84. Aschauer DF, Kreuz S, Rumpel S (2013) Analysis of transduction efficiency, tropism and axonal transport of AAV serotypes 1, 2, 5, 6, 8 and 9 in the mouse brain. PLoS One 8(9):e76310
85. Liu G et al (2005) Adeno-associated virus type 4 (AAV4) targets ependyma and astrocytes in the subventricular zone and RMS. Gene Ther 12(20):1503–1508
86. Joglekar AV, Sandoval S (2017) Pseudotyped lentiviral vectors: one vector, many guises. Hum Gene Ther Methods 28(6):291–301

87. Watson DJ et al (2002) Targeted transduction patterns in the mouse brain by lentivirus vectors pseudotyped with VSV, Ebola, Mokola, LCMV, or MuLV envelope proteins. Mol Ther 5(5 Pt 1):528–537
88. Liehl B et al (2007) Simian immunodeficiency virus vector pseudotypes differ in transduction efficiency and target cell specificity in brain. Gene Ther 14(18):1330–1343
89. Rahim AA et al (2009) Efficient gene delivery to the adult and fetal CNS using pseudotyped non-integrating lentiviral vectors. Gene Ther 16(4):509–520
90. Oliver KR, Fazakerley JK (1998) Transneuronal spread of Semliki Forest virus in the developing mouse olfactory system is determined by neuronal maturity. Neuroscience 82(3): 867–877
91. Poluri A et al (2008) Functional pseudotyping of human immunodeficiency virus type 1 vectors by Western equine encephalitis virus envelope glycoprotein. J Virol 82(24): 12580–12584

Chapter 7

Viruses for Systemic Delivery

Zikai Wang, Alon Greenbaum, and Jennifer B. Treweek

Abstract

Adeno-associated virus (AAV) is a nonpathogenic virus that has been leveraged as a gene-delivery vector due to its high transduction efficiency in many cell types and its capacity to mediate nontoxic and persistent transgene expression in infected cells. Given the demonstrated utility of AAVs in both basic scientific discovery and in treating human diseases, there is a growing interest in engineering novel traits into the recombinant AAV (rAAV) vector and refining the large-scale production of rAAVs to advance their clinical and research use. Large-scale production is required as the systemic delivery of these novel rAAVs has been envisioned as a route to treating monogenic disorders that affect whole organ physiology or function (e.g., muscle, including the heart, adipose tissue, cell types that harbor persistent viral reservoirs, globally in CNS or PNS); therefore, in contrast to local injection, a considerably higher vector dose would be required to achieve therapeutic effects. To address the production challenge, this chapter provides an experimental overview for generating recombinant single-stranded AAV at high titer and of sufficient purity for in vivo research. The methods discussed within the center on the production of rAAV vectors based on the AAV-PHP virus family. The AAV-PHP family capsid variants have been engineered to grant enhanced cell-type tropism as well as highly efficient cell transduction across the nervous system and throughout peripheral and sensory organs upon systemic delivery. By including additional levels of gene expression control through the inclusion of regulatory elements (e.g., promoters/enhancers) in the rAAV genome, researchers can now achieve either widespread or targeted expression of genetically encoded labels, sensors, and actuators in mammalian subjects for circuit-mapping and circuit manipulation endeavors. To advance the use of these optimized viral vectors, this chapter details a protocol for the high-quality manufacture and accurate titering of rAAVs using readily available reagents and standard laboratory equipment.

Key words Chemogenetics, Optogenetics, Viral vector, Adeno-associated virus, Gene therapy, Transgene expression, Capsid engineering, Tropism, Cre-recombination-based AAV targeted evolution (CREATE), Intersectional-genetic

1 Introduction: Adeno-associated Viral Vectors for Gene Transfer to the Nervous System

Recombinant viruses have been used extensively in research as vehicles for gene transfer to specific cell populations throughout the body, including the central and peripheral nervous systems (CNS and PNS). The adeno-associated virus (AAV) represents

Mark A. G. Eldridge and Adriana Galvan (eds.), *Vectorology for Optogenetics and Chemogenetics*, Neuromethods, vol. 195, https://doi.org/10.1007/978-1-0716-2918-5_7,

one of the most commonly used viruses for neuronal circuit interrogation, given the capacity of different AAV serotypes and engineered recombinant AAV vectors (rAAVs) to transduce both dividing and nondividing cells and to confer stable, long-term gene expression with minimal toxicity. In addition to serving as a nonintegrating viral vector with low immunogenicity, AAVs possess inherent cell-type or tissue-specific tropism and trafficking, both of which can be directly harnessed to limit off-target gene expression, or refined even further through capsid engineering [1–9]. Capsid engineering can be pursued via rational design [8] or directed evolution [2–5, 7, 10–12] approaches; both intend to isolate novel rAAV capsids that show enhanced transduction efficiency and/or cell-type selectivity. Intersectional-genetic methods may also be incorporated into the AAV-mediated gene transfer strategy, either to titrate transgene expression, as is exemplified by the vector-assisted spectral tracing (VAST) approach [3], or to institute additional gene-regulatory or recombinase-dependent restrictions on transgene expression (*see* Fig. 1) [3, 6, 14]. It follows that the size of the rAAV genome establishes the upper limit on the number of layers of intersectional-genetic control that can be implemented. With a capsid carrying capacity of ~5 kb, rAAVs can package most genetic cargo that is required for optogenetic and chemogenetic experiments in animals, including: regulatory elements (e.g., promoters and enhancers) for articulating transgene expression; genetically encoded sensors (e.g., genetically encoded calcium indicators (GECIs) [15–18], G-protein-coupled receptors (GPCR)-based neurotransmitter [19, 20] and neuromodulator [21–25] (GRABs) sensors [25, 26]) and actuators (e.g., optogenetic constructs such as excitatory and inhibitory opsins [27]; designer receptors exclusively activated by designer drugs (DREADD)-based chemogenetic tools [28]) for recording and manipulating neuronal activity; and small peptide or fluorescent protein tags for validating transgene expression, defining cellular niches, and/or tracing neuronal pathways. Expanding the AAV capsid carrying capacity represents an area of active research; however, currently, the ~5 kb limit encumbers the application of rAAVs to various proposed human gene therapies [11].

1.1 Engineered AAVs for Neuroscience Applications

AAV capsid engineering has yielded scientifically useful rAAV vectors with unique trafficking properties, such as directional transneuronal spread or transport across cell or organ boundaries [29, 30]. Of particular importance to basic and preclinical neuroscience research was the identification of AAV9 variants with the ability to efficiently cross the blood–brain barrier (BBB) [2]. The BBB, which is formed by the endothelial cells of the brain capillaries, normally restricts the access of blood-borne microbial and viral agents to brain cells. While a few naturally occurring AAVs, such as AAV9 and AAVrh.10 [31], can cross the BBB with exceedingly low efficiency, a family of BBB-crossing AAV9 capsid variants

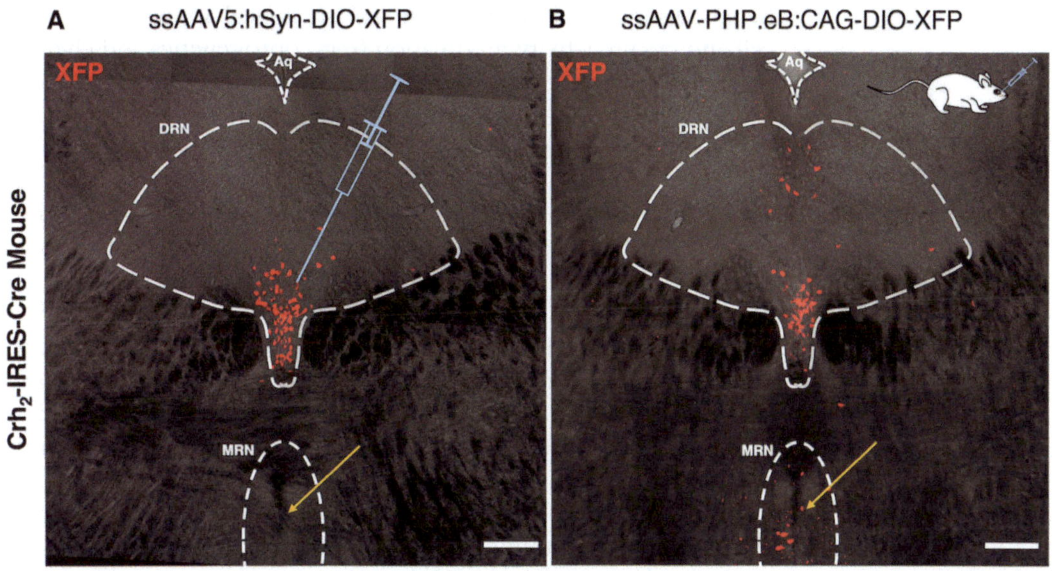

Fig. 1 Comparison of intracranial (IC) administration versus retro-orbital (RO) administration of AAVs in Crh_2-IRES-Cre transgenic mice. (**a**) AAV:hSyn-DIO-XFP was administrated by IC injection (1.5×10^{10} vg, in a total volume of 500 nL) into the DRN of a Crh2-IRES-Cre mouse (8 weeks old, JAX strain: B6.129S4(SJL)-Crhr2tm1 (cre)Lbrl/J). DRN injection coordinates with respect to Bregma: anterior-posterior ~ -4.4 mm, medial-lateral $\sim \pm 1.3$ mm, dorsal-ventral ~ -3.2 mm, with a 20° angle. (**b**) AAV-PHP.eB:CAG-DIO-XFP was RO-injected into age-matched Crh2-IRES-Cre mice at a total dose of 1.5×10^{11} viral genomes per mouse. Two weeks after injection, transgene expression was evaluated in coronal brain sections via confocal microscopy. Specifically, paraformaldehyde-fixed brains were sectioned by vibratome (Leica), and the 50 μm slices were mounted in refractive index matching solution (RIMS) [13] for imaging of the transgene product (i.e., fluorescent protein, XFP, under excitation wavelength: 561 nm). Orange arrows indicate differences in AAV distribution, in which direct brain injection of a naturally occurring serotype: AAV5, confers local transduction of Cre + neurons in the ventromedial DRN (**a**), while systemic delivery of AAV-PHP.eB grants broader coverage of Cre + cells throughout the raphe nuclei, both the DRN and MRN (**b**). Experiments on mice conformed to all relevant governmental and institutional regulations and were approved by the Institutional Animal Care and Use Committee (IACUC) and the Department of Animal Resources at the University of Southern California. Aq: aqueduct; DRN: dorsal raphe nucleus; MRN: median raphe nucleus; XFP: fluorescent protein transgene. Scale bar, 200 μm

have been identified through the Cre-recombination-based AAV targeted evolution (CREATE) and Multiplexed-CREATE (M-CREATE) platforms [2–6]. These variants, most notably AAV-PHP.B [2], AAV-PHP.eB [3], AAV-PHP.N [4], AAV-PHP.V1 [4], and AAV.CAP-B10 [5] grant noninvasive access to cells of the mammalian central nervous system (CNS) through simple systemic injection. Complementing this targeted access to neuronal and non-neuronal cells of the CNS are rAAV variants, such as AAV-PHP.S [3], that allow efficient body-wide transduction of the peripheral nervous system (PNS) [4, 32–36] and of cell classes that compose many peripheral organs and tissues (heart [1, 10, 37, 38], adipose tissue [39], liver [32], lung [12]). These vectors will benefit a range of basic research and preclinical applications, from cell and circuit activity modulation (e.g., optogenetics and chemogenetics) and in vivo imaging in biological research [11, 38, 40–44] to the bidirectional modification of gene expression and gene editing [45, 46] for the treatment of disease [11, 33, 47–50].

1.2 Methodological Considerations for AAV Production and Use in Research

Despite the tremendous promise of designer AAVs, such as those listed earlier, in both clinical therapies and basic neuroscience research, a few major hurdles continue to obstruct their use in certain applications. In addition to the constraints related to the small capsid packaging capacity, natural and designer AAVs, including the AAV-PHP capsids, typically show enhanced cell and tissue tropism in the animal model or specific strain used in their selection (e.g., AAV capsid selection in C57BL/6J mice, but with tropism that extends to non-human primates [5, 51]; AAV capsid selection in C57BL/6J mice [2, 3], but with tropism that does not extend to all inbred mouse strains or nonhuman primates [52–54]). Congruently, these designer AAVs have historically not translated well from species to species and, indeed, in some cases, even within different strains of the same species. For example, AAV-PHP.eB is capable of transducing neurons in C57BL/6NCrl and 129T2/SvEmsJ mice and in Fischer and Long-Evans rats [6]. By comparison, transcytosis of AAV-PHP.eB and other AAV-PHP.B-derived capsid variants across the BBB of BALB/cJ mice is less efficient, resulting in significantly lower CNS transduction; this trait is likely to be replicated in other species that lack specific receptors necessary for efficient BBB transcytosis (e.g., the LY6A protein or a *Ly6a* homolog), and hence should be factored into the AAV capsid selection process when designing experiments [52, 53].

When combined with intersectional-genetic strategies, the wealth of transgenic mouse lines, and the growing number of transgenic rat lines, designer AAVs permit effective targeting of distinct cell subpopulations according to cell-type and spatial constraints [3, 11, 55]. However, targeting cells according to their connectivity remains challenging. Unambiguous mapping and functional profiling of discrete cell circuits would require both exclusively retrograde-trafficking and exclusively anterograde-trafficking vector tools; ideally, vector spread would proceed through a trans-synaptic rather than transneuronal mechanism. Although many neurotropic virus strains or their engineered variants (e.g., pseudotyped and replication-defective variants of rabies virus [56], vesicular stomatitis virus [57], yellow fever virus vaccine strain 17D (YFV-17D) [58], herpes simplex virus strain H129 [59]) demonstrate efficient transneuronal spread and even preferential trafficking in either the anterograde or retrograde direction, they each possess an Achilles' heel that thwarts their ability to mediate selective, long-term transgene expression in neurons according to input–output connectivity [60].

Naturally occurring AAVs exhibit varying degrees of low-level anterograde or retrograde transport, the latter entailing virus uptake at the axon terminal and retrograde transport back to the soma. Although scientists have engineered rAAV variants with amplified anterograde [29] or retrograde [61] trafficking properties through positive selection, these rAAV variants still transduce local cells at the administration site, and most AAV serotypes, including the anterograde variant, still exhibit some capacity for retrograde transduction, particularly at high titers. With respect to intravenously administered vectors that can efficiently cross the BBB and transduce brain cells with minimal cytotoxicity, an AAV-PHP variant capable of efficient, unidirectional, transsynaptic spread upon systemic administration has yet to be identified. Thus, additional levels of gene-regulatory or anatomic control must be implemented for accurate AAV-based circuit mapping or gene delivery according to connectivity.

Nevertheless, the ability to modulate gene expression in distributed cell populations through systemic AAV injection unlocks new arenas of research into noninvasive neuromodulation and gene therapy for neurological diseases [47]. To advance the pursuit of these topics, this chapter provides a protocol for efficient and accurate rAAV production at titers sufficient for systemic administration. Systemic AAV delivery incurs the need to substantially increase the vector genome (vg) dose per subject so as to ensure broad coverage and efficient transduction of large cell populations. It is worth mentioning that systemic administration of high AAV doses is not without risk. First, it increases the likelihood of off-target transduction of peripheral organs due to the inherent promiscuity of AAVs, which is compounded by the imperfect production of high titer rAAV comprised solely of a single AAV capsid variant. Second, the more widespread biodistribution and gene transfer that results from systemic vector delivery provide additional opportunities for immunosurveillance of the rAAV capsid or transgene product, raising the potential for either to become immunogenic [2, 62]. Thus, researchers should remain vigilant of these possibilities when employing systemic, AAV-based gene delivery strategies.

1.3 Procedural Overview

The following protocol details five main stages for the production of AAV vectors for optogenetic, chemogenetic, and/or in vivo imaging experiments in rodents. These stages include (Fig. 2): AAV production via triple transfection (Stage 1), AAV harvest (Stage 2), AAV purification (Stage 3), AAV titering (Stage 4), and systemic AAV delivery (Stage 5). Users may wish to validate their systemic delivery method (e.g., intravenous (IV) or retro-orbital (RO); Fig. 1b), alongside direct brain injection (e.g., intracranial

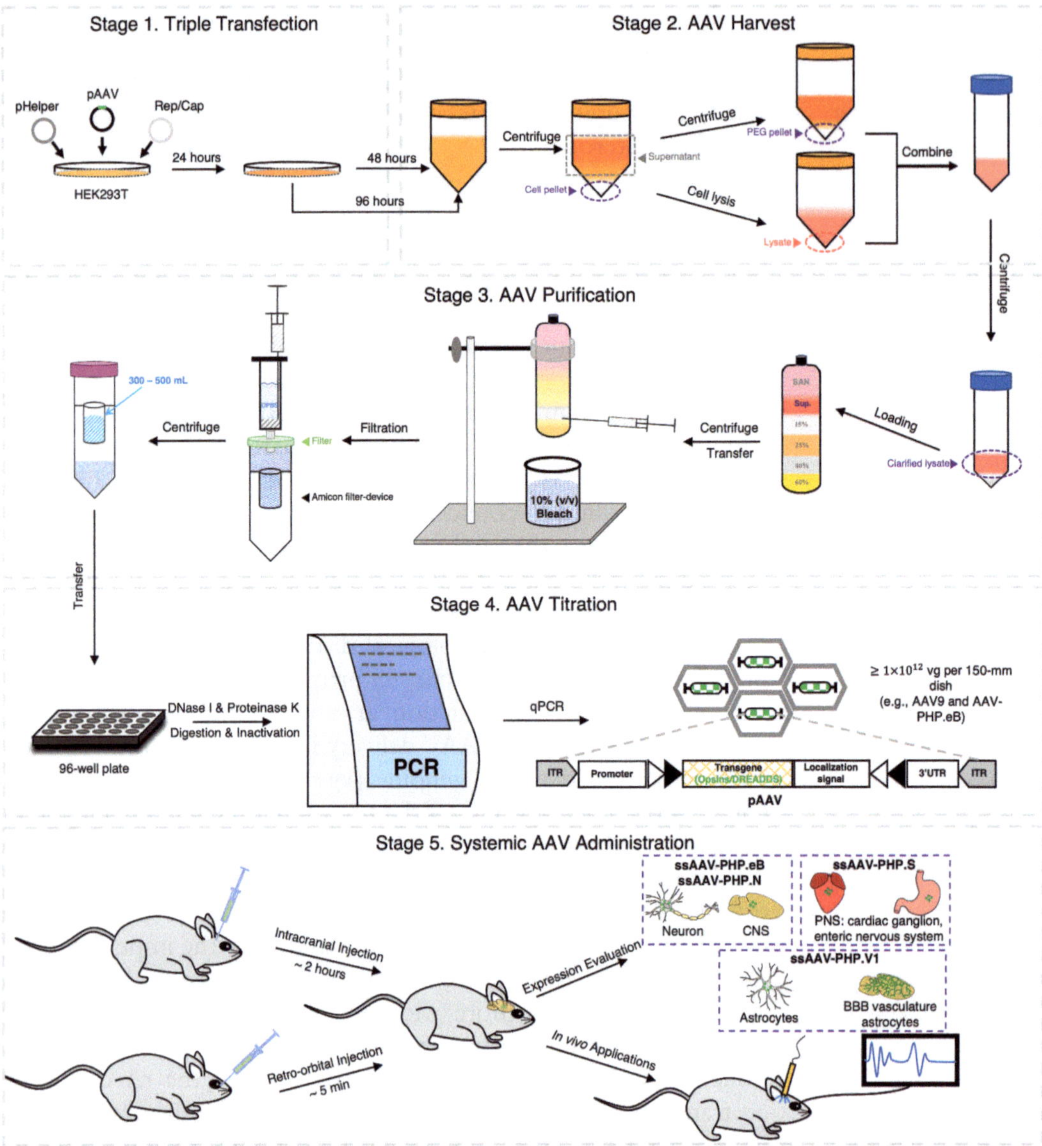

Fig. 2 Timeline and overview of the protocol. The procedure comprises five main stages: (1) triple transfection, (2) AAV harvest, (3) AAV purification, (4) AAV titration, and (5) systemic AAV administration. The entire protocol spans approximately 8 days, excluding AAV administration (Stage 5); however, there are optional pause points between certain stages, as noted in the Methods text. The pAAV plasmid contains the rAAV genome, which can be engineered by the researcher to contain an interchangeable assortment of gene regulatory elements (e.g., promoters, terminator/polyadenylation signals, and messenger RNA stability elements), transgenes that encode an opsin or DREADD (shown in green, Stage 1), localization signals, and recombinase recognition sequences. Filled green circles represent transduced cells (Stage 5)

(IC) or intracerebroventricular (ICV) injection; Fig. 1a) of the rAAV variant or parent AAV, so as to compare transduction spread and specificity, and to evaluate the level of transgene expression according to the delivery route.

rAVV production starts with polyethylenimine (PEI-based co-transfection of human embryonic kidney (HEK) cells with three plasmids:

(i) The *Rep/Cap* plasmid, which encodes the rAAV capsid variant proteins (Cap) as well as AAV proteins necessary for viral genome replication (Rep) and packaging.

(ii) The "cargo" or rAAV genome of interest (pAAV), including for instance the opsin or DREADD for activity modulation.

(iii) The pHelper plasmid, which encodes adenoviral helper proteins necessary for replication.

Upon packaging of the single-stranded rAAV genome into a capsid of choice (e.g., AAV-PHP) in HEK cells, the viruses are then harvested, purified, and titered by quantitative PCR (qPCR). Please note that the titering method must remain consistent between preparations, as even small changes in the titering procedure might lead to large variability in the reported yields. The PEI transfection protocol followed by iodixanol density gradient ultracentrifugation for AAV purification typically yields at least 1×10^{12} vector genomes (vg) per 150 mm dish for AAV-PHP.B or the parent AAV9, as measured via qPCR. Unlike cesium chloride (CsCl) gradient purification methods, the iodixanol gradient method does not grant sensitive fractionation of the full genome-containing particles from empty AAV capsids. And it is important to note that qPCR only provides an approximate readout of the presence of viral genomes (i.e., viral genome titer) in a sample; incomplete genomes that have been packaged into particles will also be counted in the titer. Assessing the integrity of the actually packaged genome requires the use of, for example, alkaline gel electrophoresis under denaturing conditions. Electron microscopy may be employed to directly visualize purified AAV particles, which permits the qualitative assessment of full-to-empty particle ratios in AAV production rounds. But neither qPCR nor electron microscopy will furnish information about the infectivity of a viral prep or the integrity of the packaged genome. Instead, more rigorous techniques such as the infectious center assay (i.e., the infectious center assay is similar to the plaque assay, but it works for non-plaque-forming viruses such as wild-type AAV and replication-deficient viral particles such as rAAV vectors) or other transduction assay must be used to verify the correct assembly and production of functional AAV particles that are capable of entering a cell and delivering the genome to the host cell nucleus [63]. Finally, various qualitative and quantitative assays may be used to assess the transduction of the target cell population; often, this is achieved through observing the expression of a transgene reporter, such as a fluorescent protein, in the infected cells [2].

When considering an appropriate scale of virus production, users must estimate the rAAV vector dose that, when administered to the subject, will mediate the desired transgene expression level in the target cell or tissue population. The optimal rAAV vector dose range for a particular experiment must be determined empirically, as the vector dose depends on the subject (i.e., species, strain, age), vector design parameters (e.g., the use of intersectional-genetic strategies or recombinase control systems), and experimental constraints. Generally, higher vector doses mediate stronger expression (i.e., high copy number) across a greater proportion of the target cell population, while lower doses result in weaker expression (i.e., low copy number) in proportionally fewer cells. As a starting point, a typical AAV-PHP.eB dose that achieves 50% neuronal transduction upon systemic administration to an adult mouse (6–8 weeks old) is 1×10^{11} to 5×10^{11} vg; AAV-PHP.S is often administered at a slightly higher dose: between 3×10^{11} and 1×10^{12} vg [3, 6]. Generally, 3–4 weeks are allowed for in vivo expression before evaluation of transduction efficiency and coverage. Any additional surgical preparation (e.g., installation of an optical fiber, recording electrodes, or cannulae) of subjects is typically performed during this period such that the timing of surgical recovery aligns with the timing of optimal transgene expression.

2 Materials

2.1 Triple Transient Transfection of HEK293T Cells

Plasmid DNA

1. Three plasmids are required to produce rAAV virions by transient transfection: (i) the Rep/Cap plasmid (e.g., pUCmini-iCAP-PHP), (ii) the pAAV, and (iii) the helper plasmid (pHelper). A brief description of how to obtain and/or prepare these plasmids follows below.
   (i) *Selecting an rAAV capsid variant for Rep/Cap.* We recommend choosing an AAV-PHP capsid based on its tropism (*see* Table 1) and production efficiency. AAV-PHP.eB

Table 1
Parameters for rAAV capsid selection

rAAV capsid variant	PHP.B [2]; PHP.eB [3]; PHP.B4, B5, B6, B7, B8 [4]	PHP.**N** [4], CAP-B10 [4]	PHP.**V1**, V2 [11]	PHP.S [8]
Tropism	CNS	Neurons	Astrocytes, BBB vasculature	Sensory neurons

(Addgene, plasmid no. 103005) would subserve experiments requiring noninvasive gene transfer to the CNS, with broad neuronal and astrocytic cell tropism, while AAV-PHP.N would enable enhanced brain-wide targeting of neuronal (NeuN+) populations with minimal transduction of astrocytes and oligodendrocyte lineage cells, even when packaged with a ubiquitous CAG promoter. AAV-PHP.S (Addgene, plasmid no. 103006) would better support systemic delivery of genes to the PNS and visceral organs, while AAV-PHP.V1 would allow targeting of BBB vasculature and endothelial cell populations, particularly when the transgene is delivered with an endothelial cell-type-specific MiniPromoter (Ple261; Addgene, plasmid no. 82563), or expressed under endothelial-cell-restricted recombinase control in Tek-cre mice [4]. AAV-PHP.eB, AAV-PHP.N, AAV-PHP.V1, and AAV-PHP.S can be produced in similar viral yields as the naturally occurring parent serotype: AAV9 or AAV-PHP.B.

(ii) *Constructing the transgene cargo, pAAV.* The rAAV genome, which must be packaged into the AAV capsid, includes both the desired genetic components (e.g., transcriptional and translational regulatory elements, transgene (s) of interest, localization signals, and intersectional-genetic components such as recombinase recognition sites) as well as two 145 bp inverted terminal repeats (ITRs), each of which flank this expression cassette. These ITRs must match the AAV serotype of the *rep* gene contained in the Rep/Cap plasmid. Most pAAV plasmids available from Addgene use AAV2 ITRs, which are compatible with the suggested pUCmini-iCAP-PHP plasmids that contain the AAV2 *rep* gene. Although there has been much effort to expand the packaging capacity of rAAVs, it is highly recommended that most users endeavor to restrict their rAAV genome size to <4.7–5 kb (inclusive of the ITRs).

(iii) *Purchasing the pHelper plasmid.* The pHelper plasmid, which contains genes (E4, E2a, and VA) from adenovirus that aid in AAV replication, is available in Agilent's AAV helper-free kit (Agilent, cat. no. 240071).

For use in rAAV production, the Rep/Cap, pAAV cargo, and pHelper plasmids, which can often be purchased from commercial vendors or procured from research labs or institutional viral vector cores as bacterial stabs, must be prepared on a large scale (i.e., 22.8 μg pUCmini-iCAP-PHP and 11.4 μg pHelper per 150 mm dish for triple transient transfection). Grow bacterial stocks (LB with agar, Sigma-Aldrich, cat. no. L3147-1KG) for the three plasmids in LB (Amresco, cat. no. J106-1KG) or Plasmid+ (Thomson Instrument, cat. no. 446300) media containing the appropriate selection antibiotic (*see* **Note 1**). Use a large-scale endotoxin-free plasmid purification kit to isolate plasmids (e.g., NucleoBond Xtra Maxi or Giga endotoxin-free (EF) plasmid purification kit, Macherey-Nagel, cat. no. 740424.50 or 740548); and elute plasmid DNA with Tris-acetate-EDTA (TAE) buffer (50×; Invitrogen, cat. no. B49) (*see* **Note 2**). Measure the DNA purity and concentration using a Qubit. These plasmid stocks may be stored frozen at −20 °C or −80 °C for several years.

Biological Materials and General Laboratory Chemicals.

2. *Cells.* The triple transfection is performed in human embryonic kidney (HEK) cells (293 or 293T; ATCC, cat. no. CRL 1573 or CRL 3216, respectively), passaged using either TrypLE Express enzyme (1×; phenol red; Gibco, cat. no. 12605-036), or a standard trypsinization protocol for adherent cultures, and grown in cell culture media (20 mL of media per 150 mm petri dish).
3. *Cell culture media.* Add 25 mL of FBS (GE Healthcare, cat. no. SH30070.03), 5 mL of MEM nonessential amino acids (NEAA) solution (100×; Gibco, cat. no. 11140-050), and 5 mL of penicillin–streptomycin (pen–strep; 5000 U/mL; Gibco, cat. no. 15070-063) to a 500 mL bottle of DMEM (high glucose, GlutaMAX supplement, pyruvate; Gibco, cat. no. 10569-044). Invert to mix and store at 4 °C for up to several months; warm to 37 °C before use. The resulting cell culture media should have a final concentration of 5% (vol/vol) FBS, 1× NEAA, and 50 U/mL pen–strep.
4. *Polyethylenimine (PEI) stock solution.* Pipette 50 mL of water-for-injection (WFI, Gibco, cat. no. A1287304) water into a 50 mL conical centrifuge tube for later use. Add 323 mg of PEI (linear, 25,000 molecular weight (MW); Polysciences, cat. no. 23966-1) to the remaining 950 mL bottle of WFI water and adjust the pH to 2–3 by adding 1 N HCl suitable for cell culture, keeping track of the volume of HCl added. Heat in a 37 °C water bath for several hours (or overnight) and occasionally shake to mix. Once dissolved, add reserved WFI water to a total volume of 1 L. Filter-sterilize, make aliquots in 50 mL conical centrifuge tubes, and store at −20 °C for up to 1 year.
5. *1 N Hydrochloric acid (HCl) solution.* HCl suitable for cell culture (Sigma-Aldrich, cat. no. H9892).
6. *Dulbecco's PBS (DPBS).* 1× DPBS with no calcium and no magnesium (Gibco, cat. no. 14190-250).
7. *Water for injection (WFI).* Water for cell culture (Gibco, cat. no. A1287304).

8. *UltraPure DNase/RNase-free distilled water* (Invitrogen, cat. no. 10977-023).
9. *$MgCl_2$-$6H_2O$* (Macron Fine Chemicals, cat. no. 5958-04).
10. *NaCl* (Millipore, cat. no. SX0420-3).
11. *KCl* (Macron Fine Chemicals, cat. no. 6858-06).
12. *N-lauroylsarcosine sodium salt* (Sigma-Aldrich, cat. no. L9150-50G).
13. *$CaCl_2$* (anhydrous; J.T. Baker, cat. no. 1311-01).
14. *Tris, ultrapure* (MP Biomedicals, cat. no. 819620).
15. *UltraPure Tris–HCl* (1 M, pH 7.5; Invitrogen, cat. no. 15567-027).
16. *UltraPure EDTA* (0.5 M, pH 8.0; Invitrogen, cat. no. 15575-020).
17. *Tris-acetate-EDTA (TAE) buffer* (50×; Invitrogen, cat. no. B49).
18. *70% (vol/vol) Ethanol.* Prepare from absolute ethanol (J.T. Baker, cat. no. 8025).
19. *10% (vol/vol) Bleach.* Prepare fresh from concentrated liquid bleach (e.g., Clorox).

2.2 AAV Harvest

1. *Salt-active nuclease* (SAN; 25 U/μl; ArcticZymes, cat. no. 70910-202).
2. *Polyethylene glycol* (PEG; MW 8000; Sigma-Aldrich, cat. no. 89510-1KG-F).
3. *SAN digestion buffer.* Add 29.22 g of NaCl, 4.85 g of Tris base, and 2.03 g of $MgCl_2 \cdot 6H_2O$ to a 1 L bottle of WFI water and shake to mix. Filter-sterilize and store at RT for up to several months. The resulting SAN digestion buffer should have a final pH of ~10.0 and a final concentration of 500 mM NaCl, 40 mM Tris base, and 10 mM $MgCl_2$.
4. *SAN + SAN digestion buffer.* Add 100 U of SAN (4 μL of 25 U/μL SAN) per milliliter of SAN digestion buffer; pipette to mix.
5. *40% (wt/vol) PEG stock solution.* Decant ~500 mL of WFI water into a 500 mL sterile bottle for later use. Add 146.1 g of NaCl to the remaining 500 mL (in the 1 L bottle of WFI water) and shake/heat until dissolved. Once completely dissolved, add 400 g PEG and heat at 37 °C overnight for up to 2 nights. Add reserved WFI water to a total volume of 1 L. Prewet the filter surface with WFI water, and then filter-sterilize the solution; allow 1–2 h for the solution to filter. This stock solution should have a final concentration of 2.5 M NaCl and 40% (wt/vol) PEG, and it may be stored at RT for several months.

2.3 AAV Purification

1. *OptiPrep.* 60% (wt/vol) iodixanol density gradient medium (Cosmo Bio USA, cat. no. AXS-1114542-5).
2. *DPBS + high salt.* To prepare a DPBS + high salt stock solution, add 29.22 g of NaCl (Millipore, cat. no. SX0420-3), 93.2 mg of KCl, and 101.7 mg of $MgCl_2 \cdot 6H2O$ to a 500 mL bottle of DPBS and shake to mix. Filter-sterilize (e.g., Stericup sterile vacuum filtration system, 0.22 μm; 1 L; Millipore, cat. no. S2G-PU11RE) and store at RT for up to several months. The resulting buffer should have a final concentration of 1 M NaCl, 2.5 mM KCl, and 1 mM $MgCl_2$ (in addition to the salts in the DPBS).
3. *DPBS + low salt.* To prepare a DPBS + low salt stock solution, add 2.92 g of NaCl, 93.2 mg of KCl, and 101.7 mg of $MgCl_2 \cdot 6H_2O$ to a 500 mL bottle of DPBS and shake to mix. Filter-sterilize and store at RT for up to several months. The resulting buffer should have a final concentration of 100 mM NaCl, 2.5 mM KCl, and 1 mM $MgCl_2$ (in addition to the salts in the DPBS).
4. *Iodixanol density gradient solutions: 15%, 25%, 40%, and 60% (wt/vol) iodixanol.* Determine the number of gradients needed and prepare the iodixanol density gradient solutions (*see* Table 2). For each layer, add OptiPrep (iodixanol), DPBS + high salt or DPBS + low salt, and phenol red (if applicable) to a 50 mL conical centrifuge tube. Invert or briefly vortex to mix. The 25% and 60% layers contain Phenol Red (Millipore, cat. no. 1072420100), which turns the solutions red and yellow, respectively, and facilitates clear demarcation of the gradient boundaries.

Table 2
Composition of iodixanol density gradients

			Number of gradients	
Layer volume mL	**Iodixanol %**	**Layer recipe mL**	**One**	**Four**
6	15	DPBS + high salt	5.0	19.8
		60% iodixanol	1.7	6.6
6	25	DPBS + low salt	3.9	15.4
		60% iodixanol	2.8	11.0
		Phenol red	0.1	0.2
5	40	DPBS + low salt	1.8	7.3
		60% iodixanol	3.7	14.7
5	60	60% iodixanol	5.5	22.0
		Phenol red	0.1	0.2

2.4 AAV Titration

1. *DNase digestion buffer.* Use a 50 mL serological pipette to measure 247.5 mL of UltraPure water into a 250 mL sterile bottle. Add 55.5 mg of $CaCl_2$, 2.5 mL of 1 M Tris–HCl, and 508 mg of $MgCl_2 \cdot 6H_2O$ and shake to mix. Filter-sterilize (e.g., Stericup sterile vacuum filtration system, 0.22 μm, 250 mL; Millipore, cat. no. SCGPU02RE) and store at RT for up to several months. The resulting buffer should have a final concentration of 2 mM CaCl2, 10 mM Tris–HCl, and 10 mM $MgCl_2$.
2. *DNase I recombinant* (RNase-free; 10 U/μL; Roche Diagnostics, cat. no. 4716728001).
3. *DNase I + DNase digestion buffer.* Add 50 U of DNase I per milliliter of digestion buffer (a 1:200 dilution of 10 U/μL DNase); pipette to mix.
4. *Proteinase K* (recombinant, PCR grade; 50 U/mL (2.5 U/mg); Roche Diagnostics, cat. no. 03115828001).
5. *Proteinase K solution.* Use a 50 mL serological pipette to measure 250 mL of UltraPure water into a 250 mL sterile bottle. Add 14.61 g of NaCl and shake to mix. Add 2.5 g of N-lauroylsarcosine sodium salt (Sigma-Aldrich, cat. no. L9150-50G) to the mixture and gently swirl to mix; N-lauroylsarcosine sodium salt is a surfactant and will generate bubbles if vortexed or mixed vigorously. Filter-sterilize (e.g., Stericup sterile vacuum filtration system, 0.22 μm, 250 mL; Millipore, cat. no. SCGPU02RE) and store at RT for up to several months. The resulting solution should have a final concentration of 1 M NaCl and 1% (wt/vol) N-lauroylsarcosine sodium salt.
6. *Proteinase K + proteinase K solution.* Add 100 μg of proteinase K per milliliter of solution (a 1:200 dilution of 50 U/mL (2.5 U/mg) proteinase K); pipette to mix.
7. *Qubit dsDNA HS Assay Kit* (Invitrogen, cat. no. Q32854), to be used with a Qubit 3.0 fluorometer (Invitrogen, cat. no. Q33216).
8. *DNA standard stock.* Set up a single 50 μL restriction digest reaction to linearize 20 μg of the plasmid DNA containing the target sequence (e.g, for plasmids with Amp resistance, this might entail using 60–80 U (3–4 μL) of ScaI). Run a small amount of the reaction on an agarose gel to ensure complete digestion. Purify the reaction using two DNA clean-up columns (DNA Clean & Concentrator kit, Zymo Research, cat. no. D4033 (DCC-25)). Measure the DNA concentration (ng/μl) using a Qubit. Dilute to ~5–10 × 10^9 single-stranded (ss) DNA molecules/μL and use the Qubit assay to verify the

concentration (ng/μl). Divide into 20 μL aliquots in DNA/RNA LoBind microcentrifuge tubes and freeze at − 20 °C for up to 1 year.

9. *DNA standard dilutions.* Create eight serial dilutions (1:10) of the DNA standard stock in triplicate (*see* **Note 3**). For each set, begin by pipetting 5 μL of the DNA standard into 45 μL of UltraPure water and then vortexing for 3 s to mix; this represents standard #8. Serially dilute standard #8 in UltraPure water and vortex to create standard #7. Repeat this process (diluting 1:10 in Ultrapure water and vortexing to mix) six remaining times to create standard #6 to #1. The final concentrations of the standard dilutions should range from $5–10 \times 10^8$ (#8) to $5–10 \times 10^1$ (# 1) ssDNA molecules per microliter.
10. *SYBR Green master mix* (Roche Diagnostics, cat. no. 04913850001).
11. *ScaI-HF restriction enzyme* (New England Biolabs, cat. no. R3122S) or other enzymes that cuts outside of the rAAV genome and within the pAAV backbone.
12. *qPCR master mix.* Immediately prior to use, prepare a qPCR master mix for the total number of reactions (i.e., wells) needed. One reaction requires 12.5 μL of SYBR Green master mix, 9.5 μL of UltraPure water, and 0.5 μL of each primer (from a 2.5 μM stock concentration), for a total of 23 μl/well. Pipette or vortex for 1–2 s to mix.
13. *qPCR Primers.* Obtain primers (Integrated DNA Technologies) corresponding to the target sequence to be amplified during qPCR (e.g., WPRE-forward: GGCTGTTGGGCACT GACAAT, WPRE-reverse: CCGAAGGGACGTAGCAGAAG, hGH polyA-forward: GTGCCCACCAGCCTTGTC, hGH polyA-reverse: TGTCTTCCCAACTTGCCCCTT).

2.5 Systemic AAV Administration

1. Insulin syringes with permanently attached needles (31 gauge × 5/16 inches; BD, cat. no. 328438).
2. Ophthalmic anesthetic (0.5% proparacaine hydrochloride ophthalmic solution, Alcon Laboratories, Inc.).
3. Materials and supplies for rodent anesthesia (e.g., Isoflurane, USP (Piramal Critical Care, cat. no. 66794-017-25) and vaporizer (V-1 Tabletop with Active Scavenging (VetEquip); medical grade oxygen (Airgas), charcoal canisters (VaporGuard Charcoal Filters, VetEquip) and/or vacuum source for scavenging anesthetic gas).
4. Warming device or heating pad (Far Infrared Warming Pad, Kent Scientific Corporation; if required by Institutional Care and Use Animal Committee for short-term rodent anesthesia).

3 Methods

Stage 1. Triple Transient Transfection of HEK293T Cells

Day 0

1. Seed HEK293T cells (approximately 2×10^7 cells) in 150 mm petri dishes (150 mm × 25 mm; Corning, cat. no. 430599) to attain 80–90% confluency on the day of transfection (i.e., transfection is typically performed after 24 h incubation of cells at 37 °C with 5% CO_2) (*see* **Note 4**).

Day 1

2. Prepare PEI + DPBS master mix. Namely, thaw PEI stock solution in a 37 °C water bath. Bring the PEI to room temperature (RT; 23 °C) and vortex to mix. Add PEI and DPBS in appropriate proportions to a 50 mL conical centrifuge tube (*see* Table 3 for per-dish quantities), and vortex again to mix.
3. Prepare DNA + DPBS solution for each viral prep. Bring plasmid DNA to RT and briefly vortex to mix. For each viral prep, calculate the quantities of plasmid DNA and DPBS (*see* Table 3 for per-dish quantities) to add to a 50 mL conical centrifuge tube; vortex the prepared DNA + DPBS solution to mix.
4. In a dropwise fashion (using a 5- or 10 mL serological pipette) and while agitating the conical tube, add the calculated volume of PEI + DPBS master mix to the DNA + DPBS solution. Cap the tube and thoroughly vortex for 10 s to mix, and then allow this tube of transfection solution to sit at RT for 2–10 min.

Table 3
Triple transfection plasmid amount guidelines for one 150 mm dish with a surface area = 152 cm^2. The μg of DNA for each plasmid (1.00 μg/μL) are calculated according to a pAAV:Rep/Cap:pHelper ratio of 1:4:2 and a total DNA amount of approximately 40 μg per dish

	DNA mass (μg)	DPBS volume (μL) for DNA + DPBS	Total volume (μL) of media	Total volume (μL) of PEI + DPBS master mix
pAAV	5.70	–	–	*(Volumes of components)*
Rep/Cap[a]	22.82	–	–	*439.23 μL PEI*
pHelper	11.41	–	–	*560.77 μL DPBS*
Total	39.93 μg	960.07 μL	20,000 μL	1000 μL

[a] *Rep/Cap* plasmid, e.g., pUCmini-iCAP-PHP

5. Add 2 mL of the transfection solution dropwise to each dish and swirl gently to mix before returning the dishes to the cell culture incubator (37 °C with 5% CO_2).

Day 2

6. Change the media 12–24 h post-transfection by aspirating the old media into 10% (vol/vol) bleach and replacing it with 20 mL of fresh, warmed media (*see* **Note 5**).

Stage 2. AAV Harvest

Day 4

1. Pipette off the cell culture media 72 h post transfection and store at 4 °C (*see* **Note 6**).
2. Add 20 mL of fresh, prewarmed media to the dishes and incubate for an additional 48 h.

Day 6

3. At 120 h post-transfection, harvest the media and cells from each petri dish by gently scraping the cells into media and transferring this media-cell suspension and the stored media from **step 1** into a 250 mL conical centrifuge tube. For multiple dishes of the same virus prep, the media-cell suspension from each dish for up to five dishes may be pooled in one tube.
4. Centrifuge the media and cells at 2000 g for 15 min at room temperature (RT).
5. Pour off the supernatant (i.e., the clarified media) into the dish's (or dishes') corresponding storage bottle from **step 1** or into a new empty media bottle, and store at 4 °C.
6. Prepare 5 mL of SAN + SAN digestion buffer (Reagent setup) per viral prep (up to five dishes).
7. Using a 5 mL serological pipette, gently resuspend the cell pellet from **step 4** in 5 mL of SAN + SAN digestion buffer; transfer suspension into a 50 mL conical tube (*see* **Note 7**) and continue to pipette or gently tap on a vortexer (on low setting) to finish resuspending the pellet.
8. Incubate in a 37 °C water bath for 1 h; then, store at 4 °C until **step 15** (up to 24 h).

Day 6 or Day 7

9. Pour the stored supernatant (**step 5**) into a 250 mL conical centrifuge tube (*see* **Note 7**).
10. Add enough 40% (wt/vol) PEG stock solution to the supernatant to create an 8% final (wt/vol) concentration

of PEG. Cap and invert the tube ten times; then, incubate on ice for 2 h.

11. Centrifuge the PEG-media mixture at 4000 g for 30 min at 4 °C.
12. Pour off the supernatant (*see* **Note 8**). Aspirate off excess supernatant without disturbing the pellet.
13. Prepare 1 mL of SAN + SAN digestion buffer per viral prep (up to five dishes) and add to pellet.
14. Using a P1000 pipetter, vigorously pipette (*see* **Note 9**) to resuspend the PEG pellet in the SAN buffer.
15. Add the resuspended pellet (**step 14**) to the pellet lysate from **step 8**.
16. Incubate the combined pellet lysates in a 37 °C water bath for 30 min.
17. Proceed directly to Stage 3: AAV Purification. Alternatively, the lysate may be stored overnight at 4 °C before Stage 3 or frozen in a dry ice–ethanol bath and stored for up to 1 week at −20 °C.

Stage 3. AAV Purification

Day 8

1. Prepare the iodixanol density gradient solutions (*see* Table 2).
2. Pour the density gradients in OptiSeal tubes (Beckman Coulter, cat. no. 361625; includes black caps) in the following layered order: 6 mL of 15% (wt/vol) iodixanol, 6 mL of 25% (wt/vol) iodixanol, 5 mL of 40% (wt/vol) iodixanol, and 5 mL of 60% (wt/vol) iodixanol (*see* **Note 10**).
3. Centrifuge the lysate from **step 17** at 2000 g for 10 min at RT.
4. With a 2 mL serological pipette, slowly pipette the supernatant (i.e., the clarified lysate from **step 3**) atop the density gradient by touching the pipette tip to the 15% iodixanol meniscus. Discard the pellet (in 10% bleach or in a biohazard waste collection bin).
5. Fill (but do not overfill) each OptiSeal tube up to the neck with SAN digestion buffer; without creating air bubbles at the top, seal the tube using its black stopper-cap, and cover with the spacer provided in the OptiSeal Tube kit (Beckman Coulter, cat. no. 361662; includes a tube rack, spacers, and spacer- and tube-removal tools/tweezers).

6. Weigh the prepared tubes with the caps and spacers; create balanced tubes (with caps and spacers) to within 5–10 mg of each other using SAN digestion buffer.
7. Load the balanced tubes into the ultracentrifuge rotor in the biosafety cabinet; fasten the rotor lid, and carefully transfer the rotor to the ultracentrifuge.
8. Spin the Type 70 Ti rotor at 350,000 g (58,400 r.p.m.) for 2 h and 25 min at 18 °C with slow acceleration and deceleration (*see* **Note 11**).
9. During the ultracentrifugation run, assemble the equipment necessary for collecting virus-containing iodixanol gradient fractions. For the clamp setup (*see* Stage 3 panel of Fig. 2), attach the rod to the support stand, and secure the clamp 25–30 cm above the stand. Also, retrieve one Amicon Ultra-15 centrifugal filter device (100 kDa molecular weight cutoff; Millipore, cat. no. UFC910024), one 5 mL and one 10 mL syringe (BD, cat. nos. 309646 and 309604), one 0.22 μm syringe filter unit (Millipore, cat. no. SLGP033RS), and one 16-gauge needle per gradient.
10. Once the rotor has come to a complete stop in the ultracentrifuge, carefully transfer the rotor back to the biosafety cabinet, unscrew the rotor lid, and remove the spacer (with the tweezers provided in the kit). Slowly remove each tube from the rotor with the OptiSeal kit-provided tool (large tweezers), and clamp the tube in-place above a 500 mL waste beaker that contains 10% (vol/vol) bleach. Wipe down the sides of the tube with 70% (vol/vol) ethanol.
11. Remove the plunger from a 10 mL syringe and set it aside; attach a 0.22 μm syringe filter-unit to the syringe tip, and balance this syringe + filter-unit atop an Amicon filter device (*see* **Stage 3** panel of Fig. 2).
12. Add 10 mL DPBS to the syringe barrel and allow the solution to begin to drip by gravity through the filter-unit and into the Amicon device while proceeding to **step 12**.
13. Using a 5 mL syringe with 16-gauge needle attached, collect virus from the 40/60% interface and 40% iodixanol layer by puncturing the OptiSeal tube ~3–5 mm below the 40/60% interface (at the tube curvature), removing the black cap with one hand, and slowly drawing back on the 5 mL syringe plunger with the opposite hand so as to collect ~4.0–4.5 mL virus/gradient (*see* **Note 12**).
14. After replacing the black stopper-cap on the OptiSeal tube, remove the 5 mL syringe/needle from the OptiSeal tube, and transfer the 4.0–4.5 mL of virus/gradient to below

the remaining milliliters of DPBS in the 10 mL syringe barrel (*see* **step 11**).

15. Insert the 10 mL syringe plunger into the syringe barrel and push the virus/gradient/DPBS mixture through the syringe filter unit and into the Amicon filter device (*see* **Note 13**).
16. Centrifuge the virus/gradient/DPBS mixture for ~5–8 min at 3000 g (RT), or until the volume of the solution in the Amicon filter-device cartridge has been reduced to ~500–1500 μL. Bleach and discard the flow-through.
17. Add 13 mL DPBS to the virus in Amicon cartridge, mix with a P1000 pipette, and centrifuge again according to the procedure in **step 15**.
18. Discard the flow-through and repeat **step 16** two additional times for a total of four buffer exchanges with DPBS (*see* **Note 14**). On the last spin, retain 300–500 μL of virus/DPBS solution in the Amicon cartridge (*see* **Note 15**).
19. Use a P200 pipette to transfer the virus from the Amicon cartridge to a 1.6 mL screw-cap vial (e.g., 1.6 mL vials; National Scientific Supply, cat. no. BC16NA-PS); store at 4 °C for up to 3 months.

Stage 4. AAV Titration

Day 8 or 9

1. Plan out the PCR 96-well plate set-up (*see* **Note 16**). Namely, reserve the first 24 wells (A1–B12) for DNA standards (*see* **Note 17**), and allocate virus samples to the remaining wells (*see* **Note 17**). Include DPBS as a negative control and a virus sample of known concentration as a positive control.
2. Digest away DNA that was not packaged into the viral capsid using DNase 1. Specifically, add 100 μL DNase I + DNase digestion buffer (*see* Subheading 2) to each 1.5 mL tube (prepare triplicate tubes for each virus sample). Vortex each virus for 1–2 s (*see* **Note 18**), and add 2 μL virus to each of three tubes. Vortex briefly (1–2 s) to mix, spin down (2000 g, RT, 10 s), and incubate in a 37 °C water bath for 1 h.
3. Inactivate the DNase by adding 5 μL EDTA to each tube, vortexing for 1–2 s to mix, spinning down the reaction mix (2000 g, RT, 10 s), and incubating in a 70 °C dry bath for 10 min.

4. Digest away the viral capsid with proteinase K. Specifically, add 120 μl proteinase K + proteinase K solution to each tube, vortex for 1–2 s to mix, and spin down (2000 g, RT, 10 s). Incubate in a 50 °C dry bath for 2 h or up to overnight.
5. During the last 20 min of the proteinase K digestion, prepare the DNA standard dilutions and use the Qubit assay ((Invitrogen, cat. no. Q32854) to measure the concentration (ng/μL) of the DNA standard stock.
6. Incubate the tubes in a 95 °C dry bath for 10 min to inactivate the proteinase K.
7. Allow the tubes to cool for 5 min. Vortex each sample for 1–2 s to mix and then add 3 μL to a new tube that contains 897 μL of UltraPure water (1:300 dilution). Vortex the diluted samples for 3 s to mix.
8. Prepare the qPCR master mix.
9. Load the PCR 96-well plate based on the experimental plan from **step 1**. Namely, first, pipette 23 μL of qPCR master mix into each designated well. Next, pipette 2 μL of each standard into wells A1–B12. Last, pipette 2 μL of each diluted sample from **step** 7 into wells C1 and onward. Seal the dish with sealing film (e.g., Genesee Scientific, cat. no. 12-529), and briefly spin down (500 g, RT, 10 s) in a PCR plate spinner (e.g., VWR International, cat. no. 89184).
10. Place the PCR plate into the qPCR machine (e.g., Analytik Jena, model no. qTOWER 2.2), and run qPCR program (*see* Table 4).
11. When the qPCR run is complete, export the cycle threshold (CT) values and enter these values into a qPCR data processing pipeline (e.g., custom Matlab script) or Excel spreadsheet modeled after Table 5. Generate a standard curve (*see* **Note 19**, *see* Table 5 for calculations) and

Table 4
qPCR cycling parameters for AAV titration

Step 1	95 °C	10 min
Step 2	95 °C	15 s
Step 3	60 °C	20 s
Step 4	60 °C	40 s
Repeat steps 2–4 40×		

Table 5
AAV titration calculator. Blue cells represent virus prep-specific values to be entered by the researcher; CT values are produced by the qPCR run. Orange cells represent values to be calculated according to the equation shown in the table. Green cells represent values to be generated from the DNA standard curve, with sample excel formulas shown in the table. If the virus prep involves pooling multiple 150 mm dishes, one should calculate the "total vector genomes per dish"

DNA standard				
Name of DNA standard: *(for record-keeping)*				
Size of DNA standard (bp)				(a)
Molecular weight of DNA standard (g/mole)[a]				(b)
DNA standard stock concentration (ng/μL)				(c)
Volume of protease-treated virus assayed in qPCR (mL)		*(Stage 4, Step 9) =*		(d)
AAV genome used in qPCR (g)		*((c) × (d)/1,000,000) =*		(e)
ds AAV genomic DNA molecules in qPCR		*((e)/(b) × 6.0221E+23) =*		(f)
ss AAV genomic DNA molecules in qPCR		*((f)/2) =*		(g)
Dilution factor calculation				
100 μL DNAse + 5 μL EDTA + 120 μL proteinase K + 2 μL virus			227 μL =	(h)
Protease treated virus dilution (e.g., 1:300)		*300 =*		(i)
Virus used in the assay (μL)		*(Stage 4, Step 2) =*		(j)
Dilution factor = final volume/initial volume		*((h) × (i)/(j)) =*		(k)
DNA standards[b]	*CT value (x-axis)*	*ssAAV molecules (y-axis)*		
1	(p_1) (p_2) (p_3)	$((t_1) - 1) = (s_1)$ $((t_2) - 1) = (s_2)$ $((t_3) - 1) = (s_3)$		
2	(p_4) (p_5) (p_6)	$((u_1) - 1) = (t_1)$ $((u_2) - 1) = (t_2)$ $((u_3) - 1) = (t_3)$		
3	(p_7) (p_8) (p_9)	$((v_1) - 1) = (u_1)$ $((v_2) - 1) = (u_2)$ $((v_3) - 1) = (u_3)$		
4	(p_{10}) (p_{11}) (p_{12})	$((w_1) - 1) = (v_1)$ $((w_2) - 1) = (v_2)$ $((w_3) - 1) = (v_3)$		
5	(p_{13}) (p_{14}) (p_{15})	$((x_1) - 1) = (w_1)$ $((x_2) - 1) = (w_2)$ $((x_3) - 1) = (w_3)$		
6	(p_{16}) (p_{17}) (p_{18})	$((y_1) - 1) = (x_1)$ $((y_2) - 1) = (x_2)$ $((y_3) - 1) = (x_3)$		

(continued)

Table 5 (continued)

7		(p_{19}) (p_{20}) (p_{21})	$((z_1) - 1) = (y_1)$ $((z_2) - 1) = (y_2)$ $((z_3) - 1) = (y_3)$		
8		(p_{22}) (p_{23}) (p_{24})	$(\text{Log}\ (g)/10) = (z_1)$ $(\text{Log}\ (g)/10) = (z_2)$ $(\text{Log}\ (g)/10) = (z_3)$		
Excel graph parameters: DNA standard CT values versus DNA standard concentrations (ssAAV molecules)					
Slope of the standard curve			*Slope* $(s_1{:}z_3, p_1{:}p_{24}) =$		(m)
Intercept of the standard curve			*Intercept* $(s_1{:}z_3, p_1{:}p_{24}) =$		(n)
R^2 value			*RSQ* $(s_1{:}z_3, p_1{:}p_{24}) =$		(r)
AAV genomes (in triplicate)					
AAV genome	CT	ssAAV molecules	Titer (vg/mL)	Volume (mL)	Total vector genomes (vg)
Virus prep #1: ________	CT_1 CT_2 CT_3	(q_1)[c] (q_2) (q_3)	$\text{avg} = \frac{(q_1+q_2+q_3)}{3}$	(mL # 1)	$\text{VG\#1} = \frac{\text{titer (vg/mL)}}{\text{volume (mL)}}$
Virus prep #2: ________	CT_4 CT_5 CT_6	(q_4) (q_5) (q_6)	$\text{avg} = \frac{(q_4+q_5+q_6)}{3}$	(mL # 2)	$\text{VG\#2} = \frac{\text{titer (vg/mL)}}{\text{volume (mL)}}$

[a]To calculate the molecular weight of a DNA sequence, use an online calculator (e.g., http://www.bioinformatics.org/sms2/dna_mw.html). Or, multiply the length (in bp) by 650 (1 bp ~ 650 g/mole)

[b]8 dilutions in triplicate

[c]$(q_1) = \left(10^{(m)\times CT_1+(n)}\right) \times (k)/(d)$

compute the titer (vg/ml) of each virus prep. If the virus is harvested from multiple dishes for a single prep, calculate the per-dish production efficiency (vg/dish).

Stage 5. Systemic AAV Administration

1. Calculate the AAV dose (vg) to administer per animal (*see* **Note 20**). Divide the dose (vg) by the titer (vg/ml) to calculate the volume of virus needed to inject one animal (*see* **Note 21**).
2. In a screw-cap vial, prepare a master mix of viruses based on the number of animals to be injected; briefly vortex each virus and master mix for 1–2 s before use. Transport and/or store the virus aliquots on the ice leading up to injection.
3. Anesthetize the first subject in a biosafety cabinet according to IACUC, governmental and institutional guidelines, and approved lab protocols.

4. Load an insulin syringe with the appropriate volume of the virus, expelling any air bubbles and syringe dead space.
5. After ensuring that the subject is fully anesthetized, place the mouse in a prone position on a small stack of paper towels on a heating pad, with the head facing the researcher's dominant hand. Place the nose cone on the mouse to maintain anesthesia, if applicable.
6. Perform the retro-orbital injection. Specifically, grasp and gently pull back the skin above and below the subject's eye with the thumb and index finger (of the nondominant hand) in a scissored position, causing the eye to protrude slightly from the socket. With the dominant hand, insert the needle and bevel down at a 30–45° angle into the medial canthus and through the conjunctival membrane. The needle should be positioned behind the eye globe in the retro-orbital sinus. Slowly release the virus into the sinus and gently remove the needle.
7. Apply mild pressure to the eyelid, and dispense 1–2 drops of proparacaine to the corneal surface to provide local analgesia.
8. Follow institutional guidelines for subject recovery. Generally, the subject is placed in a clean, warmed cage for initial recovery from anesthesia and then transferred back to a clean home cage.
9. Allow 3–4 weeks for evaluation of transgene expression, although as little as 7–10 days may be adequate for sufficient transduction and transgene expression. The appropriate timeline for a given experiment must be determined on an empirical basis.
10. Proceed with lab-specific protocol for downstream applications, such as optogenetic or chemogenetic manipulations.

4 Notes

1. Refer to the Addgene catalog or other suppliers for suggested growth conditions and selection of antibiotics. Typically, Addgene pAAV plasmids will carry antibiotic resistance to ampicillin/carbenicillin (e.g., carbenicillin disodium salt; Alfa Aesar, cat. no. J61949-06); and ScaI-HF restriction enzyme may be used to linearize this plasmid DNA.
2. pAAVs should be propagated in recombination-deficient strains such as NEB Stable, Stbl3, or SURE 2 competent cells to prevent unwanted recombination. Always verify the integrity of purified plasmids by Sanger sequencing (avoid ITRs when designing sequencing primers (Integrated DNA

Technologies)) and restriction digestion (confirm the presence of ITRs through digesting pAAVs with SmaI (New England Biolabs, cat. no. R0141S); https://www.neb.com/tools-and-resources) before proceeding with downstream applications.

3. At low concentrations, linearized DNA is prone both to degradation and to sticking to the tube's inner walls. Thus, use DNA/RNA LoBind microfuge tubes for DNA standard solutions and for AAV samples and dilutions used in AAV titration (Stage 4).
4. Passage cells at a ratio of 1:3 (i.e., divide one dish of cells into three new dishes of cells) every other day when expanding cells for viral production; and split cells at a 1:2 ratio (or 6×10^4 cells/cm^2) 24 h before transfection.
5. Replace media in an expedited fashion to minimize cell stress. Likewise, because PEI is moderately cytotoxic, with cell death of up to 20% being common, do not allow the transfection solution + media to remain unchanged for more than 24 h post-transfection as this could adversely affect titers.
6. Empty, sterile media bottles are a convenient temporary storage solution for media collected at 72 h. Alternatively, media from up to five 150 mm dishes of the same virus prep can be stored in a 250 mL centrifuge tube in preparation for **step 3** of Stage 2.
7. Save this 250 mL conical centrifuge tube for **step 9**. In **step 9**, transfer stored supernatant into this 250 mL tube.
8. Pour the supernatant into a collection flask or used media bottle and decontaminate with bleach such that the final volume of decontamination solution contains 10% bleach.
9. Pipette the pellet vigorously for ~5 min to resuspend, but avoid adding air bubbles.
10. *See* Table 2 for iodixanol gradient formulations. Use a 2 mL serological pipette (Corning brand 2 mL serological pipettes fit into the narrow OptiSeal tube opening, and they have markings for 3 mL) to add 6 mL (fill pipette twice) of 15% (wt/vol) iodixanol to each tube. Then add 6 mL of 25% (wt/vol) iodixanol under the 15% layer by carefully lowering the pipette tip to the very bottom of the tube and slowly dispensing the solution from the pipette. Avoid agitating the upper layer of 15% iodixanol. Repeat this process for subsequent layers of increasing iodixanol density, taking care not to disturb each upper layer when inserting/removing the serological pipette from the tube. To avoid generating air bubbles when removing the pipette, do NOT dispense the last ~2–3 mm of iodixanol from the pipette tip.

11. For a Beckman Coulter Optima XE-9 with Type 70Ti rotor, use profile no. 3 (accelerate across 3 min to 500 r.p.m., followed by maximum acceleration) and profile no. 9 (decelerate at maximum speed to 500 r.p.m., then decelerate across 6 min to stop).
12. Position the needle bevel facing upward while collecting virus/gradient, and slowly twist the bevel to face downward as the 25/40% interface is approached. Stop collecting virus/gradient just before this 25/40% interface (the white protein layer). Replace the black cap before removing the needle from the tube.
13. Mix the virus/gradient/DPBS liquid collected in the Amicon filter cartridge with a P1000 pipette before centrifuging.
14. Only a 2–3-min spin will be necessary for the latter washes as progressively less iodixanol gradient remains.
15. Assuming a production efficiency of at least 1×10^{12} vg/dish, the 300–500 μL final volume will permit a dose and injection volume of $\leq 1 \times 10^{12}$ vg and 100 μL, respectively.
16. Additional information on titration may be found in ref no. [64].
17. Regarding set-up, the first 24 wells (A1–B12) are reserved for standards in triplicate, such that standard no. 1 occupies wells A1–A3, standard no. 2 occupies wells A4–A6, and so on. Virus samples are assigned to the remaining wells, again in triplicate fashion, with the first virus sample occupying wells C1–C3, the second occupying wells C4–C6, etc. Pipette virus samples very carefully to minimize small volume differences, which would spoil accurate qPCR readings.
18. Do not vortex the virus vigorously or vortex longer than 1–2 s; exposure to excess force could disrupt capsid structure.
19. If the titer is $\geq 1 \times 10^{14}$ vg/mL, the virus may aggregate during 4 °C storage. Dilute the virus with DPBS to between 2×10^{13} and 5×10^{13} vg/mL; retiter the new diluted stock.
20. Brief guidelines are provided for intravenous (retro-orbital) injection in adult mice; more detailed instructions and pictures can be found here [65]. However, researchers should select a route of systemic delivery and the corresponding virus dose that is most appropriate for the specific experimental parameters and animal species. Experiments on vertebrates must conform to all relevant governmental and institutional regulations. Animal husbandry and experimental procedures involving mice were approved by the Institutional Animal Care and Use Committee (IACUC) and the Department of Animal Resources at the University of Southern California.

21. For RO injection of the mouse venous sinus, it is easier to stay within injection volumes of 40–80 μL for adult mice and ≤10 μL for neonatal mice. If <40 μL/mouse is required, use DPBS or saline to dilute the virus such that a larger volume is injected. If more than 80 μL/mouse is required, it may be more convenient to reconcentrate the virus or perform two separate injections.

Acknowledgments

We thank Dr. Nicholas C. Flytzanis for his critical reading of the manuscript. This work was supported by NIH award U01 NS120824 and the Brain and Behavior Research Foundation NARSAD Young Investigator award (to J.B.T.).

Competing Interests The author declares no competing interests.

References

1. Choudhury SR, Harris AF, Cabral DJ et al (2016) Widespread central nervous system gene transfer and silencing after systemic delivery of novel AAV-AS vector. Mol Ther 24:726–735
2. Deverman BE, Pravdo PL, Simpson BP et al (2016) Cre-dependent selection yields AAV variants for widespread gene transfer to the adult brain. Nat Biotechnol 34:204–209
3. Chan KY, Jang MJ, Yoo BB et al (2017) Engineered AAVs for efficient noninvasive gene delivery to the central and peripheral nervous systems. Nat Neurosci 20:1172–1179
4. Ravindra Kumar S, Miles TF, Chen X et al (2020) Multiplexed Cre-dependent selection yields systemic AAVs for targeting distinct brain cell types. Nat Methods 17:541–550
5. Goertsen D, Flytzanis NC, Goeden N et al (2022) AAV capsid variants with brain-wide transgene expression and decreased liver targeting after intravenous delivery in mouse and marmoset. Nat Neurosci 25:106–115
6. Challis RC, Ravindra Kumar S, Chan KY et al (2019) Systemic AAV vectors for widespread and targeted gene delivery in rodents. Nat Protoc 14:379–414
7. Davidsson M, Wang G, Aldrin-Kirk P et al (2019) A systematic capsid evolution approach performed in vivo for the design of AAV vectors with tailored properties and tropism. Proc Natl Acad Sci 116:27053–27062
8. Lee EJ, Guenther CM, Suh J (2018) Adeno-associated virus (AAV) vectors: rational design strategies for capsid engineering. Curr Opin Biomed Eng 7:58–63
9. Bey K, Deniaud J, Dubreil L et al (2020) Intra-CSF AAV9 and AAVrh10 administration in nonhuman primates: promising routes and vectors for which neurological diseases? Mol Ther Methods Clin Dev 17:771–784
10. Tabebordbar M, Lagerborg KA, Stanton A et al (2021) Directed evolution of a family of AAV capsid variants enabling potent muscle-directed gene delivery across species. Cell 184:4919–4938.e22
11. Bedbrook CN, Deverman BE, Gradinaru V (2018) Viral strategies for targeting the central and peripheral nervous systems. Annu Rev Neurosci 41:323–348
12. Goertsen D, Goeden N, Flytzanis NC et al (2022) Targeting the lung epithelium after intravenous delivery by directed evolution of underexplored sites on the AAV capsid. Mol Ther Methods Clin Dev 26:331–342. S2329050122001024. https://doi.org/10.1016/j.omtm.2022.07.010
13. Treweek JB, Chan KY, Flytzanis NC et al (2015) Whole-body tissue stabilization and selective extractions via tissue-hydrogel hybrids for high-resolution intact circuit mapping and phenotyping. Nat Protoc 10:1860–1896

14. Fenno LE, Mattis J, Ramakrishnan C et al (2014) INTRSECT: single-component targeting of cells using multiple-feature Boolean logic. Nat Methods 11:763–772
15. Chen T-W, Wardill TJ, Sun Y et al (2013) Ultrasensitive fluorescent proteins for imaging neuronal activity. Nature 499:295–300
16. Akerboom J, Chen T-W, Wardill TJ et al (2012) Optimization of a GCaMP calcium indicator for neural activity imaging. J Neurosci 32:13819–13840
17. Dana H, Mohar B, Sun Y et al (2016) Sensitive red protein calcium indicators for imaging neural activity. elife 5:e12727
18. Tian L, Hires SA, Looger LL (2012) Imaging neuronal activity with genetically encoded calcium indicators. Cold Spring Harb Protoc 2012:647–656
19. Jing M, Li Y, Zeng J et al (2020) An optimized acetylcholine sensor for monitoring in vivo cholinergic activity. Nat Methods 17:1139–1146
20. Marvin JS, Borghuis BG, Tian L et al (2013) An optimized fluorescent probe for visualizing glutamate neurotransmission. Nat Methods 10:162–170
21. Sun F, Zhou J, Dai B et al (2020) Next-generation GRAB sensors for monitoring dopaminergic activity in vivo. Nat Methods 17:1156–1166
22. Patriarchi T, Cho JR, Merten K et al (2018) Ultrafast neuronal imaging of dopamine dynamics with designed genetically encoded sensors. Science 360:eaat4422
23. Feng J, Zhang C, Lischinsky JE et al (2019) A genetically encoded fluorescent sensor for rapid and specific in vivo detection of norepinephrine. Neuron 102:745–761.e8
24. Unger EK, Keller JP, Altermatt M et al (2020) Directed evolution of a selective and sensitive serotonin sensor via machine learning. Cell 183:1986–2002.e26
25. Jing M, Zhang Y, Wang H et al (2019) G-protein-coupled receptor-based sensors for imaging neurochemicals with high sensitivity and specificity. J Neurochem 151:279–288
26. Chen J, Cho KE, Skwarzynska D et al (2021) The property-based practical applications and solutions of genetically encoded acetylcholine and monoamine sensors. J Neurosci 41:2318–2328
27. Mattis J, Tye KM, Ferenczi EA et al (2012) Principles for applying optogenetic tools derived from direct comparative analysis of microbial opsins. Nat Methods 9:159–172
28. Sternson SM, Roth BL (2014) Chemogenetic tools to interrogate brain functions. Annu Rev Neurosci 37:387–407
29. Zingg B, Chou X, Zhang Z et al (2017) AAV-mediated anterograde transsynaptic tagging: mapping input-defined functional neural pathways for defense behavior. Neuron 93:33–47
30. Zingg B, Peng B, Huang J et al (2020) Synaptic specificity and application of anterograde transsynaptic AAV for probing neural circuitry. J Neurosci 40:3250–3267
31. Zhang H, Yang B, Mu X et al (2011) Several rAAV vectors efficiently cross the blood-brain barrier and transduce neurons and astrocytes in the neonatal mouse central nervous system. Mol Ther 19:1440–1448
32. Colón-Thillet R, Jerome KR, Stone D (2021) Optimization of AAV vectors to target persistent viral reservoirs. Virol J 18:85
33. Challis C, Hori A, Sampson TR et al (2020) Gut-seeded α-synuclein fibrils promote gut dysfunction and brain pathology specifically in aged mice. Nat Neurosci 23:327–336
34. Chakrabarti S, Pattison LA, Doleschall B et al (2020) Intraarticular adeno-associated virus serotype AAV-PHP.S-mediated chemogenetic targeting of knee-innervating dorsal root ganglion neurons alleviates inflammatory pain in mice. Arthritis Rheumatol 72:1749–1758
35. Ivanchenko MV, Hanlon KS, Hathaway DM et al (2021) AAV-S: a versatile capsid variant for transduction of mouse and primate inner ear. Mol Ther Methods Clin Dev 21:382–398
36. Palfi A, Chadderton N, Millington-Ward S et al (2022) AAV-PHP.eB transduces both the inner and outer retina with high efficacy in mice. Mol Ther Methods Clin Dev 25:236–249
37. Yang L, Jiang J, Drouin LM et al (2009) A myocardium tropic adeno-associated virus (AAV) evolved by DNA shuffling and in vivo selection. Proc Natl Acad Sci 106:3946–3951
38. Rajendran PS, Challis RC, Fowlkes CC et al (2019) Identification of peripheral neural circuits that regulate heart rate using optogenetic and viral vector strategies. Nat Commun 10:1944
39. Bates R, Huang W, Cao L (2020) Adipose tissue: an emerging target for adeno-associated viral vectors. Mol Ther Methods Clin Dev 19:236–249
40. Graybuck LT, Daigle TL, Sedeño-Cortés AE et al (2021) Enhancer viruses for combinatorial cell-subclass-specific labeling. Neuron 109:1449–1464.e13
41. Mich JK, Graybuck LT, Hess EE et al (2021) Functional enhancer elements drive subclass-selective expression from mouse to primate neocortex. Cell Rep 34:108754
42. Zelikowsky M, Hui M, Karigo T et al (2018) The neuropeptide Tac2 controls a distributed

brain state induced by chronic social isolation stress. Cell 173:1265–1279.e19

43. Gradinaru V (2020) Expanding the brain researcher's toolkit. Science 369:637
44. Treweek JB, Gradinaru V (2016) Extracting structural and functional features of widely distributed biological circuits with single cell resolution via tissue clearing and delivery vectors. Curr Opin Biotechnol 40:193–207
45. Yang Y, Wang L, Bell P et al (2016) A dual AAV system enables the Cas9-mediated correction of a metabolic liver disease in newborn mice. Nat Biotechnol 34:334–338
46. Suzuki K, Tsunekawa Y, Hernandez-Benitez R et al (2016) In vivo genome editing via CRISPR/Cas9 mediated homology-independent targeted integration. Nature 540:144–149
47. Deverman BE, Ravina BM, Bankiewicz KS et al (2018) Gene therapy for neurological disorders: progress and prospects. Nat Rev Drug Discov 17:641–659
48. Rauch JN, Luna G, Guzman E et al (2020) LRP1 is a master regulator of tau uptake and spread. Nature 580:381–385
49. Zeng J, Wang Y, Luo Z et al (2019) TRIM9-mediated resolution of neuroinflammation confers neuroprotection upon ischemic stroke in mice. Cell Rep 27:549–560.e6
50. Sampson TR, Challis C, Jain N et al (2020) A gut bacterial amyloid promotes α-synuclein aggregation and motor impairment in mice. elife 9:e53111
51. Chen X, Ravindra Kumar S, Adams CD et al (2022) Engineered AAVs for non-invasive gene delivery to rodent and non-human primate nervous systems. Neuron 110(14): 2242–2257.e6. S0896627322004111. https://doi.org/10.1016/j.neuron.2022.05.003
52. Hordeaux J, Wang Q, Katz N et al (2018) The neurotropic properties of AAV-PHP.B are limited to C57BL/6J mice. Mol Ther 26:664–668
53. Huang Q, Chan KY, Tobey IG et al (2019) Delivering genes across the blood-brain barrier: LY6A, a novel cellular receptor for AAV-PHP.B capsids. PLoS One 14:e0225206
54. Batista AR, King OD, Reardon CP et al (2020) Ly6a differential expression in blood-brain barrier is responsible for strain specific central nervous system transduction profile of AAV-PHP.B. Hum Gene Ther 31:90–102
55. Brown D, Altermatt M, Dobreva T et al (2021) Deep parallel characterization of AAV tropism and AAV-mediated transcriptional changes via single-cell RNA sequencing. Front Immunol 12:730825
56. Wall NR, Wickersham IR, Cetin A et al (2010) Monosynaptic circuit tracing in vivo through Cre-dependent targeting and complementation of modified rabies virus. Proc Natl Acad Sci 107:21848–21853
57. Beier KT, Saunders A, Oldenburg IA et al (2011) Anterograde or retrograde transsynaptic labeling of CNS neurons with vesicular stomatitis virus vectors. Proc Natl Acad Sci U S A 108:15414–15419
58. Li E, Guo J, Oh SJ et al (2021) Anterograde transneuronal tracing and genetic control with engineered yellow fever vaccine YFV-17D. Nat Methods 18(12):1542–1551
59. Xu X, Holmes TC, Luo M-H et al (2020) Viral vectors for neural circuit mapping and recent advances in trans-synaptic anterograde tracers. Neuron 107:1029–1047
60. Treweek JB (2021) A repurposed vaccine grants durable anterograde access to neural circuits. Nat Methods 18:1459–1461
61. Tervo DGR, Hwang B-Y, Viswanathan S et al (2016) A designer AAV variant permits efficient retrograde access to projection neurons. Neuron 92:372–382
62. Montgomery KL, Iyer SM, Christensen AJ et al (2016) Beyond the brain: optogenetic control in the spinal cord and peripheral nervous system. Sci Transl Med 8:337rv5
63. Grieger JC, Choi VW, Samulski RJ (2006) Production and characterization of adeno-associated viral vectors. Nat Protoc 1:1412–1428
64. Gray SJ, Choi VW, Asokan A et al (2011) Production of recombinant adeno-associated viral vectors and use in in vitro and in vivo administration. Curr Protoc Neurosci 57: 4.17.1–4.17.30
65. Yardeni T, Eckhaus M, Morris HD et al (2011) Retro-orbital injections in mice. Lab Anim 40: 155–160

Chapter 8

Transcriptomic Definition of Neuron Types

Jing He and William R. Stauffer

Abstract

Understanding the dynamic functions of different cell types and the circuits in which they participate is one of the primary goals of neuroscience, but—as the existence of this book testifies—this goal presents special challenges for systems neuroscience research performed in nonhuman primates (NHPs). Namely, the strategies that enable cell type-specific access in rodents do not translate well, or easily, to NHPs. Here, we discuss how the application of novel molecular classification techniques has the potential to open new avenues for cell type-specific tool development in NHPs. High-throughput single-cell studies have created new ways to classify neurons by collecting and analyzing individual cells' transcriptomes, genomes, proteomes, and epigenomes. The focus of these cutting-edge studies has largely been on building databases of cell types in rodents and humans. Only a handful of transcriptomic studies have used brain tissue from NHPs, and even fewer have examined gene regulation in NHPs. This is a missed opportunity. For the same reasons that NHP research is the gold standard for studying cognition and human diseases—that NHPs like rhesus macaques have neuroanatomical, cytoarchitectonic, genetic, physiological, and behavioral characteristics that are more like the human than any other species in which invasive experiments are ethically permissible—single-cell studies in NHPs need to be pursued in parallel with similar studies in human and mouse. Doing so will give us the most precise picture of cell types relevant for higher cognitive functions, help to develop methods for cell type-specific NHP system neuroscience studies, and pave the way toward translational, circuit-targeted therapeutics.

Key words Nonhuman primates, Transcriptome, Neuron types

1 Introduction

Neuron type classification provides a foundation for functional studies of circuits. Single-cell technologies are a suite of powerful, quantitative, and high-throughput methods to accomplish neuron type classification. Commercial single-cell platforms can collect transcriptomic, genomic, epigenomic, and proteomic information alone or in tandem [1]. This chapter will focus on transcriptomes and epigenomic maps of chromatin accessibility—two techniques with the potential to transform our ability to achieve cell type-specific transgenesis in nonhuman primates (NHPs). Quantitative similarities and differences in transcriptomes can be used to define

Mark A. G. Eldridge and Adriana Galvan (eds.), *Vectorology for Optogenetics and Chemogenetics*,
Neuromethods, vol. 195, https://doi.org/10.1007/978-1-0716-2918-5_8,

cell types and subtypes. Genomic accessibility studies—especially those done at the single cell level—can be used to locate cell type-specific regulatory sequences. Both methods can examine cells at multiple levels of resolution—including cell classes, cell types, subtypes, and archetypes—and reveal the developmental, evolutionary, and functional relationships between different cell types. These techniques can be especially impactful when applied to primate tissue. For example, several recent studies have combined morphological characterization of unique cell types, including Von Economo neurons in the insula [2, 3] and Betz cells in the primary motor cortex [4], with single-cell characterization to identify evolutionarily conserved cell lineages and novel cell type-specific gene markers [5, 6]. Likewise, single-cell techniques have helped identify cell types that were previously unknown, including a novel striatal interneuron subtype [7], a novel mu-opioid expressing cell type in striatal interface islands [8], and 'rosehip' neurons [9]. Chromatin accessibility studies have recently isolated and validated novel gene regulatory sequences that can be used to drive cell type-specific transgene expression in mice, NHPs, and even in human explants [10–12]. Thus, single-cell transcriptomics and single-cell epigenomics together have the potential to define cell types in NHP and identify enhancers that can target those cell types.

Transcriptomes are composed of messenger RNA (mRNA). Ribosomes use these mRNA transcripts as a template for protein synthesis. The transcriptome is, therefore, tightly linked to cell identity and cell function, and measuring transcriptomes is a principle method to define cell types and cell states. Single-cell RNA Sequencing (scRNA-Seq) and single nucleus RNA sequencing (snRNA-Seq) are two methods commonly used to capture the transcriptomes. ScRNA-Seq collects more mature mRNA, whereas snRNA-Seq collects more premature mRNA—that is, RNA strands that still contain intronic regions. Rigorous comparisons of scRNA-Seq and snRNA-Seq have demonstrated that scRNA-Seq is likely to capture more transcripts per cell compared to snRNA-Seq, but the techniques lead to very similar classification results [13, 14]. We have applied both techniques to the NHP brain and consistently found that single-cell RNA-Seq was biased toward recovering microglia and other glial cells. This is likely because glial cells are generally smaller, more robust, and lack axons and dendrites. In contrast to cells, nuclei sizes are relatively uniform across cell types, and therefore snRNA-Seq provided a more unbiased estimate of the underlying tissue [8, 14]. Thus, snRNA-Seq is a validated method to capture the complete transcriptomes of cells in the NHP brain.

The epigenome is the set of modifications made to genetic material or histones, including DNA methylation-induced conformational changes. Epigenomic phenomena play a crucial role in the regulation of gene expression. Of particular importance to NHP

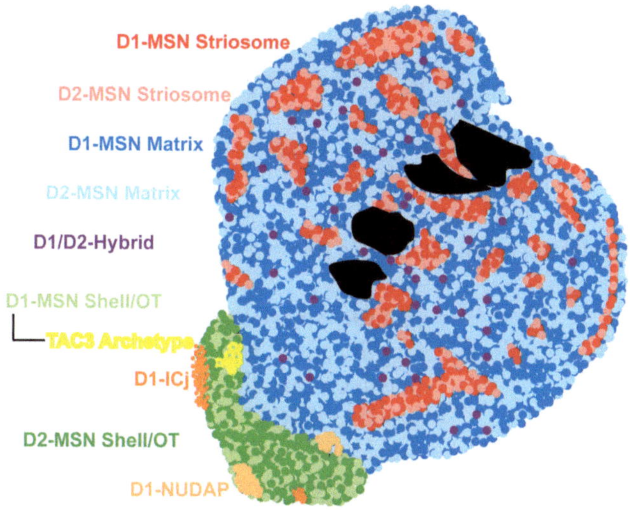

Fig. 1 Schematic view of MSN subtypes/archetypes and their spatial locations. (Adapted from He et al. [8])

studies is chromatin accessibility, which is normally evaluated via the assay for transposase-accessible chromatin using sequencing (ATAC-seq) [15, 16]. For DNA to be read and to subsequently influence cell function, the DNA must be accessible: it should not be wound tightly around histones. ATAC-Seq identifies regions of high chromatin accessibility. Single-cell ATAC-Seq [17]—which is performed on the nucleus rather than the whole cell—enables the detection of cell type-specific open chromatin regions. These regions contain putative enhancers, distal regulatory sequences that can be used to drive cell type-specific transgene activity [10–12, 18–21]. Thus, single-cell ATAC-Seq is a validated method to identify gene regulatory sequences that can drive cell type-specific expression, even in NHPs.

We recently applied snRNA-Seq and snATAC-Seq to the striatum of rhesus macaque monkeys (Fig. 1). We will use the application of snRNA-Seq to the rhesus macaque striatum to illustrate how this technique can reveal molecular insights regarding the intact NHP brain. The striatum is the major input nucleus for the basal ganglia. There are two major classes of neurons in the striatum: medium spiny neurons (MSNs) and interneurons. Despite this apparent simplicity, the striatum participates in numerous behavioral functions. We set out to characterize single cells in the NHP striatum to better understand how this structure contributes to these diverse behaviors.

1.1 Single-Cell Transcriptomics Overview

To perform scRNA-/snRNA-Seq, cells or nuclei from tissues are dissociated, and transcriptomic data are collected using plate-based methods and microfluidics-based methods. Plate-based methods

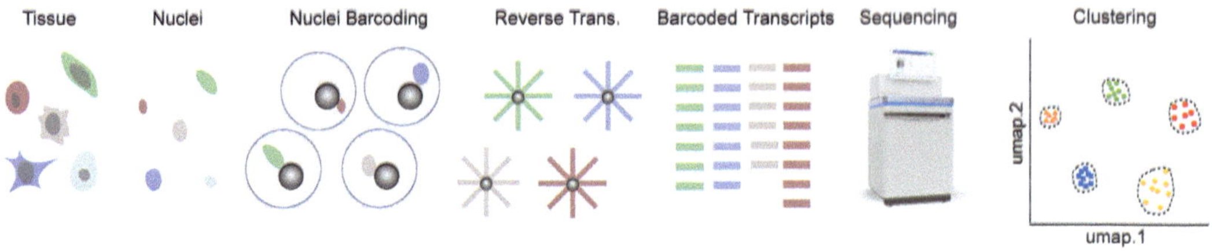

Fig. 2 Workflow of single cell/nucleus RNA sequencing

include Smart-seq and Smart-seq2 [22], cell expression by linear amplification and sequencing (CEL-seq) [23], CEL-seq2 [24], massively parallel RNA single-cell sequencing (MARS-seq) [25], single-cell RNA barcoding and sequencing (SCRB-seq) [26], and molecular crowding SCRB-seq (mcSCRB-seq) [27]. These methods are low to medium throughput and enable the parallel processing of hundreds of cells, but they enable full-length-based library construction and isoform identification. In contrast, microfluidics-based approaches, including Drop-seq [28], InDrop [29, 30], Seq-well [31], and the 10× Genomics Chromium systems, enable parallel processing of thousands, or even tens of thousands, of cells. Using these methods, RNA is barcoded and reverse transcribed to cDNA, which is then amplified and constructed into libraries for sequencing. After sequencing, the data is mapped to a species-specific reference genome, and a cell by gene (transcript count) matrix is constructed. All further data analysis, including pre-processing, principal component analysis, dimension reduction, and visualization in (Uniform Manifold Approximation and Projection) UMAP or (t-distributed stochastic neighbor embedding) TSNE space is done on the cell by gene matrix (Fig. 2).

2 Material, Methods, and Results

2.1 Imaging and Surgery

We scanned monkeys with an MRI machine (Siemens, 3T) to generate T1-weighted images of the brain and built a custom brain matrix for slicing and microdissection. For the structural MRI scan, monkeys were anesthetized with ketamine, maintained under general anesthesia with isoflurane, and head-fixed in an MRI-compatible stereotaxic frame. We generated a whole-brain model for each monkey using Brainsight (Rogue Research) based on the MRI images, and 3D printed a custom matrix for each brain with a cutting guide set every 1 mm.

For brain tissue collection, the animals were sedated with ketamine (15 mg/kg IM), ventilated, and maintained under general anesthesia with isoflurane. When we no longer detect a corneal reflex, animals are placed in a stereotaxic frame (Kopf Instruments). We used a Stryker autopsy saw and surgical drills to completely

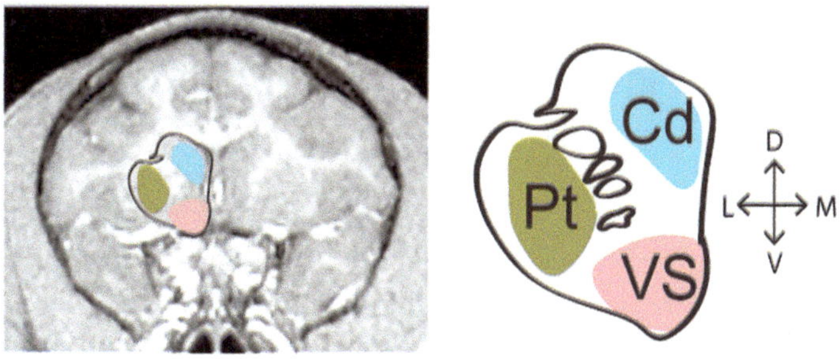

Fig. 3 MRI image (left) and schematic striatum (right) of a Rhesus macaque coronal brain section showing three striatal regions labeled with cyan (caudate nucleus), brown (putamen), and pink (ventral striatum). The right axis shows dorsal (D), ventral (V), lateral (L), and medial (M) directions. (Adapted from He et al. [8])

remove the calvarium. We perfused the circulatory system transcardially with 3–4 L of ice-cold oxygenated artificial cerebrospinal fluid (124 mM NaCl, 5 mM KCl, 2 mM $MgSO_4$, 2 mM $CaCl_2$, 23 mM $NaHCO_3$, 3 mM NaH_2PO_4, 10 mM glucose). We carefully removed the dura, cut the spinal cord and cranial nerves, and then gently removed the brain. We sliced the brain in the custom brain matrix into 4 mm slabs along the rostral-caudal axis. Brain regions of interest were isolated under a dissecting microscope for nuclei preparation.

2.2 Nuclei Isolation

We collected tissue from the caudate nucleus (Cd), putamen (Pt), and ventral striatum (VS) (Fig. 3). We isolated nuclei as previously described [32]. We put the dissected tissue (25–200 mg) in 2 mL EZ PREP buffer (Millipore Sigma, Cat# NUC-101) and crudely dissociated tissues with a loose glass Dounce homogenizer by pushing the plunger up and down 30 times. We then refined tissue dissociation with a tight glass homogenizer an additional 30 times. Next, we resuspended the nuclei with 4 mL EZ PREP buffer, centrifuged down at 500 g for 5 min, and repeated this washing step with 4 mL nuclei suspension buffer (NSB), which consists of 0.01% BSA and 0.1% RNase inhibitor in PBS. Next, we resuspended the washed nuclei in 4 mL NSB and filtered them through a 35 μm cell strainer (Corning, Cat# 352235). We labeled the nuclei by adding an equal volume of trypan blue (Thermo Fisher, Cat# 15250061), we counted the nuclei with a hemocytometer, and we diluted the nuclei with NSB to achieve a final concentration of 1000 cells/μL.

2.3 Single Nucleus RNA Sequencing Library Preparation

We used 10× Chromium Single Cell 3′ Reagent kits v3/v3.1 Chemistry (10× Genomics) for single nucleus RNA sequencing. We loaded 16,000 nuclei from each brain region onto a 10× chip with the targeted nuclei recovery of 10,000 and generated Gel

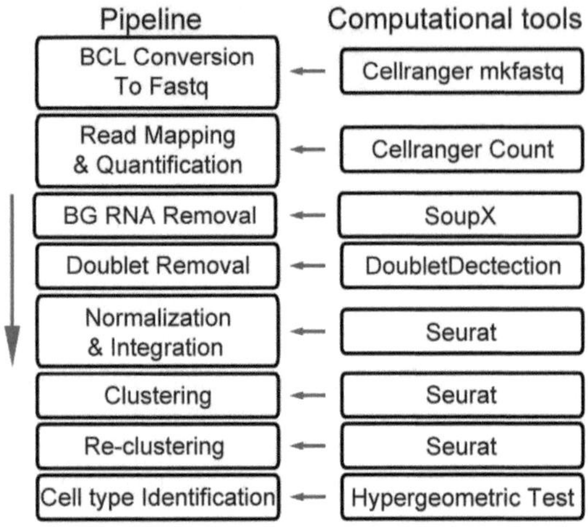

Fig. 4 Computational workflow of single cell/nuclei RNA sequencing analysis. BG: background

beads-in-emulsion (GEMs) after running through a 10× Genomics Chromium controller. We reverse-transcribed RNAs into cDNA and generated libraries according to the 10× Genomics protocol. We analyzed the quality of cDNA and final libraries using Agilent Bioanalyzer 2100 and quantified the libraries using qPCR with a KAPA Library Quantification Kit (KAPA Biosystems, Cat# KK4824). We pooled together libraries and sequenced samples to approximately 200,000 reads per nucleus with NovaSeq S4 Flow Cell Chip.

2.4 Data Processing

We used the Cell Ranger pipeline to transform raw sequence data into a cell by gene matrix. Tutorials of the Cell Ranger pipeline can be read on the 10× Genomics website: https://support.10xgenomics.com/single-cell-gene-expression/software/pipelines/latest/what-is-cell-ranger, and the Seurat pipeline for data analysis [33] (Fig. 4). Briefly, we converted the original BCL files into Fastq files using the command cell ranger mkfastq. We used the rhesus macaque reference genomes, rheMac10 [34], and the cell ranger count command to align the reads and count the transcripts. For preprocessing, we removed ambient RNA using SoupX [35], deleted ribosomal genes, and removed doublets using DoubletDetection [36]. We discarded low-quality cells by thresholding the number of genes; because neurons have more RNA than glia, we used two different thresholds to discard low-quality neurons (3000 genes) and glia (1000 genes). We normalized the data matrix and identified variable genes for each monkey's data set using Seurat's *FindVariableFeatures* function. We identified anchors using *FindIntegrationAnchors* function and used those anchors with the *IntegrateData* function to create a Seurat object with an integrated expression matrix. We scaled the integrated data, identified

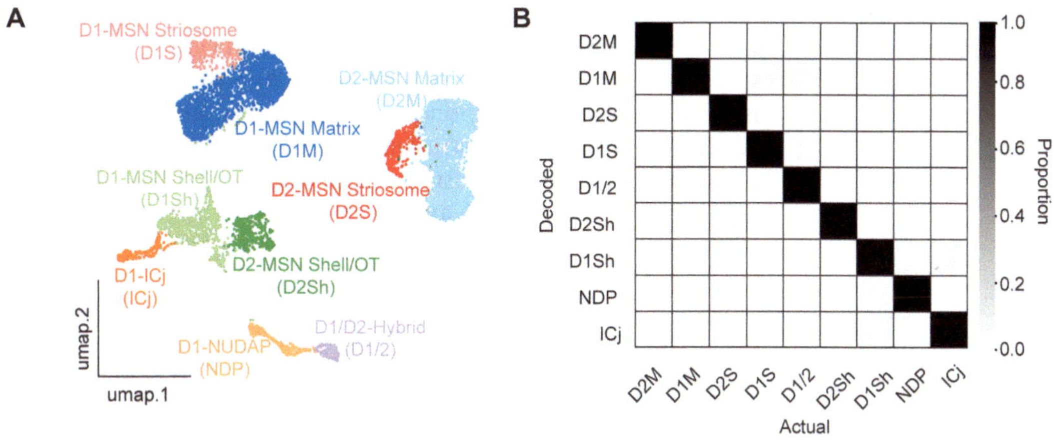

Fig. 5 Medium spiny neuron clusters in UMAP space (**a**) and the accuracy rate (**b**) between SCCAF decoded cell type and actual cell type. (Adapted from He et al. [8])

principal components (PCs), and visualized the plots in UMAP space. With Louvain clustering, we separated the nuclei into different clusters that correspond to the major cell classes, including excitatory neurons, interneurons, oligodendrocytes, microglia, oligodendrocyte precursors, astrocytes, microglia, and mural/fibroblast cells. We annotated the clusters based on well-known marker genes [37–39] and confirmed the cell cluster identities with the hypergeometric test [40–42]. All of the code we used to do this can be found on the Reward Lab GitHub (https://github.com/rewardlab/NHPstriatum).

2.5 Identification of MSN Clusters

To define the subtypes within MSNs, we first isolated the MSNs based on conventional markers, discarded other cell types, recalculated PCs, and performed UMAP dimensionality reduction on the PCs that contributed more than 80% variance. Then, we used Louvain clustering and chose a resolution that separated distinct clusters in UMAP space. Single nucleus RNA sequencing revealed nine clusters with distinct molecular signatures (Fig. 5a). We quantified the robustness of our MSN subtype clusters using the Single Cell Clustering Assessment Framework (SCCAF) [43]. SCCAF uses the area under the receiver operating characteristic (AUROC) curve to determine whether identified cell clusters correspond to different cell subtypes. SCCAF split the data into a training and testing set, built classifiers based on the gene expression in the training sets, and evaluated the performance of the classifiers using the testing sets. The recommended AUROC score is 0.9 for distinguishing distinct different cell types [43], and each MSN cluster has AUROC score of more than 0.93 (Fig. 5b). Thus, SCAAF was able to distinguish nine MSN subtypes.

2.6 MSN Subtype Annotations

We annotated the MSN subtypes based on their molecular signatures and fluorescent in situ hybridization (FISH) for differentially expressed genes. This approach demonstrated that the molecular diversity of MSNs corresponds to known anatomical or neurochemical divisions of the striatum (Fig. 5). There are five major MSN subtypes in the dorsal striatum. The larger D1- and D2-MSN clusters correspond to D1- and D2 neurons in the striatal matrix, and the smaller D1- and D2-MSN clusters are located in striosomes (patches). One small D1/D2-hybrid cluster likely corresponds to an MSN type recently described in rodents (D1H or eccentric-SPN) [38, 44]. D1/D2-hybrid cells are distributed throughout the dorsal striatum. Four MSN subtypes were identified in the ventral striatum: two D1- and D2-MSN clusters in the NAc shell and olfactory tubercle (OT) and two MSN-like neurons located in the interface islands. One type of interface island MSN-like neuron corresponds to densely packed granule cells in the islands of Calleja (ICj). Another interface island MSN-like neuron expresses high levels of the gene for the μ-opioid receptor (*OPRM1*) and is also enriched with the protein MOR encoded by the gene. These interface islands are enriched in, but not restricted to, the border regions of the VS; several interface islands in putamen can also be spotted. Based on the gene expression and distribution, these interface islands correspond to previously described Neurochemically Unique Domains in the Accumbens and Putamen (NUDAPs) [45, 46]. NUDAPs might be part of the network of "hedonic hotspots [47]," the NAc and ventral pallidum regions that are associated with the experience of pleasure in the presence of opioids [47].

2.7 Determining the Relationships Between MSN Subtypes

We first used cosine similarity based on their PCA space to assess the similarity between clusters. We found that cells within clusters were more like each other when compared with cells in other clusters. Dorsal striatum clusters were more like each other when compared to cells in ventral striatum clusters. Subtypes within anatomical structures such as D1- and D2- MSN striosome, D1- and D2-MSN matrix, and D1- and D2- MSN shell had high similarities. We then used archetype analysis further to assess the relationships between the identified MSN subtypes. Archetype analysis is a matrix decomposition technique that finds extremes—or archetypes—in continuously distributed data [48]. By using the cell types as archetypes, we assessed whether there existed discrete changes in gene expression between pairs of cell types. We found that each identified MSN subtype is characterized by discrete changes in gene expression patterns (Fig. 6a), but we also confirmed that continuous variation of gene expression between subtypes also existed. For example, archetypal analysis of single MSN subtypes revealed that two genes, *CNR1* and *CRYM*, defined a

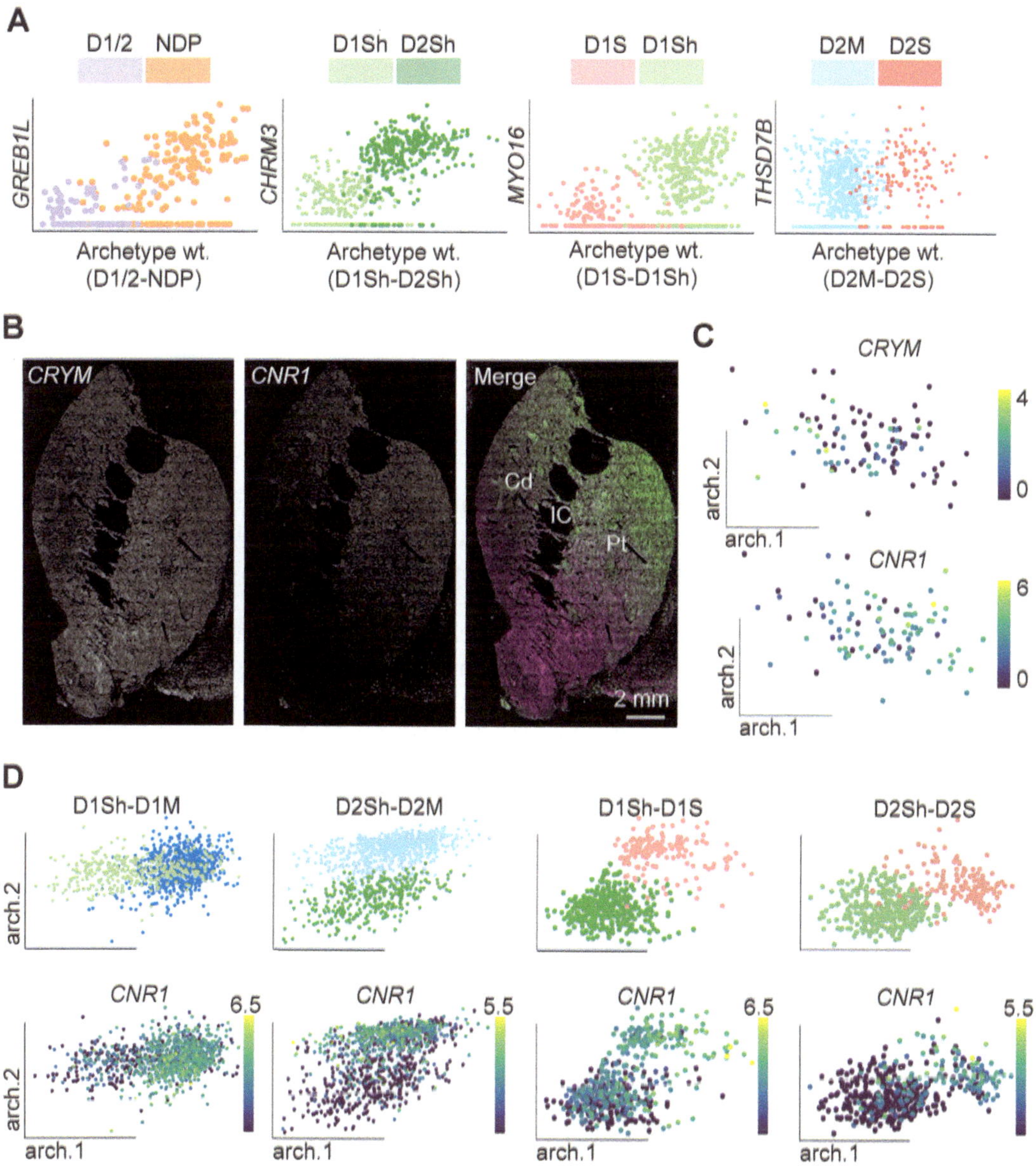

Fig. 6 Archetypal analysis of MSN subtypes. (**a**) Representative discrete genes between subtype pairs. (**b**) A continuous gradient of *CRYM* (magenta) and *CNR1* (green) on the dorsal-ventral axis. (**c**) *CRYM* and *CNR1* expressions along the D1/D2-hybrid archetype axes. (**d**) Dorsal and ventral striatum subtype pairs (top) and *CNR1* (bottom) distributions in these archetype axes. (Adapted from He et al. [8])

dorsal-ventral gradient (Fig. 6b) [44], and that *TAC3*-positive MSN archetype existed within the NAc shell (Fig. 1). The presence of these archetypes was very common and is a substantial source of variation within 'cell types.' The next phase of discovery, our next major challenge, is to define the biological functions of these archetypes.

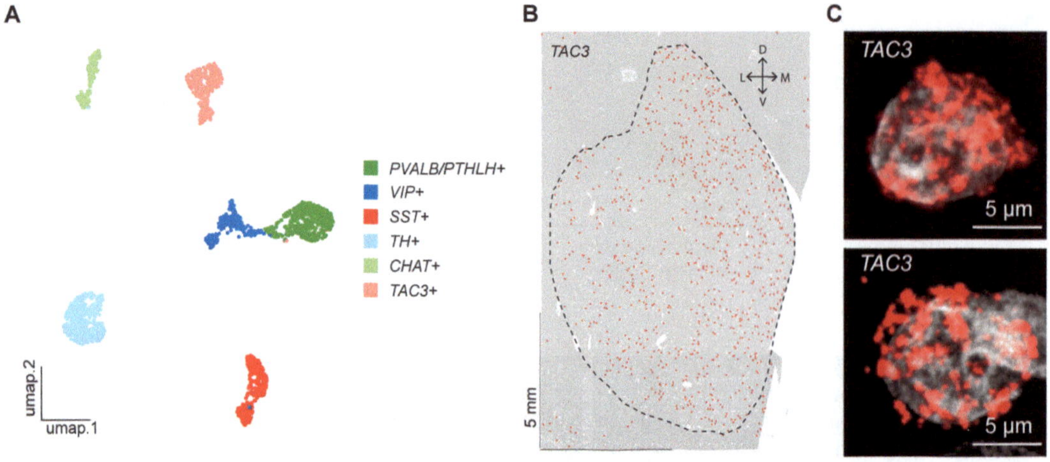

Fig. 7 Striatal interneurons in NHPs. (**a**) The clustering of interneurons revealed six populations of striatal interneurons. (**b**) FISH labeling of *TAC3* revealed this primate-specific interneuron distributed throughout the striatum. (**c**) Example *TAC3*-positive cells. (Adapted from He et al. [8])

2.8 Interneurons in Primate Striatum

To define interneurons in rhesus macaque striatum, we isolated the interneurons based on conventional markers such as LHX6, recalculated PCs for these isolated cells, performed dimensionality reduction, and visualized nuclei on UMAP space. We found that interneurons in the macaque constitute around 15% of neurons in the striatum. This is a higher proportion of interneurons than was recovered from single cell analysis of rodent striatum [38] but consistent with counts made in humans [49]. We found six distinct interneuron clusters that corresponded to known populations of striatal interneurons, including the well-known cell groups PTHLH/PVALB-, VIP-, SST-, TH-, CHAT-interneurons, and confirmed the presence of novel TAC3-positive interneurons recently discovered in the marmoset brain (Fig. 7) [7].

3 Conclusions and Discussions

The close phylogenetic relationship between humans and nonhuman primates (NHPs) makes the NHP a crucial neuroscientific model. However, the lack of cell type-specific tools prevents a deep investigation of circuit functions in NHPs. In this chapter, we demonstrated the use of snRNA-Seq to characterize cell types in the NHP striatum. snRNA-Seq revealed at least nine distinct MSN and MSN-like subtypes (Fig. 1). The analysis techniques we employed, including advanced machine learning models, identified continuous and discrete distributions of MSN subtypes. We identified five distinct MSN subtypes in the dorsal striatum and four in the ventral striatum. These MSN subtypes and archetypes provide a blueprint for studying cell type-specific functions during sophisticated primate behaviors, and the cell type-specific marker genes define potential molecular access points to enable the application of genetically coded tools in scientific or translational contexts.

Applying cell type classification to NHPs and understanding their specialized functions are crucial steps to understanding human brain function and dysfunction. However, successful targeting of neuronal cell types in the NHP brain with enhancer- or promoter-driven AAVs has been limited to a few cell types that are highly distinct from nearby cells. These include dopamine neurons in the midbrain [50, 51], Purkinje cells in the cerebellar cortex [52], Koniocellular cells in the lateral geniculate nucleus (LGN) [53], and several interneuron types ([10–12, 20, 54], and *see* Chap. 9). Applying snRNA-Seq to the monkey brain will provide a molecular definition of cell types that can be targeted based on gene expression. These molecular definitions can be used in turn—as described in the next chapter—to define cell type-specific enhancers. Thus, single-cell technologies, including transcriptomics and epigenomics, can enable cell type-specific access to behaving NHPs and pave the way to understanding sophisticated circuit functions in the human brain.

References

1. Chappell L, Russell AJC, Voet T (2018) Single-cell (multi)omics technologies. Annu Rev Genomics Hum Genet 19:15–41. https://doi.org/10.1146/annurev-genom-091416-035324
2. Nimchinsky EA, Gilissen E, Allman JM, Perl DP, Erwin JM, Hof PR (1999) A neuronal morphologic type unique to humans and great apes. Proc Natl Acad Sci U S A 96(9): 5268–5273. https://doi.org/10.1073/pnas.96.9.5268
3. Evrard HC, Forro T, Logothetis NK (2012) Von Economo neurons in the anterior insula of the macaque monkey. Neuron 74(3):482–489. https://doi.org/10.1016/j.neuron.2012.03.003
4. Braak H, Braak E (1976) The pyramidal cells of Betz within the cingulate and precentral gigantopyramidal field in the human brain. A Golgi and pigmentarchitectonic study. Cell Tissue Res 172(1):103–119. https://doi.org/10.1007/BF00226052
5. Bakken TE, Jorstad NL, Hu Q, Lake BB, Tian W, Kalmbach BE, Crow M, Hodge RD, Krienen FM, Sorensen SA, Eggermont J, Yao Z, Aevermann BD, Aldridge AI, Bartlett A, Bertagnolli D, Casper T, Castanon RG, Crichton K, Daigle TL, Dalley R, Dee N, Dembrow N, Diep D, Ding SL, Dong W, Fang R, Fischer S, Goldman M, Goldy J, Graybuck LT, Herb BR, Hou X, Kancherla J, Kroll M, Lathia K, van Lew B, Li YE, Liu CS, Liu H, Lucero JD, Mahurkar A, McMillen D, Miller JA, Moussa M, Nery JR, Nicovich PR, Niu SY, Orvis J, Osteen JK, Owen S, Palmer CR, Pham T, Plongthongkum N, Poirion O, Reed NM, Rimorin C, Rivkin A, Romanow WJ, Sedeno-Cortes AE, Siletti K, Somasundaram S, Sulc J, Tieu M, Torkelson A, Tung H, Wang X, Xie F, Yanny AM, Zhang R, Ament SA, Behrens MM, Bravo HC, Chun J, Dobin A, Gillis J, Hertzano R, Hof PR, Hollt T, Horwitz GD, Keene CD, Kharchenko PV, Ko AL, Lelieveldt BP, Luo C, Mukamel EA, Pinto-Duarte A, Preissl S, Regev A, Ren B, Scheuermann RH, Smith K, Spain WJ, White OR, Koch C, Hawrylycz M, Tasic B, Macosko EZ, McCarroll SA, Ting JT, Zeng H, Zhang K, Feng G, Ecker JR, Linnarsson S, Lein ES (2021) Comparative cellular analysis of motor cortex in human, marmoset and mouse. Nature 598(7879):111–119. https://doi.org/10.1038/s41586-021-03465-8
6. Hodge RD, Miller JA, Novotny M, Kalmbach BE, Ting JT, Bakken TE, Aevermann BD, Barkan ER, Berkowitz-Cerasano ML, Cobbs C,

Diez-Fuertes F, Ding SL, McCorrison J, Schork NJ, Shehata SI, Smith KA, Sunkin SM, Tran DN, Venepally P, Yanny AM, Steemers FJ, Phillips JW, Bernard A, Koch C, Lasken RS, Scheuermann RH, Lein ES (2020) Transcriptomic evidence that von Economo neurons are regionally specialized extratelencephalic-projecting excitatory neurons. Nat Commun 11(1):1172. https://doi.org/10.1038/s41467-020-14952-3

7. Krienen FM, Goldman M, Zhang Q, Del Rosario RCH, Florio M, Machold R, Saunders A, Levandowski K, Zaniewski H, Schuman B, Wu C, Lutservitz A, Mullally CD, Reed N, Bien E, Bortolin L, Fernandez-Otero M, Lin JD, Wysoker A, Nemesh J, Kulp D, Burns M, Tkachev V, Smith R, Walsh CA, Dimidschstein J, Rudy B, Kean LS, Berretta S, Fishell G, Feng G, McCarroll SA (2020) Innovations present in the primate interneuron repertoire. Nature 586(7828): 262–269. https://doi.org/10.1038/s41586-020-2781-z
8. He J, Kleyman M, Chen J, Alikaya A, Rothenhoefer KM, Ozturk BE, Wirthlin M, Bostan AC, Fish K, Byrne LC, Pfenning AR, Stauffer WR (2021) Transcriptional and anatomical diversity of medium spiny neurons in the primate striatum. Curr Biol 31(24):5473–5486 e5476. https://doi.org/10.1016/j.cub.2021.10.015
9. Boldog E, Bakken TE, Hodge RD, Novotny M, Aevermann BD, Baka J, Borde S, Close JL, Diez-Fuertes F, Ding SL, Farago N, Kocsis AK, Kovacs B, Maltzer Z, McCorrison JM, Miller JA, Molnar G, Olah G, Ozsvar A, Rozsa M, Shehata SI, Smith KA, Sunkin SM, Tran DN, Venepally P, Wall A, Puskas LG, Barzo P, Steemers FJ, Schork NJ, Scheuermann RH, Lasken RS, Lein ES, Tamas G (2018) Transcriptomic and morphophysiological evidence for a specialized human cortical GABAergic cell type. Nat Neurosci 21(9): 1185–1195. https://doi.org/10.1038/s41593-018-0205-2
10. Mich JK, Graybuck LT, Hess EE, Mahoney JT, Kojima Y, Ding Y, Somasundaram S, Miller JA, Kalmbach BE, Radaelli C, Gore BB, Weed N, Omstead V, Bishaw Y, Shapovalova NV, Martinez RA, Fong O, Yao S, Mortrud M, Chong P, Loftus L, Bertagnolli D, Goldy J, Casper T, Dee N, Opitz-Araya X, Cetin A, Smith KA, Gwinn RP, Cobbs C, Ko AL, Ojemann JG, Keene CD, Silbergeld DL, Sunkin SM, Gradinaru V, Horwitz GD, Zeng H, Tasic B, Lein ES, Ting JT, Levi BP (2021) Functional enhancer elements drive subclass-selective expression from mouse to primate neocortex. Cell Rep 34(13):108754. https://doi.org/10.1016/j.celrep.2021.108754
11. Dimidschstein J, Chen Q, Tremblay R, Rogers SL, Saldi GA, Guo L, Xu Q, Liu R, Lu C, Chu J, Grimley JS, Krostag AR, Kaykas A, Avery MC, Rashid MS, Baek M, Jacob AL, Smith GB, Wilson DE, Kosche G, Kruglikov I, Rusielewicz T, Kotak VC, Mowery TM, Anderson SA, Callaway EM, Dasen JS, Fitzpatrick D, Fossati V, Long MA, Noggle S, Reynolds JH, Sanes DH, Rudy B, Feng G, Fishell G (2016) A viral strategy for targeting and manipulating interneurons across vertebrate species. Nat Neurosci 19(12): 1743–1749. https://doi.org/10.1038/nn.4430
12. Vormstein-Schneider D, Lin JD, Pelkey KA, Chittajallu R, Guo B, Arias-Garcia MA, Allaway K, Sakopoulos S, Schneider G, Stevenson O, Vergara J, Sharma J, Zhang Q, Franken TP, Smith J, Ibrahim LA, Mastro KJ, Sabri E, Huang S, Favuzzi E, Burbridge T, Xu Q, Guo L, Vogel I, Sanchez V, Saldi GA, Gorissen BL, Yuan X, Zaghloul KA, Devinsky O, Sabatini BL, Batista-Brito R, Reynolds J, Feng G, Fu Z, McBain CJ, Fishell G, Dimidschstein J (2020) Viral manipulation of functionally distinct interneurons in mice, non-human primates and humans. Nat Neurosci 23(12):1629–1636. https://doi.org/10.1038/s41593-020-0692-9
13. Ding J, Adiconis X, Simmons SK, Kowalczyk MS, Hession CC, Marjanovic ND, Hughes TK, Wadsworth MH, Burks T, Nguyen LT, Kwon JYH, Barak B, Ge W, Kedaigle AJ, Carroll S, Li S, Hacohen N, Rozenblatt-Rosen O, Shalek AK, Villani AC, Regev A, Levin JZ (2020) Systematic comparison of single-cell and single-nucleus RNA-sequencing methods. Nat Biotechnol 38(6):737–746. https://doi.org/10.1038/s41587-020-0465-8
14. Bakken TE, Hodge RD, Miller JA, Yao Z, Nguyen TN, Aevermann B, Barkan E, Bertagnolli D, Casper T, Dee N, Garren E, Goldy J, Graybuck LT, Kroll M, Lasken RS, Lathia K, Parry S, Rimorin C, Scheuermann RH, Schork NJ, Shehata SI, Tieu M, Phillips JW, Bernard A, Smith KA, Zeng H, Lein ES, Tasic B (2018) Single-nucleus and single-cell transcriptomes compared in matched cortical cell types. PLoS One 13(12):e0209648. https://doi.org/10.1371/journal.pone.0209648
15. Buenrostro JD, Giresi PG, Zaba LC, Chang HY, Greenleaf WJ (2013) Transposition of native chromatin for fast and sensitive epigenomic profiling of open chromatin, DNA-binding proteins and nucleosome

position. Nat Methods 10(12):1213–1218. https://doi.org/10.1038/nmeth.2688

16. Buenrostro JD, Wu B, Chang HY, Greenleaf WJ (2015) ATAC-seq: a method for assaying chromatin accessibility genome-wide. Curr Protoc Mol Biol 109:21.29.1–21.29.9. https://doi.org/10.1002/0471142727.mb2129s109
17. Buenrostro JD, Wu B, Litzenburger UM, Ruff D, Gonzales ML, Snyder MP, Chang HY, Greenleaf WJ (2015) Single-cell chromatin accessibility reveals principles of regulatory variation. Nature 523(7561):486–490. https://doi.org/10.1038/nature14590
18. Blankvoort S, Descamps LAL, Kentros C (2020) Enhancer-Driven Gene Expression (EDGE) enables the generation of cell type specific tools for the analysis of neural circuits. Neurosci Res 152:78–86. https://doi.org/10.1016/j.neures.2020.01.009
19. Blankvoort S, Witter MP, Noonan J, Cotney J, Kentros C (2018) Marked diversity of unique cortical enhancers enables neuron-specific tools by enhancer-driven gene expression. Curr Biol 28(13):2103–2114 e2105. https://doi.org/10.1016/j.cub.2018.05.015
20. Lawler AJ, Ramamurthy E, Brown AR, Shin N, Kim Y, Toong N, Kaplow IM, Wirthlin M, Zhang X, Phan BN, Fox GA, Wade K, He J, Ozturk BE, Byrne LC, Stauffer WR, Fish KN, Pfenning AR (2022) Machine learning sequence prioritization for cell type-specific enhancer design. elife 11. https://doi.org/10.7554/eLife.69571
21. Nair RR, Blankvoort S, Lagartos MJ, Kentros C (2020) Enhancer-driven gene expression (EDGE) enables the generation of viral vectors specific to neuronal subtypes. iScience 23(3):100888. https://doi.org/10.1016/j.isci.2020.100888
22. Picelli S, Björklund AK, Faridani OR, Sagasser S, Winberg G, Sandberg R (2013) Smart-seq2 for sensitive full-length transcriptome profiling in single cells. Nat Methods 10(11):1096–1098. https://doi.org/10.1038/nmeth.2639
23. Hashimshony T, Wagner F, Sher N, Yanai I (2012) CEL-Seq: single-cell RNA-Seq by multiplexed linear amplification. Cell Rep 2(3):666–673. https://doi.org/10.1016/j.celrep.2012.08.003
24. Hashimshony T, Senderovich N, Avital G, Klochendler A, de Leeuw Y, Anavy L, Gennert D, Li S, Livak KJ, Rozenblatt-Rosen O, Dor Y, Regev A, Yanai I (2016) CEL-Seq2: sensitive highly-multiplexed single-cell RNA-Seq. Genome Biol 17:77. https://doi.org/10.1186/s13059-016-0938-8
25. Jaitin DA, Kenigsberg E, Keren-Shaul H, Elefant N, Paul F, Zaretsky I, Mildner A, Cohen N, Jung S, Tanay A, Amit I (2014) Massively parallel single-cell RNA-seq for marker-free decomposition of tissues into cell types. Science 343(6172):776–779. https://doi.org/10.1126/science.1247651
26. Soumillon M, Cacchiarelli D, Semrau S, van Oudenaarden A, Mikkelsen TS (2014) Characterization of directed differentiation by high-throughput single-cell RNA-Seq. bioRxiv:003236. https://doi.org/10.1101/003236
27. Bagnoli JW, Ziegenhain C, Janjic A, Wange LE, Vieth B, Parekh S, Geuder J, Hellmann I, Enard W (2017) mcSCRB-seq: sensitive and powerful single-cell RNA sequencing. bioRxiv:188367. https://doi.org/10.1101/188367
28. Macosko EZ, Basu A, Satija R, Nemesh J, Shekhar K, Goldman M, Tirosh I, Bialas AR, Kamitaki N, Martersteck EM, Trombetta JJ, Weitz DA, Sanes JR, Shalek AK, Regev A, McCarroll SA (2015) Highly parallel genome-wide expression profiling of individual cells using nanoliter droplets. Cell 161(5):1202–1214. https://doi.org/10.1016/j.cell.2015.05.002
29. Klein AM, Mazutis L, Akartuna I, Tallapragada N, Veres A, Li V, Peshkin L, Weitz DA, Kirschner MW (2015) Droplet barcoding for single-cell transcriptomics applied to embryonic stem cells. Cell 161(5):1187–1201. https://doi.org/10.1016/j.cell.2015.04.044
30. Zilionis R, Nainys J, Veres A, Savova V, Zemmour D, Klein AM, Mazutis L (2017) Single-cell barcoding and sequencing using droplet microfluidics. Nat Protoc 12(1):44–73. https://doi.org/10.1038/nprot.2016.154
31. Gierahn TM, Wadsworth MH 2nd, Hughes TK, Bryson BD, Butler A, Satija R, Fortune S, Love JC, Shalek AK (2017) Seq-well: portable, low-cost RNA sequencing of single cells at high throughput. Nat Methods 14(4):395–398. https://doi.org/10.1038/nmeth.4179
32. Habib N, Avraham-Davidi I, Basu A, Burks T, Shekhar K, Hofree M, Choudhury SR, Aguet F, Gelfand E, Ardlie K, Weitz DA, Rozenblatt-Rosen O, Zhang F, Regev A (2017) Massively parallel single-nucleus RNA--seq with DroNc-seq. Nat Methods 14(10):955–958. https://doi.org/10.1038/nmeth.4407
33. Satija R, Farrell JA, Gennert D, Schier AF, Regev A (2015) Spatial reconstruction of single-cell gene expression data. Nat

Biotechnol 33(5):495–502. https://doi.org/10.1038/nbt.3192
34. Warren WC, Harris RA, Haukness M, Fiddes IT, Murali SC, Fernandes J, Dishuck PC, Storer JM, Raveendran M, Hillier LW, Porubsky D, Mao Y, Gordon D, Vollger MR, Lewis AP, Munson KM, DeVogelaere E, Armstrong J, Diekhans M, Walker JA, Tomlinson C, Graves-Lindsay TA, Kremitzki M, Salama SR, Audano PA, Escalona M, Maurer NW, Antonacci F, Mercuri L, Maggiolini FAM, Catacchio CR, Underwood JG, O'Connor DH, Sanders AD, Korbel JO, Ferguson B, Kubisch HM, Picker L, Kalin NH, Rosene D, Levine J, Abbott DH, Gray SB, Sanchez MM, Kovacs-Balint ZA, Kemnitz JW, Thomasy SM, Roberts JA, Kinnally EL, Capitanio JP, Skene JHP, Platt M, Cole SA, Green RE, Ventura M, Wiseman RW, Paten B, Batzer MA, Rogers J, Eichler EE (2020) Sequence diversity analyses of an improved rhesus macaque genome enhance its biomedical utility. Science 370(6523). https://doi.org/10.1126/science.abc6617
35. Young MD, Behjati S (2020) SoupX removes ambient RNA contamination from droplet-based single-cell RNA sequencing data. Gigascience 9(12). https://doi.org/10.1093/gigascience/giaa151
36. McGinnis CS, Murrow LM, Gartner ZJ (2019) DoubletFinder: doublet detection in single-cell RNA sequencing data using artificial nearest neighbors. Cell Syst 8(4):329–337 e324. https://doi.org/10.1016/j.cels.2019.03.003
37. Gokce O, Stanley GM, Treutlein B, Neff NF, Camp JG, Malenka RC, Rothwell PE, Fuccillo MV, Sudhof TC, Quake SR (2016) Cellular taxonomy of the mouse striatum as revealed by single-cell RNA-Seq. Cell Rep 16(4): 1126–1137. https://doi.org/10.1016/j.celrep.2016.06.059
38. Saunders A, Macosko EZ, Wysoker A, Goldman M, Krienen FM, de Rivera H, Bien E, Baum M, Bortolin L, Wang S, Goeva A, Nemesh J, Kamitaki N, Brumbaugh S, Kulp D, McCarroll SA (2018) Molecular diversity and specializations among the cells of the adult mouse brain. Cell 174(4): 1015–1030 e1016. https://doi.org/10.1016/j.cell.2018.07.028
39. Zhang Y, Chen K, Sloan SA, Bennett ML, Scholze AR, O'Keeffe S, Phatnani HP, Guarnieri P, Caneda C, Ruderisch N, Deng S, Liddelow SA, Zhang C, Daneman R, Maniatis T, Barres BA, Wu JQ (2014) An RNA-sequencing transcriptome and splicing database of glia, neurons, and vascular cells of the cerebral cortex. J Neurosci 34(36): 11929–11947. https://doi.org/10.1523/JNEUROSCI.1860-14.2014
40. Grillner S, Robertson B (2016) The basal ganglia over 500 million years. Curr Biol 26(20):R1088–R1100. https://doi.org/10.1016/j.cub.2016.06.041
41. Khrameeva E, Kurochkin I, Han D, Guijarro P, Kanton S, Santel M, Qian Z, Rong S, Mazin P, Sabirov M, Bulat M, Efimova O, Tkachev A, Guo S, Sherwood CC, Camp JG, Paabo S, Treutlein B, Khaitovich P (2020) Single-cell-resolution transcriptome map of human, chimpanzee, bonobo, and macaque brains. Genome Res 30(5):776–789. https://doi.org/10.1101/gr.256958.119
42. Yin S, Lu K, Tan T, Tang J, Wei J, Liu X, Hu X, Wan H, Huang W, Fan Y, Xie D, Yu Y (2020) Transcriptomic and open chromatin atlas of high-resolution anatomical regions in the rhesus macaque brain. Nat Commun 11(1):474. https://doi.org/10.1038/s41467-020-14368-z
43. Miao Z, Moreno P, Huang N, Papatheodorou I, Brazma A, Teichmann SA (2020) Putative cell type discovery from single-cell gene expression data. Nat Methods 17(6):621–628. https://doi.org/10.1038/s41592-020-0825-9
44. Stanley G, Gokce O, Malenka RC, Sudhof TC, Quake SR (2020) Continuous and discrete neuron types of the adult murine striatum. Neuron 105(4):688–699 e688. https://doi.org/10.1016/j.neuron.2019.11.004
45. Voorn P, Brady LS, Berendse HW, Richfield EK (1996) Densitometrical analysis of opioid receptor ligand binding in the human striatum--I. Distribution of mu opioid receptor defines shell and core of the ventral striatum. Neuroscience 75(3):777–792. https://doi.org/10.1016/0306-4522(96)00271-0
46. Daunais JB, Letchworth SR, Sim-Selley LJ, Smith HR, Childers SR, Porrino LJ (2001) Functional and anatomical localization of mu opioid receptors in the striatum, amygdala, and extended amygdala of the nonhuman primate. J Comp Neurol 433(4):471–485. https://doi.org/10.1002/cne.1154
47. Berridge KC, Kringelbach ML (2015) Pleasure systems in the brain. Neuron 86(3):646–664. https://doi.org/10.1016/j.neuron.2015.02.018
48. Mohammadi S, Davila-Velderrain J, Kellis M (2020) A multiresolution framework to characterize single-cell state landscapes. Nat Commun 11(1):5399. https://doi.org/10.1038/s41467-020-18416-6

49. Lecumberri A, Lopez-Janeiro A, Corral-Domenge C, Bernacer J (2018) Neuronal density and proportion of interneurons in the associative, sensorimotor and limbic human striatum. Brain Struct Funct 223(4): 1615–1625. https://doi.org/10.1007/s00429-017-1579-8
50. Stauffer WR, Lak A, Yang A, Borel M, Paulsen O, Boyden ES, Schultz W (2016) Dopamine neuron-specific optogenetic stimulation in rhesus macaques. Cell 166(6): 1564–1571 e1566. https://doi.org/10.1016/j.cell.2016.08.024
51. Lerchner W, Corgiat B, Der Minassian V, Saunders RC, Richmond BJ (2014) Injection parameters and virus dependent choice of promoters to improve neuron targeting in the nonhuman primate brain. Gene Ther 21(3): 233–241. https://doi.org/10.1038/gt.2013.75
52. El-Shamayleh Y, Kojima Y, Soetedjo R, Horwitz GD (2017) Selective optogenetic control of Purkinje cells in monkey cerebellum. Neuron 95(1):51–62 e54. https://doi.org/10.1016/j.neuron.2017.06.002
53. Klein C, Evrard HC, Shapcott KA, Haverkamp S, Logothetis NK, Schmid MC (2016) Cell-targeted optogenetics and electrical microstimulation reveal the primate koniocellular projection to supra-granular visual cortex. Neuron 90(1):143–151. https://doi.org/10.1016/j.neuron.2016.02.036
54. Mehta P, Kreeger L, Wylie DC, Pattadkal JJ, Lusignan T, Davis MJ, Turi GF, Li WK, Whitmire MP, Chen Y, Kajs BL, Seidemann E, Priebe NJ, Losonczy A, Zemelman BV (2019) Functional access to neuron subclasses in rodent and primate forebrain. Cell Rep 26(10):2818–2832 e2818. https://doi.org/10.1016/j.celrep.2019.02.011

Chapter 9

Enhancers for Selective Targeting

Jessica Lin and Jordane Dimidschstein

Abstract

Recombinant adeno-associated viruses can be coupled with short regulatory elements to restrict viral expression to specific cellular populations. These viral vectors can be used as tools for basic research to dissect many aspects of the biology of specific cellular subtypes in health and disease, and across species. A handful of enhancers have already been described in the nervous system, and recent studies suggest that transcriptomic and epigenetic data can be leveraged to systematize the discovery of novel elements to restrict viral expression to any cell type. However, a thorough characterization of the expression profile conferred by these short sequences is required to demonstrate their utility in the experimental context in which they will be ultimately used. Here we describe a complete guide to select, screen, and validate the expression profile of enhancers to target specific subtypes of neurons.

Key words Adeno-associated viruses, Gene regulation, Gene therapy

1 Introduction

Gaining genetic access to specific neuron populations in nonhuman primates (NHPs) and humans would enable targeted circuit modulation to understand brain function and dysfunction in psychiatric disorders. In the context of basic research, the use of gene regulatory elements in transgenic mice for targeted effector and sensor expression has transformed brain circuitry studies in this model system, but technologies to conduct comparable studies in wild-type animals including NHPs are lacking. We and others have had success developing gene regulatory elements that enable cell type-restricted gene expression when used within recombinant adeno-associated virus (rAAV) vectors [1–5]. For example, we recently demonstrated the use of the Dlx5/6 [1] and the E2 enhancer [2] to drive transgene expression in forebrain GABAergic cells and parvalbumin-expressing interneurons across species, including humans. Nevertheless, identifying additional enhancer sequences that function in the context of the limited carrying capacity of AAV has been slow due to the limited success rate and low throughput

Mark A. G. Eldridge and Adriana Galvan (eds.), *Vectorology for Optogenetics and Chemogenetics*, Neuromethods, vol. 195, https://doi.org/10.1007/978-1-0716-2918-5_9,

nature of these efforts. On the therapeutic side, the importance of restricting viral expression to specific target populations is illustrated by several preclinical studies. For example, targeting hepatocytes in the liver was shown to elicit immune tolerance [6], de-targeting retinal pigmented epithelium by using photoreceptor-specific enhancers strongly reduced neuroinflammation upon injection of rAAVs in the eye [7] and selectively increasing the activity of parvalbumin-expressing interneurons had a stronger effect on seizure duration than targeting the entire GABAergic population in a mouse model of medial-temporal lobe epilepsy [8].

Locating regulatory elements that can restrict the expression of rAAVs to a desired neuronal population is a nontrivial endeavor. Sequence conservation coupled with single-cell transcriptomics and epigenomics datasets are commonly used to reduce the search space, but the complexity of gene regulation makes it impossible to accurately predict whether a given putative enhancer will retain its properties when integrated within rAAVs. As such, meticulous in vivo screening is necessary to develop and validate the tools required to precisely target and manipulate a given target cell type. Here we provide a detailed description of the screening method that was successfully implemented to nominate, test, validate, and fully characterize a series of regulatory elements capable of restricting rAAV transgene expression to parvalbumin-expressing cortical GABAergic interneurons across species (*see* Fig. 1) [2].

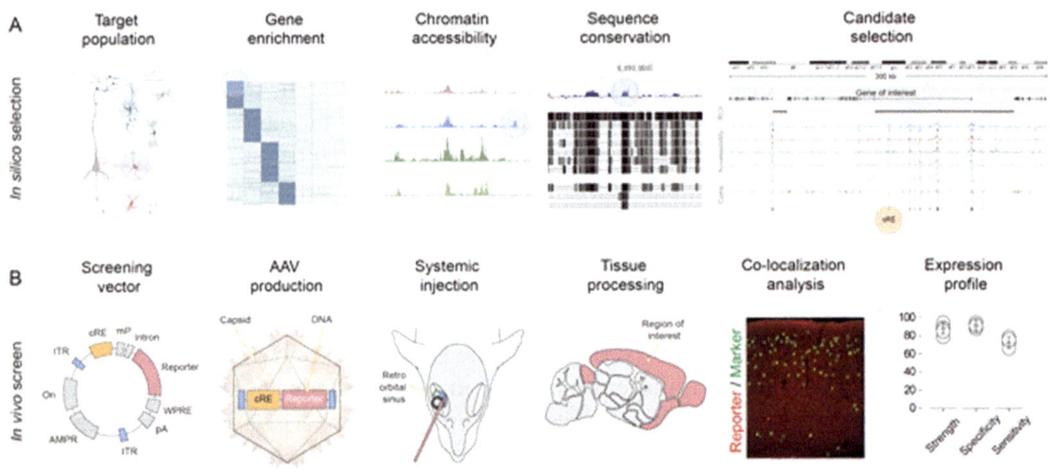

Fig. 1 Schematic representation of the in-silico selection of candidate regulatory sequences (**a**) and in vivo screening pipeline from vector design to an assessment of expression profile (**b**). ROI chromatin region of interest, cRE candidate regulatory element, ITR internal terminal repeat, mP minimal promoter, WPRE woodchuck post-regulatory element, pA poly-adenylation sequence, AMPR ampicillin resistance gene, Ori origin of replication

2 Materials

2.1 Animals

Female C57BL/6 J mice (Mus musculus; aged 10 weeks), were obtained from Jackson Labs (stock no. 000664). We used male hemizygous Dlx6a-cre mice (Mus musculus; aged 10 weeks; Jax stock no. 008199) and female homozygous INTACT mice (Mus musculus; aged 10 weeks, flox-Sun1-eGFP, Jax stock no. 021039) for the single-cell ATAC-seq. These mice were maintained at a macroenvironmental temperature and humidity ranges of 17.8–26.1 °C and 30–70%, respectively. These parameters were monitored closely and controlled within rodent colony rooms.

2.2 scATAC-Seq

The chromatin landscape of cortical interneurons was profiled using ATAC-sequencing at the single-cell resolution. Specifically, we prepared a scATAC-seq library from P28 Dlx6a-cre::Sun1-eGFP mice crossed with female homozygous INTACT mice using the 10× Genomics platform with the Chromium Single Cell ATAC Library & Gel Bead Kit v.1.0 (PN-1000111), Chromium Chip E Single Cell kit (PN-1000156) and Chromium i7 Multiplex Kit N, Set A (PN-1000084) as instructed by the manufacturer. Raw sequencing data were processed using the Cell Ranger ATAC (v.1.1.0) pipeline (10× Genomics) followed by clustering using the SnapATAC package v.1. Clusters were labeled by interneuron cardinal class and accessibility peaks were called for each class using MACS2 (v.2.2.7.1; https://github.com/taoliu/MACS). Additionally, we used bulk ATAC-seq data for cortical excitatory neurons generated by Mo et al. ([9]–accession no. GSE63137).

2.3 rAAV Production

AAVs were produced using standard methods [10]. Briefly, HEK293T cells were expanded in DMEM+ medium and subsequently transfected using polyethylenimine (PEI) combined with the three necessary plasmids (pAAV, pHelper, and pCapsid).

2.4 Systemic Delivery in Mice

Mice were sedated using an inhalant anesthetic (Isoflurane, Fisher scientific BP17541-100) and then transferred to a laminar flow hood and fit on a nose cone connected to a nonrebreathing apparatus. A 1 mL syringe (BD Medical, BD328438) is used to inject the rAAVs diluted in sterile PBS (Invitrogen, AM9625) retro-orbitally. A topical anesthetic (Proparacaine hydrochloride, Sigma-Aldrich 1571001-200MG) was applied to the eye.

2.5 Local Delivery in Mice

Mice were sedated using an inhalant anesthetic (Isoflurane, Fisher Scientific, BP17541-100) and placed in a stereotactic head frame on a temperature-controlled heating pad. The mice were injected using a sharp glass pipette (25–35 mm in diameter). The craniotomy site is covered with sterile bone wax (Fisher scientific, 50-118-0260) and the surgical site is closed with Vetbond (Fisher Scientific, NC0846393).

2.6 PFA Perfusion

Mice were anesthetized with Euthasol® administered in a 1 mL syringe (Thomas Sci., 8939D16). We diluted 32% Paraformaldehyde (PFA) Aqueous Solution (Fisher scientific, 50-980-495) to 4% with sterile PBS (Invitrogen, AM9625) and used this solution as the fixative. The mouse was placed on a dissecting pan with black wax (VWR, 470005-450). Using surgical scissors and forceps, the heart was exposed and a 27-gauge needle (BD Medical, 305109) was used to administer 4% PFA with a peristaltic pump.

2.7 Vibratome Sectioning

Sections were produced using a Leica VTS1000 vibrating microtome. For coronal sections, we embedded the tissue in 1% agarose (Lonza, 50004). We used super glue (W.B. Mason, EPIKG58548R) to secure the tissue to the vibratome platform. Sections were then placed into RNAse-free PBS (Invitrogen, AM9625) or freezing buffer made from sterile and RNAse-free reagents (PBS, Ethylene Glycol (Sigma-Aldrich, 324558-1 L), Glycerol (Sigma-Aldrich, G7893-500ML), Sodium Azide (WW Grainger, 24FC86), and water (Invitrogen, 10977015)).

2.8 Freezing Microtome Sectioning

Sections were also produced by performing freezing microtome sectioning (sliding microtome, Leica Biosystems, SM2010 R) on tissue fully saturated with 30% sucrose solution (VWR, SS1020). Then we used O.C.T. compound (Sakura, 4583) to secure the tissue to the microtome.

2.9 Immunohistochemistry

Sections were permeabilized using a blocking solution made up of 0.3% Triton X-100 (Sigma-Aldrich, ×100), 5% normal donkey (Jackson ImmunoResearch, 017-000-121) or 5% horse serum (Thermo Fisher Scientific, 26050088) and PBS (Invitrogen, AM9625). The primary antibodies were added to an antibody solution made up of 0.1% Triton X-100, 5% normal donkey or horse serum, and PBS. We used the following primary antibodies at the indicated concentrations for our PV-expressing GABAergic interneuron enhancer screen: chicken anti-GFP at 1:1000 (Abcam, catalog no. ab13970); rabbit anti-DsRed at 1:1000 (Clontech, catalog no. 632496); goat anti-PV at 1:1000 (Swant, catalog no. PVG-213); guinea-pig anti-PV at 1:2000 (Swant, catalog no. GP-72); rabbit anti-SST at 1:2000 (Peninsula, catalog no. T-4103.0050) and mouse anti-Synaptotagmin-2 at 1:250 (ZFIN, catalog no. ZDB-ATB-081002-25). We used the appropriate Alexa-Fluor conjugated secondary antibodies at 1:1000 dilution (Thermo Fisher Scientific) and DAPI (Invitrogen, D1306). The sections were mounted on glass slides using Fluoromount-G (Southern Biotech, 0100-01).

2.10 In Situ Hybridization

The following reagents from ACDBio were used for in situ hybridization: RNAscope Multiplex Fluorescent Reagent Kit v.2 (product no. 323100), RNAscope Probe Diluent (product no. 300041),

HybEZ oven (product no. 321710/321720), humidity control tray (product no. 310012), and HybEZ Humidifying Paper (product no. 310025). Additionally, we use the following ISH probes also designed by ACDBio: Gad1 (product no. 400951) Pvalb (product no. 421931), VIP (product no. 415961). TSA Plus Fluorescein, TSA Plus Cyanine 3, and TSA Plus Cyanine 5 were from PerkinElmer (nos. NEL741, NEL744, and NEL745). Of note, TSA Plus Fluorescein degrades quickly at 4 °C after reconstitution. It is best to aliquot and store it at −20 °C. Brain sections were mounted on Superfrost Plus glass slides (Thermo Fisher Scientific, 12-550-15).

2.11 Microscopy

To image the expression profile of our enhancers, we used a wide-field fluorescent microscope with tiling capabilities (Zeiss Axio Imager M2 equipped with Axiocam 512 mono) with the appropriate light source (X-Cite 120 Lamp) and filter set to distinguish between Dapi, 488 and 594 channels. Typically, 10× or 20× magnification images were acquired for further analysis. Confocal microscopy (Zeiss LSM 810) was performed using 40× magnification.

3 Methods

3.1 Selection of Candidate Regulatory Elements

Screening for regulatory elements capable of restricting viral expression to a given cellular population requires characterizing each candidate sequence in vivo. Considering the immense size of the mouse genome, the vast majority of which consists of noncoding regions, systematically tiling through all possible enhancers is currently impossible, but a number of assumptions can be combined to effectively restrict the search space to a testable number of potential regulatory elements. Among other approaches, restricting the search space to the genomic region in the immediate vicinity of genes selectively expressed in the desired population proved efficient [2, 35, 36].

Here we detail the various steps of the integrative method we developed to systematically identify regulatory elements to target specific cellular subtypes: (1) Identify a set of genes that are selectively expressed in the cellular subtypes of interest using transcriptomic analysis of single-cell RNA-seq of the relevant tissue. (2) Limit the range of the genomic region of interest to a maximum of 1 Mb away from the transcriptional start site (TSS). Because enhancers that are closer to the TSS have been shown to drive stronger expression [11], the initial search space can be reduced to a shorter window of 100 Mb up and downstream of the TSS. (3) Sequence conservation of noncoding genomic loci is suggestive of positive selection pressure and has been shown to correlate with the presence of regulatory elements [12–17]. Preferentially focusing on sequences with a high degree of cross-species conservation increases the likelihood of identifying functionally conserved enhancers that retain their properties in NHPs and humans. (4) The location of active enhancers in a given cell type correlates with chromatin accessibility (e.g., DNAseI hypersensitivity or ATACseq profiling), DNA hypomethylation, and epigenetic modifications (e.g., H3k27ac marks assessed by CHiP-seq) while promoter–enhancer interactions can be mapped using methods like Hi-C sequencing [18, 19]. Due to the complex nature of enhancer activity and the availability of these datasets for the cell type of interest, generating or integrating multiple of these datasets will increase the confidence of nominated enhancers. Additionally, the usefulness of these datasets improves with single-cell resolution by avoiding averaging effects. The number of candidate sequences identified by combining these datasets depends on the stringency of the selection criteria and the screening capability.

3.1.1 Selection of Genes Enriched in the Target Population

Identifying genes enriched in the target population can be performed by differential gene expression analysis (DGE) on relevant scRNAseq datasets. In our recent study, we performed a DGE analysis on two independent datasets [19, 20] and selected genes that were enriched in PV-interneurons in both. Importantly, defining enrichment of a gene as having a higher average expression in a cell type of interest compared to the average expression across all other cells from the dataset is prone to averaging effect (i.e., masking non-negligible expression is a subpopulation of the outgroup). To mitigate this, we assigned an identity to each cell by lumping clusters from the original studies into one of 7 “meta-clusters” that correspond to broad classes of functionally distinct cellular subtypes (i.e., the four main classes of interneurons: PV, SST, VIP, ID2–[21, 22] –two classes of pyramidal neurons: intratelencephalic, subcortical projecting neurons, and non-neuronal populations) and identified genes with an average expression in the PV cluster that was higher than the average expression in any other meta-clusters. Specifically, enrichment was defined as having an average expression of at least two times higher in PV cortical interneurons (cINs) compared with other interneurons meta-clusters and five times higher compared with pyramidal or non-neuronal cell types with an adjusted $P < 0.01$ based on pairwise comparison of each meta cluster using Student’s *t*-test and Bonferroni’s correction for multiple comparisons. We then focused on the enriched genes with the highest average expression in PV cINs and manually selected a final subset of seven genes that showed an expression pattern consistent with PV cIN enrichment using ISH data from the Allen Brain Atlas mouse brain (that is, sparse labeling in the S1 cortex with biased distribution toward the deep cortical layers).

3.1.2 Cross-Species Conservation

Cross-species conservation is a hallmark of enhancers [12–17]. Enhancers consist of relatively short genomic sequences (10–1000 bp long) interspersed with transcription-factor binding motifs. It is likely that positive selection pressure preferentially applies to these motifs. Hence, limiting the search to long and continuous stretches of highly conserved sequences will likely result in a high rate of false-negative. Here we used PhastCons analysis to get the conservation scores of sequences conserved across 60 mammalian species that can be used to guide candidate selection [23] of sequences with conservation. The "phastCons-Element60way" track was downloaded from the UCSC genome browser portal [24] with the table browser interface.

3.1.3 scATAC-Seq

Enhancers that are actively engaged in gene regulation are located in euchromatin and are bound to a complex array of transcription factors and other DNA-interacting proteins. As such, the various methods developed to assess the epigenetic landscape of specific cell types can be used to predict their location within the genome. These include the assessment of chromatin accessibility, histone modifications, transcription factor binding, and nucleosome positioning [25–31]. Among these, ATACseq [25] is well established and can be implemented at single-cell resolution (single-cell has the advantage of removing the averaging effects of bulk but suffers from a lack of depth resulting from unsequenced open chromatin regions due to damaged DNA material during the protocol). Data generated by this method can be processed with easy-to-implement analytical pipelines to resolve the chromatin accessibility landscape of any cell type of interest [32, 33]. The accessibility "peaks" can then be used as a set of candidate sequences potentially involved in the regulation of the gene of interest. To assess the chromatin landscape of GABAergic populations we profiled adult cortical tissue using scATAC-seq on interneurons isolated from a transgenic mouse line expressing a nuclear reporter exclusively in interneurons. Clusters of cells with similar accessibility profiles were resolved using the SnapATAC pipeline and annotated based on the accessibility of marker genes for cardinal classes. Individual chromatin accessibility peaks could then be isolated using the peak-caller MACS2 and were used to predict the location of candidate regulatory sequences.

3.1.4 Selection of Candidate Enhancers

Together, the genomic coordinates of the selected genes, the chromatin accessibility peaks of each cellular population, and the highly conserved regions can be converted into BED file format (a tab-separated value format that logs the chromosome, start, and stop position of each element). These files can be easily intersected and combined using the Bedtools package. We defined each candidate regulatory element as a chromatin-accessibility peak that must be located within 100 kb of the noncoding regions from the

TSS of one of the seven selected marker genes and overlap with a highly conserved region defined by the modified phastcons track. The resulting Bedfile was imported to the Integrated Genome Browser for visual inspection and a total of 35 enhancers were selected for screening [34].

3.2 AAV Design/Production

3.2.1 rAAV Genome Vector Design

The only requirement for producing functional rAAVs is to flank the recombinant DNA with internal terminal repeat sequences and limit its size to ~4.7 kb [37]. In addition to the candidate enhancer sequence, several structural elements can be incorporated to enhance and stabilize the expression of the transgene of interest. These include minimal promoter, chimeric intron, woodchuck hepatitis virus post-transcriptional regulatory element (WPRE), and a poly-adenylation signal [38]. The inclusion of each element depends on the size of the transgene and the desired level of expression of the transgene.

3.2.2 Transgene

The specificity of the experimental needs will ultimately dictate the nature of the transgene that will be incorporated into the viral vector. Fluorescent reporters or epitope tags can be targeted to different cellular compartments to visualize various aspects of the cells. Nuclear-localized reporters can facilitate colocalization studies and allow for FACS isolation of the cells. Cytoplasmic reporters are preferable for electrophysiological studies while membrane-bound reporters can be used to assess cellular morphology [39–47]. Importantly, the stability of the reporter to the various steps of the downstream tissue processing must be taken into consideration, as exemplified by the fact that the dTomato reporter can be used in combination with RNAscope while GFP is entirely degraded by the process. Reporters of neuronal activity (e.g., calcium or voltage sensors–[48–50], chemogenetic and optogenetic effectors can be used to assess and manipulate the activity of the target neuronal population [50–53]. Importantly, some recombinases (e.g., Cre-recombinase) are highly active at low dose, which can lead to nonspecific expression of the recombinant-dependent reporter. This has been observed using AAV-Dlx-Cre in R26-Ai9 mice and can be mitigated by using a recombinase with lower activity (e.g., Flp-recombinase–[54]) or combined with a genomic locus that is less ubiquitously accessible [55].

3.2.3 Capsid

Tropism for specific organs, bias for broad cellular populations and species, and strain-dependent transduction efficiency must be taken into account when selecting a capsid for rAAV production [56]. Our group uses the PHPeB capsid for systemic injection in mice [57]. This capsid efficiently crosses the blood-brain barrier in C57Bl6 and 129S mouse strains, which allows for widespread infection of neurons throughout the CNS. Although PHPeB was shown not to cross the BBB with high efficiency in NHP, it is efficient at transducing neurons upon local injections across several species including zebra-finches, mice, rats, marmosets, and macaques.

3.2.4 rAAV Production and Storage

rAAVs can be produced in-house with standardized and easy-to-implement protocols [10]. In our laboratory, we use polyethylenimine-cotransfection (PEI) of HEK293T cells followed by Iodixanol-based purification. Briefly, HEK293T cells are expanded in DMEM+ medium and subsequently transfected using PEI combined with the three necessary plasmids (pAAV, pHelper, and pCapsid). The cells producing rAAVs were then collected after ~60 h (i.e., at 90% confluence), concentrated, and purified using Iodixanol gradient. The resulting preparation should be aliquoted immediately and stored at −80 °C. Importantly, the virus should only be thawed on ice before use and never refrozen. The remaining volume can be stored for at least 1 year at 4 °C without a noticeable loss of efficiency. The titer of the preparation can be estimated by quantitative PCR using validated primers targeting the WPRE element. Notably, the resulting estimate of the number of DNA fragments per unit of volume can include unpackaged viral DNA. As such, qPCR titration is used as an estimate of the consistency of production rather than a quantification of the number of functional viral particles. To obtain a more reliable estimate, determine the multiplicity of infection (MOI – # packaged AAV genome copies / # of cells to be infected) needed for the relevant biological system.

3.3 Enhancer Screening Pipeline

To assess the expression profile of the viral transgene conferred by the candidate regulatory sequence, we injected individual rAAVs systemically in adult mice. After an incubation period of 3 weeks, the brains are harvested and the expression profile of the viral reporter is assessed using immunohistochemistry or in situ hybridization. Enhancers conferring highly specific expression in the target population (>90%) are then further validated. First, we test whether the specificity is dose-dependent by local injection (which leads to a significantly higher viral load per cell). We then test their ability to drive functional expression of effector transgenes (i.e., chemogenetic and optogenetic effectors). Finally, we assess whether the targeting specificity is maintained across species. We strongly believe that these three additional validation steps are essential to fully characterize the expression profile of the enhancers and constitute a prerequisite to making these tools available to the community.

3.3.1 Systemic Delivery

When screening enhancers, we typically perform systemic injections in adult mice (5–6 weeks) to assess the global expression profile within the brain and across organs. For systemic injection in adult mice (~35 g), we inject a mixture of 1 × 10^11 viral particles produced with the PHPeB capsid and diluted in sterile PBS into the retro-orbital sinus of each animal. First, mice were sedated using an inhalant anesthetic. After being anesthetized, the animals were transferred to a laminar flow hood and fit on a nose cone connected to a nonrebreathing apparatus. Then, the injection is performed with a 1 mL syringe and the volume of the virus should be above 50 μL and below 200 μL to maximize precision and limit blood dilution. One eye was gently protruded by applying gentle pressure above and below the eye. Next, the syringe filled with the viral mixture was inserted into the medial canthus at a slight angle. After injecting completely, the needle was slowly withdrawn to prevent backflow. A topical anesthetic was applied to the eye. Postoperative monitoring was performed for 5 days post-injection.

The incubation period depends on the experimental design and is typically 2–3 weeks for reporters and 5–6 weeks for effectors. Importantly, other protocols are described for systemic delivery (i.e., delivery via the tail and heart) and can be adapted for neonatal animals.

3.3.2 Local Delivery

Animals were anesthetized under isoflurane (1–3% in oxygen) and placed in a stereotactic head frame on a temperature-controlled heating pad. After a craniotomy and durotomy were performed above the region of interest, the animals were injected with 50–500 nL of the desired virus at a rate of 10–25 nL/min using a sharp glass pipette (25–35 mm in diameter). The glass pipette is left in place for 5–15 min after the injection to minimize backflow. The craniotomy site is covered with sterile bone wax and the surgical opening is closed with Vetbond. The animals were returned to their home cages for an incubation period of at least 1 week with 5 days of postoperative monitoring. In our case, to inject into mice we used the following coordinates: somatosensory cortex S1: 1.0 mm posterior, 3.0 mm lateral, 0.7 /0.4 mm ventral relative to bregma; hippocampus CA1: 1.6 mm posterior, 1.8 mm lateral, 1.2 mm ventral relative to bregma; striatum: 0.5 mm posterior, 2.0 mm lateral, 3.2 mm ventral relative to bregma.

Direct parenchymal injections can be used when the experiment requires viral delivery restricted to specific brain regions. Importantly, local injections typically lead to a higher viral load compared to systemic injections. This results in significantly higher transgene expression and possible alteration of expression specificity. However, we showed that the E2 retained its specificity upon local injection at various titers and could drive detectable expression within 48 h.

3.3.3 PFA Perfusion and Brain Dissection

Mice were anesthetized by intraperitoneal injection of Euthasol® using a 1 mL syringe, and then transcardially perfused with 4% paraformaldehyde (PFA). After exposing the beating heart, a small incision was made in the right atrium, and a 27-gauge needle (or smaller) connected to a peristaltic pump was inserted into the left ventricle. An initial infusion of PBS was performed to flush out the blood until the liver turned white, followed by an infusion of 20–30 mL of the 4% PFA (until stiff). After perfusion, the brain was dissected out of the skull using forceps and placed in 4% PFA overnight.

Notably, some protein markers such as PV are sensitive to tissue damage sustained during the dissection of the brain from the skull. For example, PV immunostaining will be depleted in regions directly underneath damaged pia. To mitigate this, we allow 2–3 h after perfusion before extracting the brain. This gives more time for the formaldehyde crosslinking to stabilize the proteins, which stiffens the brain and facilitates its dissection. It is important to adapt the post-fixation time to the downstream processing step (e.g., post-fixation time exceeding 3 h will result in complete loss of signal for VIP immunohistochemistry).

3.3.4 Vibratome Sectioning

Tissue sections were produced using a Leica VTS1000 vibrating microtome. The brains are sectioned into 40 μm slices either coronally or sagittally, where sagittal sections are used to assess global expression across brain regions and coronal sections are preferred to resolve specific structures (e.g., amygdala). For coronal sections, we used a razor blade to cut through the cerebellum to create a flat surface. We then embedded the tissue in 1% agarose and super glued the block onto the vibratome platform for sectioning. For sagittal sections, we removed the spinal cord and cut the brain down the midline. The tissue was then super glued directly to the platform and sectioned without agarose embedding. Sections were then placed into RNAse-free PBS or freezing buffer made from sterile and RNAse-free reagents. Before downstream tissue processing, the freezing buffer can be removed with 2–3 washes with PBS.

3.3.5 Freezing Microtome Sectioning

For nonhuman primate tissue, after perfusing with 4% PFA, we performed freezing microtome sectioning (sliding microtome, Leica Biosystems, SM2010 R). After postfixation, the tissue was placed in a container filled with a 30% sucrose solution for approximately 3 days. We waited until the tissue was fully saturated and sank to the bottom of the container. Then we filled the microtome with dry ice, secured the tissue to the platform of the microtome using O.C.T. compound (Sakura, 4583), and started sectioning after the tissue was completely frozen.

3.3.6 Immunohistochemistry

After the appropriate post-fixation period following perfusion, we often performed immunohistochemistry (IHC) to assess the performance of the rAAVs by identifying neuronal cell types of interest with marker proteins and amplifying the virally delivered fluorescent reporters.

Floating tissue sections taken out of the freezing buffer were washed three times with PBS to remove the freezing buffer. These wash steps can be skipped if the sections were stored in PBS (short-term storage). Then the floating sections were permeabilized using a blocking solution for 30 min. However, for this protocol, some antibodies might work better without detergent, such as the mouse anti-Gad67 (Milipore, catalog no. MAB5406). Sections were incubated 2–3 overnights in antibody solution with the appropriate concentrations of the primary antibodies according to the manufacturer at 4 °C. Next, the sections were washed at least three times with PBS. Then the sections were incubated with Alexa-Fluor-conjugated secondary antibodies at 1:1000 dilution in antibody solution for one overnight. After at least three washes with PBS, the sections were counterstained with DAPI, washes at least one time with PBS, and mounted on glass slides using Fluoromount-G.

3.3.7 In Situ Hybridization

Brain sections were washed once in PBS, followed by three washes in 0.1% Triton X-100 and PBS, mounted on Superfrost Plus glass slides, and baked at 60 °C in the HybEZ oven for 25 min. It is important to make sure the tissue is adhered well to the slide as well as having no folds in the tissue as this will cause the section to fall off the slide over the course of the protocol. The slides were then submerged in 4% PFA for 30 min and washed three times in H_2O. RNAscope H_2O_2 was applied to each section for 5 min at room temperature. The slides were then washed three times in H_2O before being submerged in prewarmed 90 °C H_2O for 15 s, followed by prewarmed 90 °C RNAscope Target Retrieval for 15 min. Slides were washed three times in H_2O before RNAscope Protease III was applied to each section, and then incubated for 15 min at 40 °C in the HybEZ oven. Slides were washed three times in H_2O and then incubated with probe solution diluted to 1:50 with probe diluent for 2 h at 40 °C in the HybEZ oven. Next, the sections were washed three times in the RNAscope wash buffer followed by fluorescence amplification. Of note, probes against the RNA of the reporter revealed nonspecific staining that we speculate comes from the viral DNA. To reveal the viral reporter, we followed the RNAscope protocol with an IHC amplification of the dTomato viral reporter. (GFP viral reporters seemed to degrade during the RNAscope protocol). In short, the sections mounted on the slides were incubated in blocking solution (0.3% Triton X-100 + 5% normal horse serum in PBS) for 30 min. Then sections were incubated in antibody solution (0.1% Triton X-100 + 5% normal horse serum in PBS) with rabbit anti-DsRed at 1:250 at 4 °C overnight. The sections were then washed three times with PBS, incubated with Alexa-Fluor-conjugated secondary antibodies at 1:500 for 2 h, counterstained with DAPI, and mounted on glass slides using Fluoromount-G.

3.4 Evaluation of Enhancer Expression Profile

The expression profile conferred by a regulatory element can be assessed using various metrics. The transgene is expressed at a certain level (i.e., magnitude of viral transgene expression in each cell), with a given specificity for the target population (i.e., percentage of cells co-expressing the viral transgene and a marker of the target population) and a given sensitivity (i.e., the percentage of the target population that express the viral transgene). Depending on the method of injection, cells outside the target organ can also be exposed to the virus, which can also be investigated if relevant to the experimental design. Immunohistochemistry (IHC) and in-situ hybridization (ISH) protocols should be carefully optimized to assess the expression of both the transgene and marker genes for the target population. IHC is easier to implement but fluorescent ISH can be used to reveal virtually any gene with a high signal-to-noise ratio. We often use both methods as a means to cross-validate our results.

3.4.1 Strength of Expression

The nature of the transgene dictates the optimal range in which it should be expressed. High expression of recombinase can lead to off-target expression while low expression of chemogenetic effector leads to insufficient cellular response upon stimulation. The strength of expression can be modulated by adapting the viral dose and incubation time, but it ultimately depends on the intrinsic strength of the regulatory element incorporated into the rAAV. To assess the strength of expression of each enhancer, we standardized the viral backbone, age of the animal at injection, viral dose, incubation time, tissue processing and imaging parameters. We then average pixel intensity of the cell bodies of each cell expressing the viral reporter and report the average overall fluorescently labeled cells per enhancer. Because we demonstrated that the level of expression conferred by the Dlx enhancer is sufficient for chemogenetic effectors, we thus use it as a reference against which we test all other candidate enhancers.

3.4.2 Sensitivity and Specificity

For the quantification, we first mark all cells expressing the viral reporter and all cells expressing the marker gene for the target population independently. We then overlay the markings of cells from each color channel to calculate ratios corresponding to specificity and sensitivity. This analysis is performed for specific brain regions (i.e., somatosensory cortex) and can be reported as bulk or broken down into specific subregions (i.e., cortical layers). We distinguish between cortical layers by looking at a DAPI staining from which we can infer layers based on cellular density. To systematize the image analysis, we used Cell Profiler (https://cellprofiler.org) to detect cell boundaries in each channel and report strength, specificity, and sensitivity per region and cortical layers (automatically detected based on differential density using the Dapi staining). Altogether, these quantifications were assessed from at least two animals and at least three brain sections per animal to ensure reproducibility. Additionally, we assess these criteria over different brain regions because successful enhancers for the same cell type can have interesting regional differences as shown in our publication. For promising therapeutic candidates, we also look at the expression across organs to get a sense of potential off-target effects.

4 Conclusion

Recent technological advances allow for querying the transcriptomic and epigenetic landscape of functionally distinct subtypes of cells across all tissue with unprecedented depth and resolution. These insights can be combined to identify endless lists of candidate regulatory sequences. Transforming these candidates into effective tools is achievable but far from obvious, and requires a thorough in vivo characterization of the properties they confer to viral vectors. rAAV equipped with selective enhancers will likely constitute a revolution for both basic research and gene therapy as they will open the door to understanding the role of specific cell types in normal and pathological conditions. Because this is achievable across species, it is likely to constitute the backbone of novel therapeutic interventions with unprecedented specificity.

References

1. Dimidschstein J et al (2016) A viral strategy for targeting and manipulating interneurons across vertebrate species. Nat Neurosci 19:1743–1749
2. Vormstein-Schneider D, Lin JD et al (2020) Viral manipulation of functionally distinct interneurons in mice, non-human primates and humans. Nat Neurosci 23:1629–1636
3. Bedbrook CN, Deverman BE, Gradinaru V (2018) Viral strategies for targeting the central and peripheral nervous systems. Annu Rev Neurosci 41:323–348
4. Hrvatin S et al (2019) A scalable platform for the development of cell-type-specific viral drivers. elife 8:e48089
5. Mich JK et al (2021) Functional enhancer elements drive subclass-selective expression from mouse to primate neocortex. Cell Rep 34(13): 108754
6. Mingozzi F et al (2003) Induction of immune tolerance to coagulation factor IX antigen by in vivo hepatic gene transfer. J Clin Invest 11(9):1347–1356
7. Xiong W et al (2019) AAV cis-regulatory sequences are correlated with ocular toxicity. Proc Natl Acad Sci U S A 116(12):5785–5794
8. Vesuna S et al (2020) Deep posteromedial cortical rhythm in dissociation. Nature 586:87–94
9. Mo A et al (2015) Epigenomic signatures of neuronal diversity in the mammalian brain. Neuron 86:1369–1384
10. Gray SJ et al (2011) Production of recombinant adeno-associated viral vectors and use in in vitro and in vivo administration. Curr Protoc Neurosci 4:4.17
11. Fulco CP et al (2016) Systematic mapping of functional enhancer-promoter connections with CRISPR interference. Science 354(6313):769–773
12. Visel A et al (2007) VISTA enhancer browser–a database of tissue-specific human enhancers. Nucleic Acids Res 35:D88–D92
13. Bejerano G et al (2004) Ultraconserved elements in the human genome. Science 304: 1321–1325

14. Dimitrieva S, Bucher P (2013) UCNEbase—a database of ultraconserved non-coding elements and genomic regulatory blocks. Nucleic Acids Res 41:D101–D109
15. Andersson R et al (2014) An atlas of active enhancers across human cell types and tissues. Nature 507:455–461
16. Dousse A, Junier T, Zdobnov EM (2016) CEGA—a catalog of conserved elements from genomic alignments. Nucleic Acids Res 44: D96–D100
17. Dickel DE et al (2018) Ultraconserved enhancers are required for normal development. Cell 172:491–499
18. Ron G et al (2017) Promoter-enhancer interactions identified from Hi-C data using probabilistic models and hierarchical topological domains. Nat Commun 8:2237
19. Huang J et al (2018) Dissecting super-enhancer hierarchy based on chromatin interactions. Nat Commun 9(1):943
20. Tasic B et al (2016) Adult mouse cortical cell taxonomy revealed by single cell transcriptomics. Nat Neurosci 19:335–346
21. Saunders A et al (2018) Molecular diversity and specializations among the cells of the adult mouse brain. Cell 174:1015–1030.e16
22. Tremblay R, Lee S, Rudy B (2016) GABAergic interneurons in the neocortex: from cellular properties to circuits. Neuron 91:260–292
23. Siepel A et al (2005) Evolutionarily conserved elements in vertebrate, insect, worm, and yeast genomes. Genome Res 15(8):1034–1050
24. Kent WJ et al (2002) The human genome browser at UCSC. Genome Res 12(6): 996–1006
25. Buenrostro J et al (2013) Transposition of native chromatin for fast and sensitive epigenomic profiling of open chromatin, DNA-binding proteins and nucleosome position. Nat Methods 10:1213–1218
26. Barski A et al (2007) High-resolution profiling of histone methylations in the human genome. Cell 129:823–837
27. Johnson DS et al (2007) Genome-wide mapping of in vivo protein-DNA interactions. Science 316:1497–1502
28. Mikkelsen T et al (2007) Genome-wide maps of chromatin state in pluripotent and lineage-committed cells. Nature 448:553–560
29. Spitz F, Furlong E (2012) Transcription factors: from enhancer binding to developmental control. Nat Rev Genet 13:613–626
30. Tuteja G et al (2014) Automated discovery of tissue-targeting enhancers and transcription factors from binding motif and gene function data. PLoS Comput Biol 10(1):e1003449
31. He HH et al (2010) Nucleosome dynamics define transcriptional enhancers. Nat Genet 42(4):343–347
32. Fang R et al (2021) Comprehensive analysis of single cell ATAC-seq data with SnapATAC. Nat Commun 12:1337
33. Feng J et al (2012) Identifying ChIP-seq enrichment using MACS. Nat Protoc 7(9): 1728–1740
34. Freese NH et al (2016) Integrated genome browser: visual analytics platform for genomics. Bioinformatics 32(14):2089–2095
35. Jüttner J et al (2019) Targeting neuronal and glial cell types with synthetic promoter AAVs in mice, non-human primates and humans. Nat Neurosci 22:1345–1356
36. Blankvoort S, Witter MP, Noonan J, Cotney J, Kentros C (2018) Marked diversity of unique cortical enhancers enables neuron-specific tools by enhancer-driven gene expression. Curr Biol 28(13):2103–2114
37. Wu Z, Yang H, Colosi P (2010) Effect of genome size on AAV vector packaging. Mol Ther 18(1):80–86
38. Powell SK, Rivera-Soto R, Gray SJ (2015) Viral expression cassette elements to enhance transgene target specificity and expression in gene therapy. Discov Med 19(102):49–57
39. Chalfie M et al (1994) Green fluorescent protein as a marker for gene-expression. Science 263:802–805
40. Heim R, Tsien RY (1996) Engineering green fluorescent protein for improved brightness, longer wavelengths and fluorescence resonance energy transfer. Curr Biol 6:178–182
41. Matz MV et al (1999) Fluorescent proteins from nonbioluminescent Anthozoa species. Nat Biotechnol 17:969–973
42. Shaner NC et al (2004) Improved monomeric red, orange and yellow fluorescent proteins derived from Discosoma sp. red fluorescent protein. Nat Biotechnol 22:1567–1572
43. Strack RL et al (2009) A rapidly maturing far-red derivative of DsRed-Express2 for whole-cell labeling. Biochemistry 48:8279–8281
44. Griesbeck O et al (2001) Reducing the environmental sensitivity of yellow fluorescent protein. J Biol Chem 276:29188–29194
45. Kalderon D, Roberts BL, Richardson WD, Smith AE (1984) A short amino acid sequence able to specify nuclear location. Cell 39:499–509

46. Robbins J, Dilworth SM, Laskey RA, Dingwall C (1991) Two interdependent basic domains in nucleoplasmin nuclear targeting sequence: identification of a class of bipartite nuclear targeting sequence. Cell 64:615–623
47. Fowler DK et al (2016) A MultiSite gateway toolkit for rapid cloning of vertebrate expression constructs with diverse research applications. PloS one 11(8):e0159277
48. Chen TW et al (2013) Ultrasensitive fluorescent proteins for imaging neuronal activity. Nature 499(7458):295–300
49. Piatkevich KD et al (2018) A robotic multidimensional directed evolution approach applied to fluorescent voltage reporters. Nat Chem Biol 14(4):352–360
50. Marvin JS et al (2019) A genetically encoded fluorescent sensor for in vivo imaging of GABA. Nat Methods 16:763–770
51. Magnus CJ et al (2019) Ultrapotent chemogenetics for research and potential clinical applications. Science 364:6436
52. Boyden ES et al (2005) Millisecond-timescale, genetically targeted optical control of neural activity. Nat Neurosci 8:1263–1268
53. Yizhar O et al (2011) Neocortical excitation/inhibition balance in information processing and social dysfunction. Nature 477(7363):171–178
54. Graybuck LT et al (2021) Enhancer viruses for combinatorial cell-subclass-specific labeling. Neuron 109(9):1449–1464.e13
55. Chen Q et al (2020) Dysfunction of cortical GABAergic neurons leads to sensory hyperreactivity in a Shank3 mouse model of ASD. Nat Neurosci 23:520–532
56. Watakabe A et al (2015) Comparative analyses of adeno-associated viral vector serotypes 1, 2, 5, 8 and 9 in marmoset, mouse and macaque cerebral cortex. Neurosci Res 93:144–157
57. Chan KY et al (2017) Engineered AAVs for efficient noninvasive gene delivery to the central and peripheral nervous systems. Nat Neurosci 20:1172–1179

Chapter 10

Pathway-Selective Reversible Perturbations Using a Double-Infection Technique in the Macaque Brain

Wim Vanduffel and Tadashi Isa

Abstract

Brain research during the last two decades has been characterized by extraordinary tool developments, which allowed researchers to unravel brain mechanisms underlying motor and cognitive functions at unprecedented spatiotemporal resolution. Inspired by genetic-based tools that became mainstream for invertebrate and rodent studies, nonhuman primate researchers started to embrace the power of such techniques. Primates, however, are much less genetically tractable, which imposes limitations on the transfer of genetic tools to monkeys. Despite this constraint, researchers have successfully introduced opto- and chemogenetics in nonhuman primate studies, mainly through injection of viral vector constructs which allow body-foreign genes to be expressed in cells of the nervous system. One particularly fascinating leap forward in this domain pertains to the reversible manipulation of specific brain pathways. Researchers are now able to causally link properties of very specific brain pathways to behavior and brain-wide activity in primates. Besides the extra dimension this brings to basic brain research, it also holds great therapeutic promise. For the first time, it has become feasible to manipulate the function of one specific efferent pathway while leaving other functions and pathways originating or arriving in that area largely unaffected. This chapter summarizes the progress that has been made in the field of pathway-selective perturbations in nonhuman primates.

Key words Macaque, Viral vector, Pathway-selective, Inactivation, Activation, Reversible, Causal

1 Introduction

Although Ernest Auburtin did not rise to scientific prominence, this physician left an enduring mark on systems neuroscience more than 150 years ago. When he softly poked specific parts of the frontal cortex of someone surviving a suicide attempt, this patient temporarily lost his ability to speak while leaving other cognitive functions unaffected. Instead, another physician became famous as he discovered that a lesion in the left frontal cortex of another patient could be linked to his inability to produce speech [1, 2]. Auburtin's poking experiments were performed 10 years earlier and were more telling than Broca's lesion study, because of

Mark A. G. Eldridge and Adriana Galvan (eds.), *Vectorology for Optogenetics and Chemogenetics*, Neuromethods, vol. 195, https://doi.org/10.1007/978-1-0716-2918-5_10,

the reversible nature of his "pressure" perturbations on an otherwise intact brain. Nevertheless, this brain region, which is critical for producing speech, finally became known as Broca's area. Both studies showed the exquisite power of perturbations to causally link regional brain activity, or the lack thereof, to specific behaviors.

Inspired by such findings, researchers started to develop tools that allowed them to artificially alter the activity of specific areas, which enabled them to causally link regional brain activity to behavior under experimentally controlled conditions [3]. In addition to their level of invasiveness, these procedures can be subdivided into *reversible* (e.g., Auburtin's poking experiments) and *irreversible* perturbation methods (e.g., Broca's lesioned patient), and in tools that either increase (e.g., by electrical microstimulation) or decrease normal activity in a targeted region (e.g., by injection of a GABA agonist such as muscimol) [4, 5]. A third important dimension concerns the *spatiotemporal resolution* of these perturbation tools, spanning the entire spectrum from ultrafast optogenetic stimulation of specific neurons in a small part of the brain [6–9], to patient or animal lesions lacking any temporal resolution and which often encompass more than one area [10–12]. For the interpretation of perturbation results, one always has to consider the possibility of functional reorganizations [13]. In other words, any read-out of a brain perturbation reflects behavior or activity of an altered abnormal brain. Molecular, anatomical, and functional changes may be triggered at different and possibly unknown spatiotemporal scales [14]. Such anatomo-functional reorganizations are apparent after permanent lesions, for example, as evidenced by functional recovery after a stroke [15]. However, they likely exist also at short or even ultrashort timescales, via latent connections which may be of little or unknown relevance in an otherwise intact brain. Another important consideration when using perturbation tools is that altered activity in a specific brain region alters activity locally, as well as in down- and upstream-connected regions [16, 17]. Since a single brain region may serve as a hub for multiple independent functional networks, all of them will be influenced when perturbing that hub. Moreover, depending on the specificity of the perturbation method, axons just passing through the targeted area can be affected. In the latter case, functional networks unrelated to the targeted region will also be altered.

To mitigate these shortcomings and to bring circuit analyses to macroscale levels, tools were recently developed to precisely manipulate the activity of specific pathways connecting two separate brain regions, while keeping other pathways originating or terminating in one of these two regions largely untouched. Such pathway-selective perturbation tools finally provided researchers with the opportunity to investigate how a pathway, rather than an area, contributes to behavior and activity across a functional network (*see* Fig. 1). The latest genetic-based tools supersede heroic past attempts to isolate

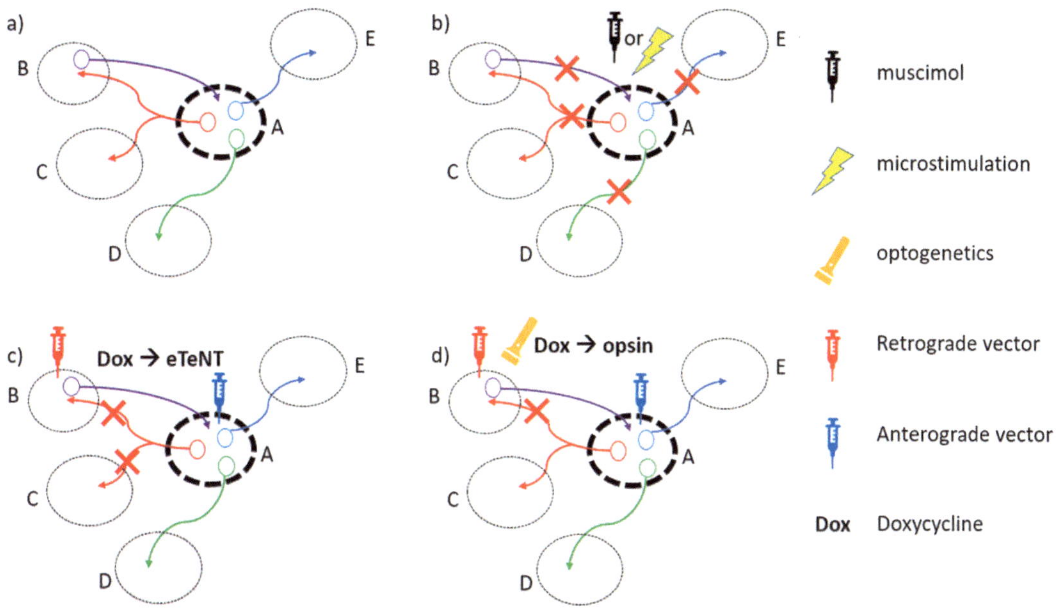

Fig. 1 Summary of possible effects of pathway selective perturbations and comparison with standard perturbations. (**a**) Area A (source area) is connected with areas B-E via different efferent projection neurons (green, red, and blue). Only for one area B, reciprocal connections with A are indicated by the purple afferent neuron, but they may exist between all pairs of areas. (**b**) Traditional areal perturbation methods, such as pharmacological inactivation using the GABA agonist muscimol, or electrical stimulation will affect all efferent and afferent connections. Note that in case of microstimulation this can be enhanced or suppressed activity, depending on stimulation parameters (e.g., [18]). (**c**) In case of pathway-selective perturbations using a double infection approach, one only affects the efferent connections from the source area A to target area B, with the caveat that also bifurcated axons can target another area (area C in this case). (**d**) Pathway selectivity can be achieved by stimulating the axon terminals in target area B, either by local illumination in case of an optogenetic experiment, or by local infusion of a ligand in case of a chemogenetic experiment. However, in the former case, we should be careful about the possible antidromic propagation of action potentials and effects on other target(s) of the stimulated neurons through bifurcating collaterals

the activity of specific connections between two regions which relied on antidromic microstimulation using the so-called collision approach [19].

Exactly as for the aforementioned "areal" perturbation methods, pathway-selective manipulations can be categorized as reversible or irreversible, and activity-enhancing or activity-suppressing tools with varying spatiotemporal characteristics. Genetic-based pathway-selective perturbation tools were first developed in invertebrates and rodents and have been adopted only slowly by nonhuman primate researchers. Several research groups played a prominent role in the development and use of these highly promising vector-based pathway-selective perturbation methods in primates (including T. Isa, K. Kobayashi, K. Kobayashi, K. Inoue, M. Takada, O. Hikosaka, Y. Miyashita, T. Minamimoto, M. Sakagami, and M. Oguchi). This chapter aims to summarize and describe the current state-of-the-art for performing pathway-selective perturbation experiments in nonhuman primates. We believe that these tools are crucial assets for providing a mechanistic understanding of information flow and processing in a normal brain and how these pathways are affected by pathology. Ultimately, they may become clinically relevant by selectively correcting malfunctioning circuits in neurological and psychiatric diseases.

One possibility to achieve pathway selectivity is to insert a gene into the region of DNA coding for circuit-specific molecules. When a transgene is introduced into the germline, genetic manipulation is noninvasive. It needs to be noted, however, that a genetic "signature" of a pathway may not be unique, or neurons may bifurcate and connect to multiple targets, which may impact the specificity of this approach. More restrictive for widespread use, however, is that this option requires transgenic animals, the development of which is still very challenging in primates. The second option is to exploit the spatial properties of a circuit: one genetic element is introduced at the origin of a pathway (source area) and the other at its termination (target area). When the two genetic elements combine within a neuron, they can alter the function of the pathway [20–26]. Note that pathway selectivity also implies direction selectivity. Most inter-areal connections are bidirectional, and most often, researchers aim to study the functional impact of a source area upon a target area, while leaving the reverse connections intact. It needs to be remarked, however, that a reverse connection may be affected indirectly by altered intra-areal activity in the target area that triggers downstream effects in projection neurons located within the target region itself.

Before discussing the different strategies that can be followed to achieve pathway selectivity in perturbation experiments, we will briefly discuss some relevant methodological issues.

1.1 Injection of Viral Vectors

1.1.1 Targeting

Due to the paucity of transgenic nonhuman primate lines, one has to inject at least one or two viral vectors. To identify source and target regions one can rely on atlases. However, inter-subject variability cannot be accounted for by most atlases. Therefore, subject-specific anatomical knowledge obtained using magnetic resonance imaging (MRI) improves the identification of injection targets. For example, not only T1- and T2-weighted imaging but also diffusion imaging and blood vessel imaging can be effective tools to identify targets and avoid major blood vessels during the injections. To achieve more precise targeting in the context of functional rather than anatomical networks, one should rely on specific response characteristics of areas, especially if one aims to target sub-compartments. Functionally guided injections can be based on fMRI and electrophysiology, but potentially also on optical imaging, focal ultrasound imaging, or the combination of any of these methods [3, 27–29]. Injectrodes offer exquisite resolution as they allow the recording of electrophysiological activity just prior to the injections [30].

1.1.2 Injections

Once the target(s) are defined, either based on prior anatomical or functional information, one needs to prepare for injections. This can be achieved using stereotactic techniques, preferentially with the addition of fixed external markers that are MR visible [31]. An important issue is that the MR space, which can be deformed, needs to be registered to the real anatomy [32, 33]. One way is to identify anatomical features in MR and real-life space (e.g., specific features of the skull). Once the head of the subject is placed in a stereotactic (or non-stereotactic) device, one can triangulate the targets based on these external markers. Alternatively, one calculates the targets based on stereotactic coordinates only, with the caveat that the animal needs to be positioned in exactly the same manner inside the MR scanner as the stereotactic frame. This means that an MR-compatible stereotactic frame has to be used during surgery, which may not be as stable and reliable as metal stereotactic frames. Several pre-surgery planning methods have been developed to visualize the registration between function/anatomical and blood vessel maps and "real" space in real-time [31].

Once the target is precisely localized, one can administer the vector using pressure injection with small-volume Hamilton syringes (e.g., 1–100 μL, depending on the final volume to be injected, *see* below) attached to infusion pumps that precisely control the duration and flow rate of the injections. These can be connected to Hamilton needles (30–31 ga) attached to polytetrafluoroethylene tubing. For superficial structures one can also use pulled glass micropipettes with very small tips attached to, for example, Nanoject Injectors. Typically, very small volumes are injected: for example, 0.5–0.75 μL (@ 0.1 μL/min), with multiple injections at different depths within a specific target region. One should wait 2–5 min between subsequent injections and lower the injection needle in steps of 100 or 200 μm. Afterward, the process can be repeated in nearby trajectories (e.g., 1 mm apart) to load a structure of interest with a sufficient viral vector. Recently, Fredericks and colleagues developed arrays of multichannel injection devices that allow faster and more uniform vector delivery compared to single-channel methods in primates [31 and *see* Chap. 13]. To reduce backflow through the injection track, which can be problematic when targeting deep and small brain structures, convection-enhanced delivery (CED) methods can be considered an excellent alternative [27, 34]. This injection technique generates a pressure gradient at the tip of an infusion device to deliver vectors directly through the interstitial spaces of the central nervous system. Large brain volumes (50 μL per site) can be infused with sufficient pressure to drive advection and hence achieve a wider and more even distribution of the vector.

1.1.3 Assessing Successful Delivery

Total volume per target can be in the order of 10 μL, although much larger volumes can be useful if one aims to transduce large structures such as some basal ganglia structures (e.g., the caudate nucleus or the putamen). Co-infusion of MR contrast agents such as $MnCl_2$-$4H_2O$ or gadolinium, together with the viral vectors, allows to verify whether the infusions were successful. Low concentrations (e.g., 0.1 mM Mn^{2+}) are sufficient for MRI-based visualization of the injected area, although higher concentrations, up to 10 mM have been used without side effects [35, *see* Chap. 13]. Although these contrast agents can indicate whether the injection was accurately targeted and provide information about the center of mass of the injected area, it does not reveal the total extent of vector diffusion or transduced brain volumes.

In-vivo fluorescence imaging either of exposed cortical regions [27, 36] or using optical probes to measure fluorescence in deep regions [30] are excellent alternatives to check whether viral vector injections resulted in effectively transduced brain tissue. These options are commonly employed in conjunction with optogenetic techniques, for which access to the tissue is required for the delivery of light. Likewise, positron emission tomography (PET) imaging can provide in-vivo information critical for the successful expression of designer receptors in pathway-selective chemogenetic experiments [37, 38].

1.2 Stimulation of Axon Terminals: Use of Anterograde Vectors

The most straightforward way to perturb activity in a pathway-selective manner is to transduce neurons in a source area, using a viral vector containing the genetic material for light-sensitive ion channels [39, 40], pumps [41], enzymes, or DREADDs (designer receptor exclusively activated by a designer drug) [38, 42–44]. After expression of these genes in transduced projection neurons, one can stimulate or inhibit the axon terminals in the target region with light (in the case of optogenetics), or by injection of a ligand in the target area that selectively binds to the exogenous receptor (in case of a chemogenetic experiment). Recently, a novel high-affinity and selective agonist ligand, deschloroclozapine (DCZ), has been described that will most likely replace clozapine-*N*-oxide (CNO) in chemogenetic experiments because of CNO's sluggish kinetics, metabolic liabilities and potential off-target effects [45, 46]. This axonal pathway-selective perturbation method has the advantage that one can stimulate or inhibit precisely the connection between source and target region, leaving unaffected possible bifurcations of axons that connect the source region with other, nontarget, regions. The disadvantage is that one has to rely on viral vectors with exclusively anterograde and no retrograde transduction properties–note that vectors with exclusive anterograde transduction properties are very rare. In the case that even a few neurons are transduced retrogradely in the target region, one risks stimulating the pathway in the reverse direction. It is possible that secondary target regions may also be impacted if these retrogradely transduced neurons bifurcate (*see* Fig. 1).

Several studies have successfully employed this method. For example, Ken-Ichi Inoue and his colleagues injected an AAV2 vector carrying ChR2 with a CMV promoter (AAV2-CMV-ChR2-EYFP) unilaterally in the frontal eye fields (FEF) of rhesus monkeys and stimulated the axon terminals in the superior colliculus (SC), a tectal structure that is known to receive direct input from the FEF. It was previously known that signals from the FEF to the SC causally contribute to the generation of saccadic eye movements [47]. However, optogenetic stimulation of the FEF's projection neurons targeting the SC showed that these effects are most likely mediated via direct FEF-SC connections, although alternative indirect pathways via the basal ganglia cannot be completely discarded [48]. By illuminating the FEF terminals in the ipsilateral SC, many of the SC neurons were excited by the blue laser light, with only a few showing inhibited responses. This means that axons from FEF projection neurons to the SC can directly modulate activity in the latter structure. In addition, optical stimulation of these fibers triggered saccadic eye movements toward the response fields of neurons residing in the stimulation sites within the SC–in fact with stronger effects than prior optogenetic approaches targeting the FEF directly [49, 50]. The Inoue et al. [48] study clearly showed a combined neuronal and behavioral effect of stimulating transduced axon terminals remote from the injection site.

In another study, Nurminen and colleagues aimed to inactivate feedback projections from visual area V2 to primary visual cortex by injecting a viral vector cocktail containing AAV9.CaMKII.Cre and a second vector carrying the gene for an outward proton pump Archaerhodopsin-T (AAV9.Flex.CAG.ArchT-GFP) in the second visual area of marmosets [51]. Expression of ArchT is Cre-dependent in this case, hence only neurons transduced by both vectors will express ArchT. This viral vector combination shows negligible retrogradely labeled neurons in area V1. Hence optical stimulation of target zones within area V1 will stimulate primarily axon terminals from projection neurons with their cell bodies in area V2. These neurons carry feedback from V2 to V1 and optogenetic inactivation decreased responses of V1 neurons, but increased their receptive field (RF) size. Thus, this pathway-selective manipulation of feedback from V2 to V1 modulates RF size, surround suppression and response amplitude of postsynaptic V1 neurons.

Pathway-selective optogenetic stimulation of axon terminals was also performed in a recent macaque study by Amita et al. [29]. Using an injectrode, by which the MRI-guided target was verified based on its neuronal activity, these authors injected AAV2-CMV-ChR2-EYFP in the tail of the caudate nucleus and optically stimulated axon terminals in a subsector of the substantia nigra reticularis (SNr) and the external globus pallidus (GPe). This experiment showed that optical stimulation of striatal terminals in SNr causes prolonged excitation of visual-saccadic neurons in the superior colliculus (SC), and also induces contralateral saccades.

To study the functional role of corticothalamic projections, Galvan and coworkers infused primary motor cortex (M1) and premotor cortex (PM) of rhesus monkeys with either AAV2/5-CaMKIIα-hChR2-EYFP or AAV2/5- CaMKIIα-C1V1-EYFP [52]. They relied on an injectrode for identifying the relevant injection zones, based on electrophysiological properties and behavioral effects induced by electrical microstimulation. Next, they optically stimulated either the injection zones in motor cortex or the corticothalamic axon terminals while recording single unit activity in the vicinity of the stimulation sites. In contrast to short-latency increases in firing observed in transduced and photo-stimulated cortical neurons, photo-activation of corticothalamic terminals yielded long-latency increases or decreases in firing of thalamic neurons. In combination with electron microscopic results, these data suggested the recruitment of intra-thalamic GABAergic circuits by the corticothalamic afferents. Also, this study did not find evidence for retrogradely labeled neurons in the thalamus.

These aforementioned studies showed that selective optical stimulation of axon terminals can be a highly effective approach to investigate brain pathways. A major issue is that highly selective anterograde viral vectors need to be injected in the source area without causing any retrograde labeling of neurons in the target region. Even a small number of retrogradely neurons in the intended target area will affect the interpretation of results generated by such axonal stimulation experiments.

1.3 Stimulation of Retrogradely Transduced Cell Bodies: Use of Retrograde Vectors

Pathway-selective manipulation can also be achieved by optically stimulating retrogradely transduced cell bodies. In this approach, a viral vector with strong retrograde gene transfer properties has to be injected into the target region at the site of the axon terminals. This vector transduces remote cell bodies in the source area, and optical stimulation targets the transduced soma. The retrograde stimulation approach is appealing since one can alter the activity of the projection neurons while recording from them at the same time. Moreover, this approach allows the identification of these neurons based on their activity induced by the photostimulation or based on their histochemical and anatomical properties. For transducing neurons retrogradely, several AAV, CAV, and lentiviral vectors have been developed (e.g., *see* Chap. 2). Compared to the axonal stimulation approach, one needs to achieve high transduction rates, requiring efficient vectors and usually relatively high titers. Note that the retrograde-stimulation strategy suffers from a similar "selectivity" issue as mentioned in the previous paragraph for the anterograde strategy. Transduction of anterograde axons in the intended target area will affect the interpretation of results. Hence, one should rely on vectors that exclusively transduce neurons retrogradely, which are also exceedingly rare. These issues can be overcome by using the double infection strategy with combined retro- and anterograde vectors (*see* below).

1.3.1 Efficient Retrograde Vectors

Highly efficient lentiviral vectors that have been used for pathway-selective perturbations are the HiRet [53] and NeuRet vectors [54], which are pseudo-typed with fusion glycoprotein type B2 and type E, respectively (*see* Chap. 2). HiRet transduces both neurons and glial cells, while the NeuRet vector is largely specific for neurons. Moreover, the capacity for retrograde gene delivery is better for NeuRet compared to HiRet, and NeuRet induces fewer inflammatory reactions. This may be because transgene delivery into glial cells by the HiRet vector around the injection site may induce an immune response which can result in damage [21]. The neuronal specificity of the NeuRet vector reduces the risk of tumorigenesis due to the prevention of transgene insertion into the genome of dividing cells in the brain, which would be beneficial for potential therapeutic purposes. Thus, NeuRet is a more suitable vector for retrograde perturbation experiments, both in rodents and macaques [55]. Recently, it has been shown that a new retrograde gene transfer vector (AAV2-retro) is even more efficient than NeuRet, at least in the rat [23], AAV2-retro may be also very efficient in many [56] though not all [42, 57] regions of the primate brain. In general, transduction selectivity, efficiency, and spread of viral vectors can be considerably different across animal species, but also across neuron types and areas within the same species [58, 59]. Hence, they should be tested for each model species and experimental paradigm.

1.3.2 Double-Infection Strategies

A third pathway-selective perturbation approach relies on a double-infection strategy [21–23, 26], through a combination of binary systems and efficient retrograde transduction. A binary system consists of two genes. One of them drives the expression of the other gene when both are delivered to the same neuron. The first gene usually encodes a transcription activator or a DNA recombinase that binds to a special element on DNA molecules and induces transcription. Two widely used binary systems for controlling transgene expression are Cre-recombinase and tetracycline-regulated systems [22, 23].

The double-infection system is highly selective and was pioneered in primates by the groups of Kazuto Kobayashi, Dai Watanabe and Tadashi Isa. They used an efficient retrograde viral vector (HiRet or NeuRet) carrying the gene coding for an enhanced tetanus neurotoxin (eTeNT) and a reporter gene (e.g., enhanced green fluorescent protein; EGFP) behind a tetracycline responsive element (TRE). It should be noted that the gene for the tetanus neurotoxin was transformed to have higher expression in mammalian neurons, hence the name "enhanced" tetanus neurotoxin. When the retrograde viral vector is injected in a target area, axon terminals incorporate the genetic material that is transported to the cell bodies of the projection neurons located in the source area. In that source area an anterograde vector carrying a highly sensitive Tet-ON sequence (rtTAV16) is injected, to double-transduce the cell bodies of the projection neurons. This efficient Tet-ON design enables a quick onset of gene expression. Thus, only in double-infected neurons eTeNT can be regulated under TET-on control, only when the antibiotic doxycycline is present. When doxycycline is administered to the transduced animals, it binds to rtTAV16 in the double-transduced neurons, which activates the TRE promoter triggering the expression of the eTeNT and EGFP (*see* Fig. 2). Thus, double-infected neurons showing fluorescence effectively produce the neurotoxin, which selectively suppresses synaptic

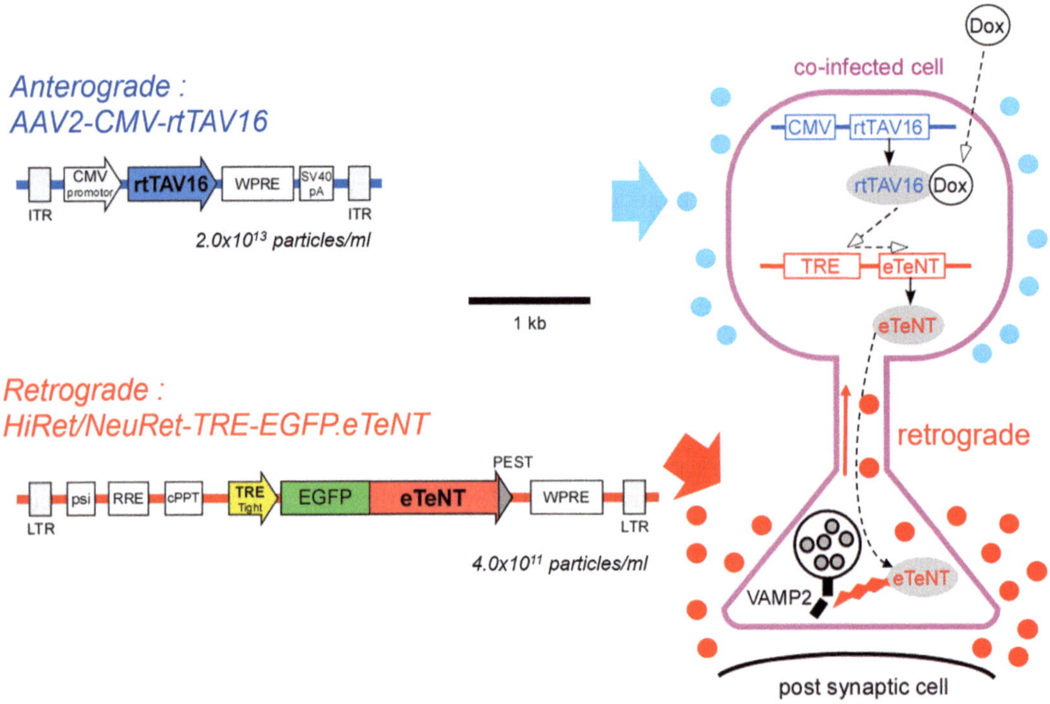

Fig. 2 Schematic diagram of the double infection method using a combination of efficient retrograde and anterograde vectors (adapted with permission from [25]). Doxycycline-controlled transcriptional activation mechanism includes (1) expression of a variant of reverse tetracycline transactivator (rtTAV16, Tet-on sequence); (2) binding of rtTAV16 to the tetracycline-responsive element (TRE) exclusively in the presence of a tetracycline derivative like doxycycline (Dox) and thereby activating transcription; (3) expression of the fluorescent protein (EGFP) and enhanced tetanus neurotoxin light chain protein (eTeNT) downstream of the TRE promoter; and (4) eTeNT cleaves the vesicle-associated membrane protein (VAMP) isoform 2 and inhibits enzyme secretion

transmission in the target area. Specifically, eTeNT blocks synaptic transmission of the double-transduced neurons after systemic administration of doxycycline. When eTeNT is transcribed and translated, it will be carried to the axon terminals where it cleaves the vesicle-associated membrane protein 2 (VAMP2) which is normally required to regulate transmitter release, thereby preventing docking and fusion of synaptic vesicles containing neurotransmitter.

Kinoshita and colleagues were the first to use the technique in the spinal cord of macaques, with the aim to transiently interrupt transmission of propriospinal neurons [25]. A retrograde vector, HiRet-TRE-EGFP.eTeNT, was injected into the region containing motor neurons targeted by the propriospinal neurons, and an anterograde vector AAV2-CMV-rtTAV16 into the intermediate zone of the cervical C2–5 segments where their cell bodies are located. The transient silencing of these propriospinal neurons showed their critical involvement in the control of hand dexterity. With the same technique, the authors showed afterward that propriospinal neurons are also involved in the recovery of hand dexterity after a corticospinal tract lesion [60]. In this follow-up study, NeuRet-TRE-EGFP.eTeNT was used as a retrograde vector, but AAV-2/DJ-CMV-rtTAV16 as an anterograde vector. In another study, Isa's group applied the same double infection method to target the connections between the superior colliculus and the pulvinar in monkeys with a lesion in striate cortex. Reversible inactivation of this pathway, using HiRet-TRE-eGFP.eTeNT as retrograde vector and AAV1-CMV-rtTAV16 as anterograde vector, selectively impaired the production of visually guided saccades in this blindsight model [61].

In a very challenging experiment, Ninomiya and colleagues succeeded in blocking a cortico-cortical pathway in macaques using the double infection technique. Specifically, they reversibly inactivated the connections between ventral premotor and medial prefrontal cortex, using AAV2-retro-CMV-rtTAV16 as retrograde vector and AAV-DJ-TRE-EGFP-eTeNT as anterograde vector. Reversible inactivation of the connections between ventral premotor and medial prefrontal cortex revealed impaired processing of observed but not executed actions in the context of social action monitoring [62]. Thus, the double-infection technique is also useful to target larger cortico-cortical pathways and investigate their causal contribution in higher cognitive functions of primates.

In the previous examples, the double-infection method was used in conjunction with either behavioral tests or electrophysiology. In a collaborative study between the Vanduffel and Isa groups, the pathway between the ventral tegmental area (VTA) and the nucleus accumbens (NAc) was reversibly inactivated using both HiRet- and NeuRet-TRE-EGFP.eTeNT as retrograde vectors and AAV2-CMV-rtTAV16 as anterograde vector [63]. The effect of

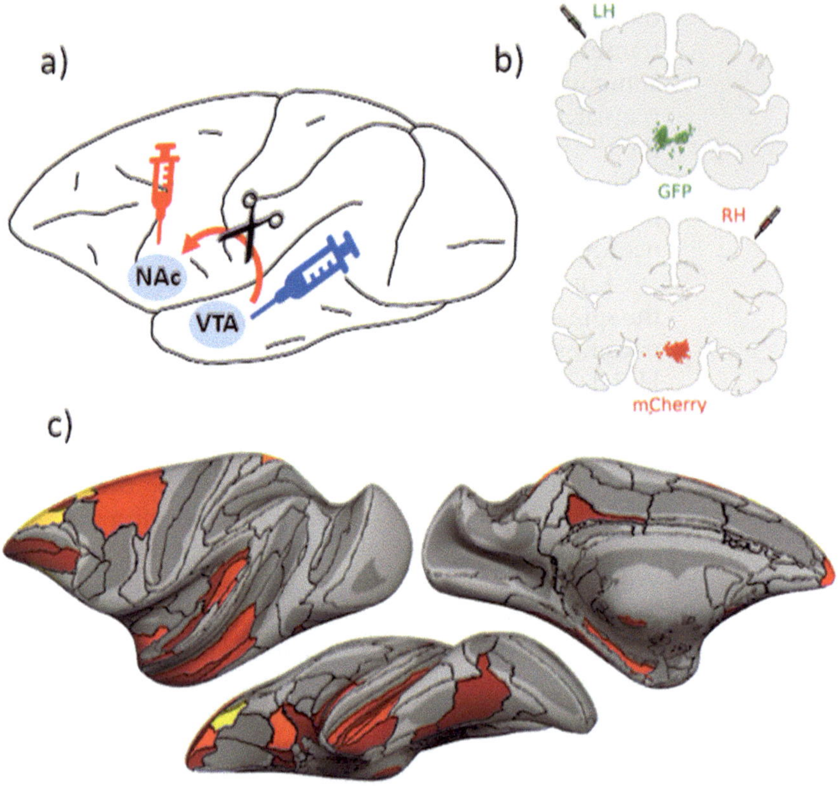

Fig. 3 Pathway-selective inactivation of the connections between the ventral tegmental area (VTA) and nucleus accumbens (NAc) in the macaque using the double-infection technique (reproduced with permission from Vancraeyenest et al. [63]). (**a**) A retrograde vector (red syringe) was injected in the NAc, and the anterograde vector in the VTA (*see* Fig. 2). (**b**) Injections were performed bilaterally with a green (Green Fluorescent Protein, GFP) and a red (mCherry) fluorescent protein in the left (LH) and right hemisphere (RH), respectively. The double-transduced neurons, as revealed by immunocytochemistry from all sections, are back-projected to a representative plane through the VTA. These fluorescent neurons effectively produced the neurotoxin eTeNT when doxycycline is given to the animals. (**c**) Overview of cortical areas showing an increase in functional connectivity (FC) induced by the pathway-selective inactivation with other brain areas that is significantly correlated with the increase in FC between the VTA and NAc. The color scale indicates for each cortical region the number of area-area pairs showing changes in FC that are significantly correlated with that of the VTA-NAc pair

this reversible inactivation was assessed on two cognitive tasks aimed to dissociate motivational from reward-based reversal learning effects and on whole-brain functional connectivity (Fig. 3). The study revealed a relatively small number of retrogradely labeled neurons in the VTA: between 650 and 1500, which is less than 5% of the total number of dopaminergic neurons in the VTA of monkeys [64]. Double-transduced neurons were largely restricted to the VTA. Moreover, most of these neurons were dopaminergic as shown by a double staining revealing the expression of the fluorescent protein in tyrosine hydroxylase (TH)-positive neurons. Likely, the immunohistochemical identification of neurons producing eTeNT in the presence of doxycycline is an underestimation of the real number of double-transduced neurons, since in-situ hybridization (hence at RNA instead of protein level) revealed an order of magnitude more labeled neurons. The histological results confirm that surprisingly good cell-type selectivity can be achieved without the use of specific promoters, by relying on the specificity of anatomical connections. The majority (~60%) of the connections between VTA and NAc are dopaminergic [65]. Despite the relatively low number of double-labeled neurons detected with immunohistochemistry, reversible inactivation of these neurons profoundly reduced the ability of these monkeys to perform trials requiring high effort to obtain a high reward. Although high-effort motivational behavior was clearly affected, the monkeys did not show a deficit in reinforcement-based learning during an object discrimination reversal task—a task in which this pathway was expected to be critical.

Although dopaminergic signaling from VTA to the NAc was impaired during the administration of doxycycline, whole-brain resting state fMRI revealed that fMRI-based functional connectivity paradoxically increased between VTA and NAc. Moreover, the functional connectivity between NAc and several other brain regions also increased. Changes in functional connectivity between the VTA and NAc, induced by the pathway-selective inactivation, correlated positively with changes in functional connectivity of areas located in the temporal pole, orbitofrontal cortex, dorso- and ventrolateral prefrontal cortex as well as anterior cingulate cortex (Fig. 3). While seemingly counterintuitive, mesoaccumbal blockage may increase functional connectivity in relevant functional networks by reducing the inhibitory output of the predominantly GABAergic NAc neurons, which also provide a feedback signal to regulate the DA output of the VTA [66–68]. In turn, this release of feedback inhibition in the VTA itself may increase the functional coupling with the latter structure through its extra-accumbal projection sites. In general, these findings are in line with the reduced functional connectivity observed in humans after pharmacologically increased dopamine levels [69]. Thus, dopamine may play an important role in regulating the functional coupling between many brain regions.

1.4 Limitations

Despite their appeal, pathway-selective perturbations have limitations. A first limitation concerns possible immune responses when multiple injections are required. Immune responses after a second injection seem to be idiosyncratic and difficult to predict. Checking blood plasma for antibodies prior to each experiment may provide indications about possible immune responses. However, even without preexisting immunity, viral vector injections can produce immune responses, which may decrease the efficacy of subsequent injections [70, 71].

Although for some pathways reasonably good cell-type specificity can be achieved without a selective regulatory element, for example in the case of the VTA to NAc pathway [63], there is significant room for improvement by incorporating cell-type specific promotors/enhancers in the anterograde vector (e.g., using a TH promoter for targeting DA neurons). Unfortunately, cell-type specific promoters for nonhuman primates are still limited, and those that have been used successfully show typically low transduction efficiencies, despite their selectivity (e.g., [72]). Transduction efficiencies of viral vectors are limiting factors in general, in part because knowledge obtained in other species may not be transferrable to primates. Moreover, knowledge about transduction efficiencies in one brain region of the same species may not even generalize to other regions of the same species [58, 59].

Another factor that needs to be considered when interpreting the results of pathway-selective manipulations is that axonal projections between a source and target area may branch to other areas (Fig. 1). Hence behavioral results may be explained by the connection between source and intended target area, but also between source and non-target area that is connected via the bifurcated axons. This is particularly relevant for these approaches whereby a ligand (e.g., for DREADDs) or drug (e.g., doxycycline for a TET-On system) is given systemically. This possible confound can be partially mitigated by local infusion of a ligand in the target region. Good anatomical knowledge about the existence or absence of branched axons is, however, very useful for the interpretation of pathway-selective manipulations. This knowledge can be partially obtained by the injected viral vectors themselves [73].

Another important consideration is the temporal resolution that one aims to achieve in a given experiment. The doxycycline TET-on strategy has a temporal resolution of days to weeks [25], and for pathway-selective chemogenetic approaches the resolution will be in the order of hours [43]. Regarding doxycycline TET-on experiments, one has indeed to consider that the antibiotic has to be given for extensive periods of time. Kinoshita et al. [25] showed that their effects became apparent after a few days, and behavioral effects in the Vancraeyenest et al. [63] study emerged only after a week. The temporal dynamics of a doxycycline TET-on experiment may depend on the specific experimental paradigm, or possibly the number of double-transduced neurons. Given the relatively long periods that doxycycline is given, potential issues related to antibiotic resistance may emerge–which have not been documented so far. If one is interested in dissecting cognitive operations at high temporal resolution, for example, during different epochs of a task trial [74], the combination of a pathway perturbation tool with optogenetics is currently the most ideal tool. Instead of a sequence coding for eTeNT in the double infection technique, one can insert a sequence for a depolarizing or hyperpolarizing opsin. This would retain the pathway selectivity as only double labeled neurons would express the opsin. The obvious drawback, however, is that an optic fiber needs to be inserted in either the source or target region in order to activate the opsins. As noted earlier, backpropagation of action potentials and activation of bifurcated axons have to be considered in the interpretation of the results.

Lastly, some of the efficient retrograde vectors are lentiviral vectors, which have a higher packing capacity than adeno-associated viral vectors, yet they may require stricter biosafety regulations in some institutions, which may limit their use [21].

1.5 Prospects

To lift circuit-level whole-brain research in primates to yet another level, one should consider tools through which activity of multiple pathways in the same brain can be selectively increased or decreased, either in combination or in isolation. Different combinations of excitatory and inhibitory DREADDs (activated by different ligands) and/or opsins (activated with light of different wavelengths), together with a double-infection approach with highly efficient retrograde and anterograde vectors may soon bring these exciting possibilities within reach of the primate systems neuroscience community. Ultimately, precise and selective manipulation of specific pathways may offer great therapeutic value to battle neurological and psychiatric diseases. The field of pathway-selective manipulations in primates is built on the vast knowledge about the use of viral vectors gathered during the past decades in the realm of gene therapy. It is to be expected that, in the future, clinicians may also be able to enhance or suppress communication along specific brain connections which are affected by pathology, while leaving others unchanged. Such an approach could mitigate the side effects often observed using more generic approaches where all connections from an area and their associated functions are disturbed using nonspecific pharmacological interventions, or electrical microstimulation (Fig. 1). For example, the study of Vancraeyenest et al. [63] revealed that reduced activity in the VTA-to-NAc pathway specifically affects motivational behavior, while leaving at least some forms of reward-based learning unaffected. Since many psychological disorders, including obsessive-compulsive behavior, depression, and addiction are characterized by malfunctioning motivational behavior, reversible up- or down-regulation of this pathway may someday improve the quality of life of such patients.

Acknowledgments

This work received funding from KU Leuven C14/21/111; Fonds Wetenschappelijk Onderzoek-Vlaanderen (FWO-Flanders) G0D5817N, G0B8617N, G0E0520N, VS02219N; the European Union's Horizon 2020 Framework Programme for

Research and Innovation under Grant Agreement No 945539 (Human Brain Project SGA3) to W.V, and funding from JSPS (19H05723, 19H01011, 19KK0192, 22H04992, 1082703), JST (JPMJCR1651) and AMED (21dm0207093h0002, 20dm0107151h0005, 21dm0307005h0004) to T.I.

Glossary

Perturbation	Artificial alteration of neuronal activity. This can be an artificial increase or decrease of activity.
Pathway	A specific connection between remote brain regions via axons originating in a source region (where the cell body is located) and projecting to a target region (where synapses are made with postsynaptic cells).
Source area	Area housing the soma of the neurons projecting to the target area.
Target area	Area where axon terminals of the projection neurons from the source area terminate.
Viral vector	A tool through which genetic material can be delivered into cells. Usually nonpathogenic, non-replicating, modified viruses are used to deliver genetic material to neurons. The viral vector construct typically comprises: (i) a viral expression system used to deliver the opsin/chemogenetic receptor gene; (ii) a promoter/enhancer fragment supporting cell type-specific targeting; (iii) genes of interest such as the genetic information of a designer receptor or light-activated protein or opsin; and (iv) a reporter gene to visualize opsin expression levels and to determine cell-type specificity.
Transduction	Is the process by which foreign genetic material is introduced into a cell by a viral vector.
AAV	Adeno-associated vector. One important advantage of AAVs compared to other vectors is that in transduced cells, the genetic material is generally not integrated into the genome of the host cell.
LV	Lentiviral vector. These can integrate genetic information into the DNA of host cells. More or longer genes can be packaged in LVs compared to AAVs, in other words, they have a better packaging capacity.
DREADD	Designer receptor exclusively activated by a designer drug [43].
Opsin	In the context of optogenetics, these light-sensitive proteins are used to activate (depolarize, e.g., using channelrhodopsin or ChR2) or inactivate (hyperpolarize, e.g., ArchT) neurons by the light of specific wavelengths [75].
Anterograde	(Transport) from soma to axon terminal

Retrograde	(Transport) from axon terminal to soma
Injectrode	A combination of an electrode and infusion device that allows to record single unit activity prior and during the injection of a substance (e.g., viral vector and drug)

References

1. Mohammed N, Narayan V, Patra DP, Nanda A (2018) Louis Victor Leborgne ("Tan"). World Neurosurg 114:121–125. https://doi.org/10.1016/j.wneu.2018.02.021
2. Stookey B (1963) Jean-Baptiste Bouillaud and Ernest AUBURTIN. Early studies on cerebral localization and the speech center. JAMA 184: 1024–1029. https://doi.org/10.1001/jama.1963.73700260007011
3. Klink PC, Aubry JF, Ferrera VP, Fox AS, Froudist-Walsh S, Jarraya B, Konofagou EE, Krauzlis RJ, Messinger A, Mitchell AS, Ortiz-Rios M, Oya H, Roberts AC, Roe AW, Rushworth MFS, Sallet J, Schmid MC, Schroeder CE, Tasserie J, Tsao DY, Uhrig L, Vanduffel W, Wilke M, Kagan I, Petkov CI (2021) Combining brain perturbation and neuroimaging in non-human primates. NeuroImage 235: 118017. https://doi.org/10.1016/j.neuroimage.2021.118017
4. Bruce CJ, Goldberg ME, Bushnell MC, Stanton GB (1985) Primate frontal eye fields. II. Physiological and anatomical correlates of electrically evoked eye movements. J Neurophysiol 54(3):714–734
5. Hikosaka O, Wurtz RH (1985) Modification of saccadic eye movements by GABA-related substances. I. Effect of muscimol and bicuculline in monkey superior colliculus. J Neurophysiol 53(1):266–291. https://doi.org/10.1152/jn.1985.53.1.266
6. Deisseroth K (2015) Optogenetics: 10 years of microbial opsins in neuroscience. Nat Neurosci 18(9):1213–1225. https://doi.org/10.1038/nn.4091
7. Deisseroth K (2012) Optogenetics and psychiatry: applications, challenges, and opportunities. Biol Psychiatry 71(12):1030–1032. https://doi.org/10.1016/j.biopsych.2011.12.021. S0006-3223(11)01271-6 [pii]
8. Yizhar O, Fenno LE, Davidson TJ, Mogri M, Deisseroth K (2011) Optogenetics in neural systems. Neuron 71(1):9–34
9. Gradinaru V, Thompson KR, Zhang F, Mogri M, Kay K, Schneider MB, Deisseroth K (2007) Targeting and readout strategies for fast optical neural control in vitro and in vivo. J Neurosci 27(52):14231–14238
10. Calautti C, Baron JC (2003) Functional neuroimaging studies of motor recovery after stroke in adults: a review. Stroke 34(6): 1553–1566
11. Keck T, Mrsic-Flogel TD, Vaz AM, Eysel UT, Bonhoeffer T, Hubener M (2008) Massive restructuring of neuronal circuits during functional reorganization of adult visual cortex. Nat Neurosci 11(10):1162–1167
12. Ward NS (2005) Plasticity and the functional reorganization of the human brain. Int J Psychophysiol 58(2–3):158–161. https://doi.org/10.1016/j.ijpsycho.2005.02.009
13. Buonomano DV, Merzenich MM (1998) Cortical plasticity: from synapses to maps. Annu Rev Neurosci 21:149–186. https://doi.org/10.1146/annurev.neuro.21.1.149
14. Balan PF, Gerits A, Zhu Q, Kolster H, Orban GA, Wardak C, Vanduffel W (2018) Fast compensatory functional network changes caused by reversible inactivation of monkey parietal cortex. Cereb Cortex 29:2588. https://doi.org/10.1093/cercor/bhy128
15. Ward NS (2005) Neural plasticity and recovery of function. Prog Brain Res 150:527–535. https://doi.org/10.1016/S0079-6123(05)50036-0
16. Ekstrom LB, Roelfsema PR, Arsenault JT, Bonmassar G, Vanduffel W (2008) Bottom-up dependent gating of frontal signals in early visual cortex. Science 321(5887):414–417. https://doi.org/10.1126/science.1153276
17. Moeller S, Freiwald WA, Tsao DY (2008) Patches with links: a unified system for processing faces in the macaque temporal lobe. Science 320(5881):1355–1359. https://doi.org/10.1126/science.1157436
18. Murris SR, Arsenault JT, Vanduffel W (2020) Frequency- and state-dependent network effects of electrical stimulation targeting the ventral tegmental area in macaques. Cereb Cortex 30(8):4281–4296. https://doi.org/10.1093/cercor/bhaa007

19. Ferraina S, Pare M, Wurtz RH (2002) Comparison of cortico-cortical and cortico-collicular signals for the generation of saccadic eye movements. J Neurophysiol 87(2): 845–858. https://doi.org/10.1152/jn.00317.2001
20. Park HG, Carmel JB (2016) Selective manipulation of neural circuits. Neurotherapeutics 13(2):311–324. https://doi.org/10.1007/s13311-016-0425-7
21. Isa T (2022) Double viral vector intersectional approaches for pathway-selective manipulation of motor functions and compensatory mechanisms. Exp Neurol 349:113959. https://doi.org/10.1016/j.expneurol.2021.113959
22. Inoue K, Matsumoto M, Takada M (2021) Nonhuman primate Optogenetics: current status and future prospects. In: Yawo H (ed) Optogenetics, advances in experimental medicine and biology 1293, vol 2., 2015/09/22 edn. Springer, Singapore, pp 345–358. https://doi.org/10.1007/978-981-15-8763-4_22
23. Koshimizu Y, Isa K, Kobayashi K, Isa T (2021) Double viral vector technology for selective manipulation of neural pathways with higher level of efficiency and safety. Gene Ther 28(6): 339–350. https://doi.org/10.1038/s41434-020-00212-y
24. Sooksawate T, Isa K, Matsui R, Kato S, Kinoshita M, Kobayashi K, Watanabe D, Kobayashi K, Isa T (2013) Viral vector-mediated selective and reversible blockade of the pathway for visual orienting in mice. Front Neural Circuits 7:162. https://doi.org/10.3389/fncir.2013.00162
25. Kinoshita M, Matsui R, Kato S, Hasegawa T, Kasahara H, Isa K, Watakabe A, Yamamori T, Nishimura Y, Alstermark B, Watanabe D, Kobayashi K, Isa T (2012) Genetic dissection of the circuit for hand dexterity in primates. Nature 487(7406):235–238. https://doi.org/10.1038/nature11206
26. Oguchi M, Tanaka S, Pan X, Kikusui T, Moriya-Ito K, Kato S, Kobayashi K, Sakagami M (2021) Chemogenetic inactivation reveals the inhibitory control function of the prefronto-striatal pathway in the macaque brain. Commun Biol 4(1):1088. https://doi.org/10.1038/s42003-021-02623-y
27. Yazdan-Shahmorad A, Diaz-Botia C, Hanson TL, Kharazia V, Ledochowitsch P, Maharbiz MM, Sabes PN (2016) A large-scale interface for optogenetic stimulation and recording in nonhuman primates. Neuron 89(5):927–939. https://doi.org/10.1016/j.neuron.2016.01.013
28. Ruiz O, Lustig BR, Nassi JJ, Cetin A, Reynolds JH, Albright TD, Callaway EM, Stoner GR, Roe AW (2013) Optogenetics through windows on the brain in the nonhuman primate. J Neurophysiol 110(6):1455–1467. https://doi.org/10.1152/jn.00153.2013
29. Amita H, Kim HF, Inoue KI, Takada M, Hikosaka O (2020) Optogenetic manipulation of a value-coding pathway from the primate caudate tail facilitates saccadic gaze shift. Nat Commun 11(1):1876. https://doi.org/10.1038/s41467-020-15802-y
30. Tamura K, Takeda M, Setsuie R, Tsubota T, Hirabayashi T, Miyamoto K, Miyashita Y (2017) Conversion of object identity to object-general semantic value in the primate temporal cortex. Science 357(6352): 687–692. https://doi.org/10.1126/science.aan4800
31. Ohayon S, Tsao DY (2012) MR-guided stereotactic navigation. J Neurosci Methods 204(2): 389–397. https://doi.org/10.1016/j.jneumeth.2011.11.031
32. Murris SR, Arsenault JT, Raman R, Vogels R, Vanduffel W (2021) Electrical stimulation of the macaque ventral tegmental area drives category-selective learning without attention. Neuron 109(8):1381–1395 e1387. https://doi.org/10.1016/j.neuron.2021.02.013
33. Arsenault JT, Vanduffel W (2019) Ventral midbrain stimulation induces perceptual learning and cortical plasticity in primates. Nat Commun 10(1):3591. https://doi.org/10.1038/s41467-019-11527-9
34. Hadaczek P, Kohutnicka M, Krauze MT, Bringas J, Pivirotto P, Cunningham J, Bankiewicz K (2006) Convection-enhanced delivery of adeno-associated virus type 2 (AAV2) into the striatum and transport of AAV2 within monkey brain. Hum Gene Ther 17(3): 291–302. https://doi.org/10.1089/hum.2006.17.291
35. Fredericks JM, Dash KE, Jaskot EM, Bennett TW, Lerchner W, Dold G, Ide D, Cummins AC, Der Minassian VH, Turchi JN, Richmond BJ, Eldridge MAG (2020) Methods for mechanical delivery of viral vectors into rhesus monkey brain. J Neurosci Methods 339: 108730. https://doi.org/10.1016/j.jneumeth.2020.108730
36. Nassi JJ, Avery MC, Cetin AH, Roe AW, Reynolds JH (2015) Optogenetic activation of normalization in alert macaque visual cortex. Neuron 86(6):1504–1517. https://doi.org/10.1016/j.neuron.2015.05.040
37. Nagai Y, Kikuchi E, Lerchner W, Inoue KI, Ji B, Eldridge MA, Kaneko H, Kimura Y,

Oh-Nishi A, Hori Y, Kato Y, Hirabayashi T, Fujimoto A, Kumata K, Zhang MR, Aoki I, Suhara T, Higuchi M, Takada M, Richmond BJ, Minamimoto T (2016) PET imaging-guided chemogenetic silencing reveals a critical role of primate rostromedial caudate in reward evaluation. Nat Commun 7:13605. https://doi.org/10.1038/ncomms13605

38. Mimura K, Nagai Y, Inoue KI, Matsumoto J, Hori Y, Sato C, Kimura K, Okauchi T, Hirabayashi T, Nishijo H, Yahata N, Takada M, Suhara T, Higuchi M, Minamimoto T (2021) Chemogenetic activation of nigrostriatal dopamine neurons in freely moving common marmosets. iScience 24(9):103066. https://doi.org/10.1016/j.isci.2021.103066
39. Diester I, Kaufman MT, Mogri M, Pashaie R, Goo W, Yizhar O, Ramakrishnan C, Deisseroth K, Shenoy KV (2011) An optogenetic toolbox designed for primates. Nat Neurosci 14(3):387–397. https://doi.org/10.1038/nn.2749. nn.2749 [pii]
40. Han X, Qian X, Bernstein JG, Zhou HH, Franzesi GT, Stern P, Bronson RT, Graybiel AM, Desimone R, Boyden ES (2009) Millisecond-timescale optical control of neural dynamics in the nonhuman primate brain. Neuron 62(2):191–198. https://doi.org/10.1016/j.neuron.2009.03.011
41. Chow BY, Han X, Dobry AS, Qian X, Chuong AS, Li M, Henninger MA, Belfort GM, Lin Y, Monahan PE, Boyden ES (2010) High-performance genetically targetable optical neural silencing by light-driven proton pumps. Nature 463(7277):98–102
42. Cushnie AK, El-Nahal HG, Bohlen MO, May PJ, Basso MA, Grimaldi P, Wang MZ, de Velasco Ezequiel MF, Sommer MA, Heilbronner SR (2020) Using rAAV2-retro in rhesus macaques: promise and caveats for circuit manipulation. J Neurosci Methods 345:108859. https://doi.org/10.1016/j.jneumeth.2020.108859
43. Roth BL (2016) DREADDs for neuroscientists. Neuron 89(4):683–694. https://doi.org/10.1016/j.neuron.2016.01.040
44. Eldridge MA, Lerchner W, Saunders RC, Kaneko H, Krausz KW, Gonzalez FJ, Ji B, Higuchi M, Minamimoto T, Richmond BJ (2016) Chemogenetic disconnection of monkey orbitofrontal and rhinal cortex reversibly disrupts reward value. Nat Neurosci 19(1):37–39. https://doi.org/10.1038/nn.4192
45. Yan X, Telu S, Dick RM, Liow JS, Zanotti-Fregonara P, Morse CL, Manly LS, Gladding RL, Shrestha S, Lerchner W, Nagai Y, Minamimoto T, Zoghbi SS, Innis RB, Pike VW, Richmond BJ, Eldridge MA (2021) [(11)C]deschloroclozapine is an improved PET radioligand for quantifying a human muscarinic DREADD expressed in monkey brain. J Cereb Blood Flow Metab 41(10):2571–2582. https://doi.org/10.1177/0271678X211007949
46. Nagai Y, Miyakawa N, Takuwa H, Hori Y, Oyama K, Ji B, Takahashi M, Huang XP, Slocum ST, DiBerto JF, Xiong Y, Urushihata T, Hirabayashi T, Fujimoto A, Mimura K, English JG, Liu J, Inoue KI, Kumata K, Seki C, Ono M, Shimojo M, Zhang MR, Tomita Y, Nakahara J, Suhara T, Takada M, Higuchi M, Jin J, Roth BL, Minamimoto T (2020) Deschloroclozapine, a potent and selective chemogenetic actuator enables rapid neuronal and behavioral modulations in mice and monkeys. Nat Neurosci 23(9):1157–1167. https://doi.org/10.1038/s41593-020-0661-3
47. Hanes DP, Wurtz RH (2001) Interaction of the frontal eye field and superior colliculus for saccade generation. J Neurophysiol 85(2):804–815. https://doi.org/10.1152/jn.2001.85.2.804
48. Inoue KI, Takada M, Matsumoto M (2015) Neuronal and behavioural modulations by pathway-selective optogenetic stimulation of the primate oculomotor system. Nat Commun 6:8378. https://doi.org/10.1038/ncomms9378
49. Gerits A, Farivar R, Rosen BR, Wald LL, Boyden ES, Vanduffel W (2012) Optogenetically induced behavioral and functional network changes in primates. Curr Biol 22(18):1722–1726. https://doi.org/10.1016/j.cub.2012.07.023
50. Ohayon S, Grimaldi P, Schweers N, Tsao DY (2013) Saccade modulation by optical and electrical stimulation in the macaque frontal eye field. J Neurosci 33(42):16684–16697. https://doi.org/10.1523/JNEUROSCI.2675-13.2013
51. Nurminen L, Merlin S, Bijanzadeh M, Federer F, Angelucci A (2018) Top-down feedback controls spatial summation and response amplitude in primate visual cortex. Nat Commun 9(1):2281. https://doi.org/10.1038/s41467-018-04500-5
52. Galvan A, Hu X, Smith Y, Wichmann T (2016) Effects of optogenetic activation of corticothalamic terminals in the motor thalamus of awake monkeys. J Neurosci 36(12):3519–3530. https://doi.org/10.1523/JNEUROSCI.4363-15.2016
53. Kato S, Kobayashi K, Inoue K, Kuramochi M, Okada T, Yaginuma H, Morimoto K, Shimada T, Takada M, Kobayashi K (2011) A lentiviral strategy for highly efficient retrograde

gene transfer by pseudotyping with fusion envelope glycoprotein. Hum Gene Ther 22(2):197–206. https://doi.org/10.1089/hum.2009.179

54. Kato S, Kuramochi M, Takasumi K, Kobayashi K, Inoue K, Takahara D, Hitoshi S, Ikenaka K, Shimada T, Takada M, Kobayashi K (2011) Neuron-specific gene transfer through retrograde transport of lentiviral vector pseudotyped with a novel type of fusion envelope glycoprotein. Hum Gene Ther 22(12):1511–1523. https://doi.org/10.1089/hum.2011.111
55. Tanabe S, Uezono S, Tsuge H, Fujiwara M, Miwa M, Kato S, Nakamura K, Kobayashi K, Inoue KI, Takada M (2019) A note on retrograde gene transfer efficiency and inflammatory response of lentiviral vectors pseudotyped with FuG-E vs. FuG-B2 glycoproteins. Sci Rep 9(1):3567. https://doi.org/10.1038/s41598-019-39535-1
56. Bandettini PA, Kwong KK, Davis TL, Tootell RB, Wong EC, Fox PT, Belliveau JW, Weisskoff RM, Rosen BR (1997) Characterization of cerebral blood oxygenation and flow changes during prolonged brain activation. Hum Brain Mapp 5(2):93–109
57. Albaugh DL, Smith Y, Galvan A (2020) Comparative analyses of transgene expression patterns after intra-striatal injections of rAAV2-retro in rats and rhesus monkeys: a light and electron microscopic study. Eur J Neurosci 52(12):4824–4839. https://doi.org/10.1111/ejn.15027
58. Watakabe A, Ohtsuka M, Kinoshita M, Takaji M, Isa K, Mizukami H, Ozawa K, Isa T, Yamamori T (2015) Comparative analyses of adeno-associated viral vector serotypes 1, 2, 5, 8 and 9 in marmoset, mouse and macaque cerebral cortex. Neurosci Res 93:144–157. https://doi.org/10.1016/j.neures.2014.09.002
59. Gerits A, Vancraeyenest P, Vreysen S, Laramee ME, Michiels A, Gijsbers R, Van den Haute C, Moons L, Debyser Z, Baekelandt V, Arckens L, Vanduffel W (2015) Serotype-dependent transduction efficiencies of recombinant adeno-associated viral vectors in monkey neocortex. Neurophotonics 2(3):031209. https://doi.org/10.1117/1.NPh.2.3.031209
60. Tohyama T, Kinoshita M, Kobayashi K, Isa K, Watanabe D, Kobayashi K, Liu M, Isa T (2017) Contribution of propriospinal neurons to recovery of hand dexterity after corticospinal tract lesions in monkeys. Proc Natl Acad Sci U S A 114(3):604–609. https://doi.org/10.1073/pnas.1610787114
61. Kinoshita M, Kato R, Isa K, Kobayashi K, Kobayashi K, Onoe H, Isa T (2019) Dissecting the circuit for blindsight to reveal the critical role of pulvinar and superior colliculus. Nat Commun 10(1):135. https://doi.org/10.1038/s41467-018-08058-0
62. Ninomiya T, Noritake A, Kobayashi K, Isoda M (2020) A causal role for frontal cortico-cortical coordination in social action monitoring. Nat Commun 11(1):5233. https://doi.org/10.1038/s41467-020-19026-y
63. Vancraeyenest P, Arsenault JT, Li X, Zhu Q, Kobayashi K, Isa K, Isa T, Vanduffel W (2020) Selective mesoaccumbal pathway inactivation affects motivation but not reinforcement-based learning in macaques. Neuron 108(3):568–581 e566. https://doi.org/10.1016/j.neuron.2020.07.013
64. German DC, Manaye K, Smith WK, Woodward DJ, Saper CB (1989) Midbrain dopaminergic cell loss in Parkinson's disease: computer visualization. Ann Neurol 26(4):507–514. https://doi.org/10.1002/ana.410260403
65. Yetnikoff L, Lavezzi HN, Reichard RA, Zahm DS (2014) An update on the connections of the ventral mesencephalic dopaminergic complex. Neuroscience 282:23–48. https://doi.org/10.1016/j.neuroscience.2014.04.010
66. Day JJ, Carelli RM (2007) The nucleus accumbens and Pavlovian reward learning. Neuroscientist 13(2):148–159. https://doi.org/10.1177/1073858406295854
67. Robison AJ, Nestler EJ (2011) Transcriptional and epigenetic mechanisms of addiction. Nat Rev Neurosci 12(11):623–637. https://doi.org/10.1038/nrn3111
68. Scofield MD, Heinsbroek JA, Gipson CD, Kupchik YM, Spencer S, Smith AC, Roberts-Wolfe D, Kalivas PW (2016) The nucleus accumbens: mechanisms of addiction across drug classes reflect the importance of glutamate homeostasis. Pharmacol Rev 68(3):816–871. https://doi.org/10.1124/pr.116.012484
69. Ramaekers JG, Evers EA, Theunissen EL, Kuypers KP, Goulas A, Stiers P (2013) Methylphenidate reduces functional connectivity of nucleus accumbens in brain reward circuit. Psychopharmacology 229(2):219–226. https://doi.org/10.1007/s00213-013-3105-x
70. El-Shamayleh Y, Horwitz GD (2019) Primate optogenetics: progress and prognosis. Proc Natl Acad Sci U S A 116(52):26195–26203. https://doi.org/10.1073/pnas.1902284116
71. El-Shamayleh Y, Ni AM, Horwitz GD (2016) Strategies for targeting primate neural circuits

with viral vectors. J Neurophysiol 116(1): 122–134. https://doi.org/10.1152/jn.00087.2016

72. Stauffer WR, Lak A, Yang A, Borel M, Paulsen O, Boyden ES, Schultz W (2016) Dopamine neuron-specific optogenetic stimulation in rhesus macaques. Cell 166(6): 1564–1571 e1566. https://doi.org/10.1016/j.cell.2016.08.024
73. Zubair M, Murris SR, Isa K, Onoe H, Koshimizu Y, Kobayashi K, Vanduffel W, Isa T (2021) Divergent whole brain projections from the ventral midbrain in macaques. Cereb Cortex 31(6):2913–2931. https://doi.org/10.1093/cercor/bhaa399
74. Acker L, Pino EN, Boyden ES, Desimone R (2016) FEF inactivation with improved optogenetic methods. Proc Natl Acad Sci U S A 113(46):E7297–E7306. https://doi.org/10.1073/pnas.1610784113
75. Gerits A, Vanduffel W (2013) Optogenetics in primates: a shining future? Trends Genet 29(7):403–411. https://doi.org/10.1016/j.tig.2013.03.004

Chapter 11

Pathway-Specific Chemogenetic Manipulation by Applying Ligand to Axonally Expressed DREADDs

Maricela X. Martinez, Mitchell R. Farrell, and Stephen V. Mahler

Abstract

New tools for controlling neural populations such as opto- and chemogenetics afford unprecedented precision in pinpointing the neural circuit substrates of behavior. In addition to the ability of these tools to allow bidirectional control of genetically defined and anatomically defined cell populations, they also offer special advantages for interrogating the behavioral functions of *neural pathways*, setting them apart from more traditional approaches. In this chapter, we outline considerations for planning chemogenetic pathway-manipulation experiments and describe our protocol for using virally expressed designer receptors exclusively activated by designer drugs (DREADDs) and axonal application of clozapine-n-oxide (CNO) to modulate neural pathways in behaving rats.

Key words DREADDs, CNO, Pathway-specific, Axonal, Microinjection, Viral vectors

1 Introduction

1.1 Why Choose DREADDs Over an Optogenetic Approach?

Optogenetic approaches employ light-sensitive channels such as channel- or halorhodopsins that are expressed in the brain via viral vectors or other transgenic strategies [1]. When a suitable wavelength of light is applied during an experiment, the result is generally strong, reversible, and exceptionally rapid excitation or inhibition of neurons and pathways. In contrast, chemogenetic approaches like DREADDs involve inducing the expression of synthetic G-protein coupled receptors (GPCRs) which engage endogenous G-protein signaling mechanisms to control cell activity [2]. DREADDs do not bind endogenous neurotransmitters but instead, respond to an experimenter-administered drug that is ideally otherwise inert. DREADDs come in several varieties, including muscarinic-like excitatory (G_s-coupled rM3Ds or G_q-coupled hM3Dq) or inhibitory ($G_{i/o}$-coupled hM4Di) DREADDs, which respond to compounds that are structurally similar to clozapine

Mark A. G. Eldridge and Adriana Galvan (eds.), *Vectorology for Optogenetics and Chemogenetics*,
Neuromethods, vol. 195, https://doi.org/10.1007/978-1-0716-2918-5_11,

such as clozapine-n-oxide (CNO; [3], though other chemogenetic strategies with different mechanisms have also been described [4–8]).

DREADDs have a less precise temporal window of control than most optogenetic approaches due to their GPCR-based mechanism, and their activation by a ligand rather than by light. Yet DREADDs also afford certain advantages over conventional optogenetics. DREADDs can be less invasive since optogenetic approaches typically involve chronically implanted optic fibers coupled that require tethered light sources, thus restricting movement, and preventing use in some experimental apparatuses (though noninvasive transcranial or wireless light delivery methods are in development) [9–11]. In contrast, DREADDs require only application of the DREADD agonist drug prior to testing, thus allowing animals to freely move through even the most complex experimental testing chambers [12, 13]. DREADDs can also be very useful for manipulating large or irregularly shaped brain regions, since when the agonist is systemically administered, all DREADD receptors in the brain will likely be influenced, whereas optogenetic experiments are typically limited by the ability of nondestructive intensities of light to spread through the brain [12]. This may be of particular use for large, primate brains [14–21], facilitating the potential translational applications of DREADDs. In addition, by virtue of functioning via self-regulating and -limiting GPCR signaling cascades rather than ion channels [22, 23], DREADDs can be described as *biasing* cells toward activity or quiescence [12, 24], rather than imposing these states, as is usually the case with optogenetics. Therefore, chemogenetic platforms like DREADDs afford the ability to experimentally (and someday maybe therapeutically) stimulate or inhibit activity in select neurons or pathways, while presumably preserving the temporal dynamics endogenously generated by the brain (Fig. 1a). Perhaps coupled with the development of noninvasive, systemically administered viral vectors [25, 26 and *see* Chap. 7], we predict the application of actuatable systems like DREADDs may be useful in treating psychiatric disorders in the future, echoing similar suggestions by others [27]. However, much work is required to know where and how to target such systems to treat disease—this will first require a deeper understanding of how neural circuits modulate normal and abnormal behavioral states.

1.2 Targeting Neural Pathways with DREADDs

Chemogenetic approaches afford a key advantage over conventional behavioral neuroscience techniques—the ability to target neural pathways. This can be achieved in two main ways. The first is by using combinations of viral vectors traveling anterogradely or retrogradely, expressing Cre or Flp recombinase or related conditional expression systems—a powerful approach with many practical considerations that are discussed in detail in Chap. 10 of this volume.

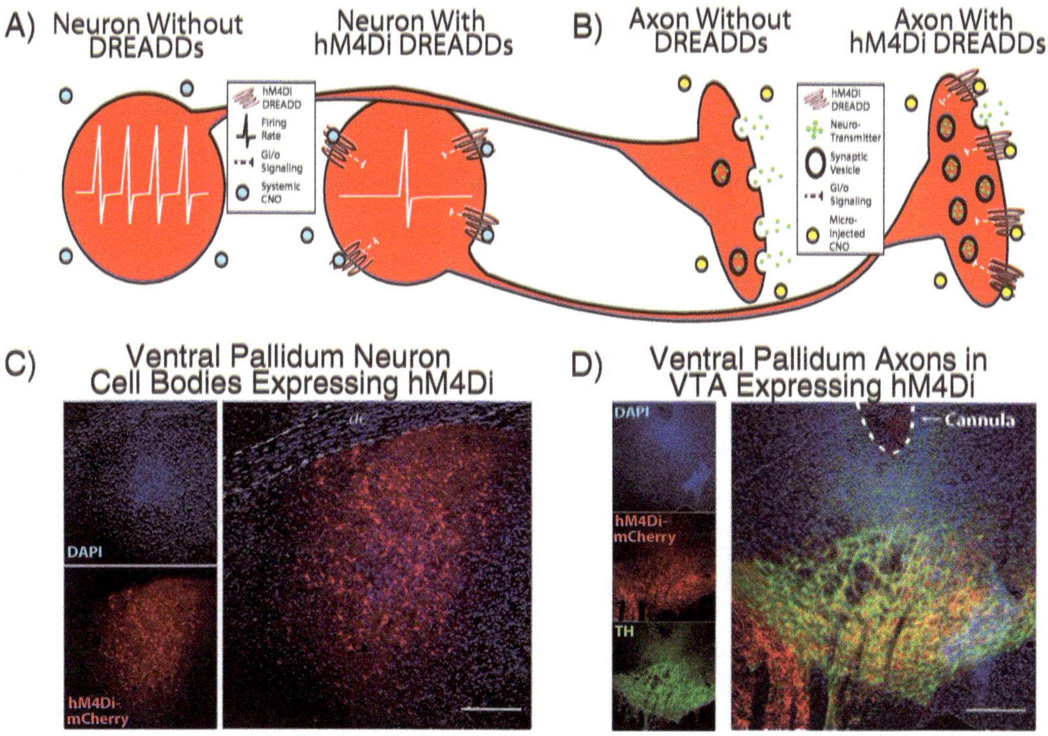

Fig. 1 DREADD actions on cell firing and neurotransmitter release. (**a**) Neurons infected with hM4Di, $G_{i/o}$-coupled inhibitory DREADDs exhibit a decrease in firing rate due to engagement of $G_{i/o}$ signaling when CNO is administered systemically. No such effects of systemic CNO are generally seen in neurons without DREADDs. (**b**) Microinjections of CNO directly upon axon terminals of hM4Di DREADD-expressing neurons modulate neurotransmitter release by engaging $G_{i/o}$ signaling presynaptically, allowing pathway-specific modulation. No such effects of microinjected CNO are generally seen in neurons without DREADDs. (**c**) Ventral pallidum neuron cell bodies expressing hM4Di DREADDs are shown, with blue DAPI stain showing cell nuclei, and the fluorescent red tag mCherry resulting from neuron-targeted viral vector injection. (AAV2-hSyn-hM4Di-mCherry). Merged image at right. *ac* anterior commissure. (**d**) DREADD-expressing axons (red) of ventral pallidum hM4Di-DREADD expressing neurons are shown in ventral tegmental area, in proximity to dopamine neurons (defined by tyrosine hydroxylase: TH staining in green). Damage from a cannula implanted above DREADD-expressing axons, allowing pathway-specific inhibition of this circuit with CNO microinjections, is visible at top. DAPI is shown in blue. Merged image at right. Scale Bars = 250 μm

Another way to target pathways with DREADDs involves microinjecting a DREADD agonist such as CNO directly upon axon terminals of DREADD-expressing neurons, thereby presynaptically modulating neurotransmitter release from the targeted neurons (Fig. 1b). This approach takes advantage of the fact that, after viral vector transduction, most neurons readily transport DREADDs (and opsins) to axonal processes, where they are inserted in presynaptic terminals. We were the first to show that applying CNO to axonally expressed $G_{i/o}$ DREADDs reduces the frequency of inhibitory postsynaptic currents [28]; specifically, in GABAergic projections from ventral pallidum (VP) to the ventral tegmental area (VTA; Fig. 1c, d). We also showed that intra-VTA CNO microinjection upon VP DREADD-expressing axons also suppressed cocaine-seeking behavior. Nearly concurrently with this report, Stachniak et al. [29] similarly demonstrated that axonally expressed inhibitory DREADDs act presynaptically via local application of CNO to inhibit pathways, using both electrophysiological and behavioral assays to confirm this. Notably, though Stachniak et al. developed an axon-selective DREADD ($hM4D^{nrxn}$), robust axonal expression is readily seen even with standard non-axon targeted DREADDs [28, 30–33].

More recently, we extended this work by showing that engaging excitatory G_q- or G_s-coupled DREADDs via axonal CNO application facilitates neurotransmitter release, highlighting the suitability of excitatory DREADD variants for pathway-specific *stimulation* of neural circuits [30, 34]. Specifically, we showed that G_q- or G_s-coupled DREADDs facilitate, while $G_{i/o}$-coupled DREADDs suppress, dopamine release from axon terminals of VTA dopamine neurons in nucleus accumbens [30]. Moreover, we found that microinjecting CNO upon G_q-coupled DREADD-expressing dopamine neuron axons in vivo causes distinct patterns of behavior depending on whether projections to medial prefrontal cortex, nucleus accumbens, or basolateral amygdala were stimulated [30]. Clearly, local microinjection of CNO upon DREADD-expressing neural compartments is a viable strategy for bidirectionally controlling anatomically defined neural pathways in vivo.

1.3 Disadvantages of Axonal DREADD Agonist Application to Control Pathways

No neuroscience method is perfect, and pathway control using DREADDs is no exception. One issue with this approach for use in behavioral experiments is the necessity of applying the DREADD agonist via intracranial microinjection. Repeated microinjections can cause damage to local tissue, so we have typically limited our experiments to a maximum of eight total microinjections [28, 30, 35–38]. In addition, microinjections can be stressful to animals (and researchers with unsteady hands!), so comparing CNO to vehicle microinjections is required to help control for this variable. Another consideration is the volume and concentration of the DREADD agonist to be microinjected. In general, larger injection volumes lead to greater spread [39, 40], which could be advantageous or disadvantageous, depending on experimental aims. We have used 1 mM CNO in 0.3 or 0.5 μL of 0.5% dimethylsulfoxide (DMSO) in artificial cerebrospinal fluid, which elicits robust, DREADD-specific effects when microinjected in vivo in several different pathways, independently validated across a number of labs in both mice and rats [28, 30–32, 41–47]. Regardless, since CNO in the brain can have DREADD-independent effects under certain conditions, potentially in part via local back metabolism to clozapine [48–51], comparing the effects of CNO in animals with or without DREADDs is essential for each new experiment.

Few studies have reported the use of DREADD agonists other than CNO (such as JHU60, C-21, or low-dose clozapine [18, 48, 52–54]) directly in the brain to manipulate neurotransmitter release, though we know of no reason why this would not be broadly effective. In fact, a recent study in nonhuman primates administered the DREADD agonist deschloroclozapine intracranially to inhibit medial prefrontal cortex projections to the lateral mediodorsal thalamus or the dorsal caudate nucleus, which revealed pathway-specific changes in working memory and decision-making [55]. Whether alternative agonists outperform CNO as a selective DREADD actuator when injected intracranially must be established in the future, perhaps using head-to-head comparisons of ligands in well-characterized behavioral tasks, neural firing, or other outcomes [56].

Next, we provide a detailed protocol for use of intracranial CNO upon DREADD-expressing axons to manipulate neural pathways in vivo.

2 Materials

2.1 Preoperative Supplies

1. *Ketamine* (MWI Animal Health; MWI 092209).
2. *Xylazine* (MWI Animal Health; MWI 510650).
3. *Hair Clippers* (Wahl Clipper Corp; Model MC3).
4. *Iodine Pads* (Fisher; C12400).

2.2 Viral Injection Surgery Supplies

5. *Stereotaxic Instrument* (KOPF; Model 942) with *Standard Holder* (KOPF; Model 1770) and *Cannula Holder* (KOPF; Model 1776-P1).
6. *Scalpel Handle* (Fine Science Tools; 91003-12) & *Scalpel Blades* (Integra Miltex; 4-111 no.11).
7. *Bulldog Clips* (Fine Science Tools; 18039-45 & 18038-45).
8. *0.9% Sodium Chloride* (ICU Medical; 07983-02).
9. *Cotton Tipped Wood Applicators Non-sterile 6″* (Dynarex; 4302).
10. *Black Food Color* (McCormick).
11. *Pipette Puller* (Narishige; Model # 1754-6).
12. *NanoFil Syringe* (World Precision Instruments; NANOFIL).
13. *Glass Pipettes* (Hirschmann; 9600105).
14. *Picospritzer III* (Parker; 052-0500-900).
15. *Hand Drill & Drill Bits* (CellPoint Scientific; Ideal Micro Drill, 1 mm).

2.3 Virus Loading Supplies

16. *Viral Vectors* (Addgene; e.g., for pan-neuronal expression: AAV2-hSyn-hM4D(Gi)-mCherry #50475; or for Cre-dependent expression: AAV2-hSyn-DIO-hM4D(Gi)-mCherry #44362).
17. *Tygon E-3603 Tubing* (Sigma Aldrich; Z765244-50FT).
18. *10 mL Syringe* (Fisher; BD309604)
19. *Vial Caps for Front Filling Virus* (DWK Life Sciences; 240,917).
20. *Parafilm* (Fisher; 13-374-12).

2.4 Cannula Implantation Supplies

21. *Guide Cannulae* (23 ga, Plastics One; C232G-2.0/SPC).
22. *Screwdriver* (Plastics One; 8J0026030X01).
23. *Skull Screws 0-80X1/8* (Fisher; NC1249270).
24. *Tweezer* (Fine Science Tools; 11052-10).
25. *Dental Cement and Solvent* (Pearson Dental; C73-00-70; C73-00-78).

2.5 Postoperative Supplies

26. *Sutures* (Esutures; D684N).
27. *Antibiotic Cream* (Globe Triple Antibiotic Ointment).
28. *Meloxicam 20 mL* (MWI Animal Health; 502,080).
29. *Cefazolin 10 mL* (MWI Animal Health; MW 008902).

2.6 Microinjection Supplies

30. *Stylets* (Plastics One; C232DC/SPC).
31. *Stainless Steel Wire* (Plastics One; 014BSH/30S).
32. *PE20 Tubing* (Instech; BPTE-20).
33. *1 μL Hamilton Syringe* (Hamilton; 80,100).
34. *Syringe Pump* (Thermo M365 Orion Sage Syringe Pump).
35. *Microinjector Tips* (28 ga; Plastics One; C232I/SPC; cut by manufactured to desired length).
36. *Clozapine N-oxide* (HelloBio; HB1807-100 mg).
37. *DMSO* (Sigma Aldrich; D5879-1 L).
38. *Artificial Cerebrospinal Fluid* (Harvard Apparatus; 59-7316).

3 Protocol

Our lab primarily studies addiction-related reward circuits, and has used DREADDs coupled with CNO microinjections to manipulate two main pathways during behavior in our work: VTA dopamine neuron projections to the forebrain (using excitatory and inhibitory DREADDs) [30, 31], and VP projections to VTA (using inhibitory

DREADDs; [28]). These general methods are more widely applicable however, and can be adapted for a wide range of pathways, as shown by our collaborators [32, 57].

3.1 Experimental Protocol for Targeting Pathways with Axon-Targeted DREADD Manipulations

We use adult male and female Long-Evans rats for such studies, pair- or trio-housed with ad libitum food and water in a reverse light-dark cycle colony room. In order to first stereotaxically target VTA or VP with a DREADD vector, rats are anesthetized with intraperitoneal injections of ketamine (65 mg/kg) and xylazine (8 mg/kg), and the fur from the top of the head is shaved, and iodine is applied to disinfect the area. After placing the rat securely into the frame of a stereotaxic instrument, an incision along the midsagittal line of the skull is made with a scalpel, then the skin is retracted using bulldog clamps. The skull is thoroughly cleaned with saline, and the skull sutures are lightly dyed with sterile food coloring to help visualize them. Ensure the skull is flat by comparing the Z coordinate of the skull at bregma and lambda, and adjust the head until these are level. Either microinjection syringes (e.g., Nanofil) or a glass pipette with a Picospritzer can be used to inject the virus.

Many labs use custom-length microinjection syringes (e.g., 23–33 ga; [58, 59]) mounted onto a stereotaxic arm to inject virus directly into the brain. This method minimizes problems with injector clogs, makes filling the injector easier, and allows use of automated injection pumps to push slowly and steadily the desired volume into the brain. However, this approach also may lead to some unwanted spread of the vector to sites dorsal to the targeted region, presumably due to back flow along the uniform gauge needle. To minimize this dorsal spread, we often use house-drawn glass pipettes instead, with a shank at least 2 mm longer than the depth of the targeted region (e.g., >10 mm for our lab's purposes). This is because we have observed that pipettes tend to restrict expression to a small, targeted region, presumably since the tip is very small (30–50 μM), and the increasing-diameter of the pipette dorsal to the tip helps limit dorsal spread of the vector. See Fig. 2 for a schematic of these dynamics. We have also found that slow injection rates, and leaving the injector in place for at least 5 min after injection are also useful for minimizing unwanted vector spread outside the targeted region.

If using a glass pipette to inject the vector, first use a stereotaxic-mounted reference glass pipette with a blunted tip to find bregma/lambda, then find the desired bregma-relative anterior-posterior (X) and medial-lateral (Y) coordinates for your region of interest. Drill a >1 mm wide hole above the intended injection site, removing bone fragments and debris that may clog the pipette, and breach the dura mater with a needle. When the skull holes above your injection site are ready, it is time to prepare your viral vector for injection.

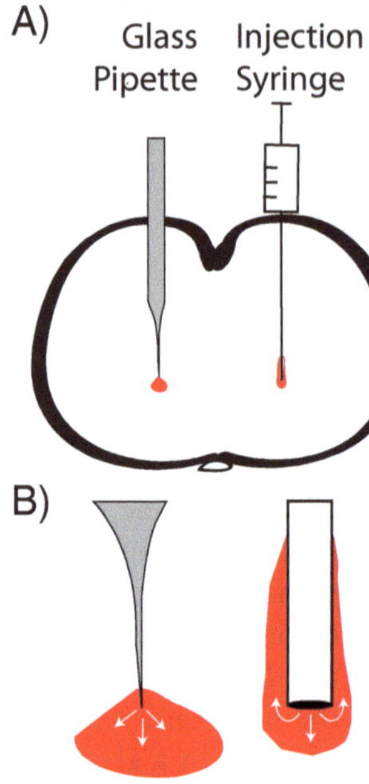

Fig. 2 Impact of injection system on viral spread in the brain. Glass pipette and injection syringe impact on dorsal spread of vector injection, based on our observations of DREADD expression patterns. (**a**) Glass pipettes (left) restrict virus expression to a small, targeted region, and limit dorsal spread due to tapered shank and small-diameter tip. Injection syringes in contrast allow less control of dorsal spread along the path of least resistance along the consistent diameter needle. (**b**) A closer view of these injection strategies is shown

Smaller injection volumes will achieve a more restricted spread. Note that different AAV serotypes will spread and express differently, and they may express quite differently in distinct brain regions and pathways, so this should always be tested with each batch of the vector before conducting behavioral experiments. AAVs should be thawed and aliquoted upon arrival from the supplier, stored at −80 °C, and kept frozen until the time of use. Once thawed, the AAV can be stored on ice for several hours at least [60], though we try to keep it at as low a temperature as possible without re-freezing.

If using a glass pipette (with a tip broken to an internal diameter of 30–100 μm), connect it with Tygon tubing to a 10 mL syringe, mount it on a manipulatable arm (we use the standard holder and cannula holder, **item 5** in the material list above), and position it above a well-lit stage such as a vial cap covered with clean Parafilm. Pipette an aliquot of a vector onto the stage and carefully maneuver the pipette tip into the drop of vector, then use the large 10 mL syringe to carefully create sufficient suction in the tube to front-fill the pipette with vector. Draw at least 200 nL greater volume than will be injected in the brain, as some vector will be lost or wasted in testing or transfer.

Connect the filled pipette to the picospritzer using Tygon tubing, and mount it on the surgical stereotaxic arm in the same position as the reference pipette, which it will replace. Drip saline into the drilled holes in the skull, until the liquid is flush with skull surface, then lower the pipette into the hole in the skull until it breaks the surface of the saline pool: this is the coordinate you will use as the skull-relative ventral (Z) coordinate. Lower the pipette into the brain to the desired coordinate at a brisk pace (for targeting VTA in adult rats we use bregma-relative coordinates (mm); AP: –5.5, ML: ±0.8, DV: −8.1, for VP we use: AP: 0.1, ML: ±2.0, DV: −8.0), then lower it another 100–200 μM, only to slowly retract it to the original position. This creates a "pocket" into which the virus can diffuse, which may help restrict/guide injection spread. We inject 0.2–1 μL of most AAV vectors (for VTA we normally inject 0.5–0.75 μL of a Cre-dependent vector in TH:Cre rats), depending on the viral serotype, size of the intended area of infection, injection method, and promoter system—again, testing your protocol in advance is essential for success! We prefer to inject most vectors slowly (e.g., <200 nL/min), with Picospritzer pressure kept under 40 psi, and pressure durations from 10 to 100 ms. Leave the injector in place as long as feasible after injection to allow diffusion from the injection site (at least 5 min). Ensure that you are actually delivering the virus, for example by watching the vector move down the pipette as the injection occurs. When the post-injection waiting period is complete, withdraw the pipette very slowly at first (e.g., 1 mm/min), then more quickly as the probe retreats from the injection site. If injecting bilaterally, test that the injector is still unclogged, verify that the skull surface is at a similar Z coordinate on both sides, then repeat the above steps. After surgery, suture the skin closed, and allow at least 4 weeks incubation time for expression and axonal transport of DREADDs before proceeding (though we sometimes start behaviorally training rats about a week after surgery while the virus is still incubating in experiments requiring weeks of training prior to test, such as drug self-administration/reinstatement).

3.1.1 Cannulae Implantation

For microinjecting CNO upon DREADD-expressing axons to manipulate specific neural pathways, intracranial cannulae must be implanted in their vicinity. This can occur either in the same surgery as viral transduction described above, or in a second surgery conducted closer to the desired test day. We have found that cannula placements are maximally accurate when they are implanted closer to the time of testing (e.g., 2–3 weeks before), compared to long before the test (e.g., >6 weeks). When planning your cannulae implantation surgery, you will want to carefully consider the trajectory employed to target the region of interest and avoid allowing cannulae to penetrate the ventricles, sinuses, or other regions of importance to the neural circuits under investigation. For example, we use a 6° lateral angle to target VTA, to avoid the midsagittal sinus.

After anesthetizing the animal as described above, mount the cannulae to be implanted on the stereotaxic arm. Find bregma or lambda in the X, Y, and Z axes using the tips of both cannulae, and move the cannulae to the desired X and Y coordinates relative to this landmark. Mark the skull with ink or a marker, and drill ~1.5–2 mm diameter holes at these sites. Next, move the stereotaxic arms aside, and create burr holes at least 5 mm away from cannulae holes, and not upon skull sutures, using a hand drill. Screw 3–4 jewelers' screws into these holes with a fine-tip tweezer and screwdriver. These screws will serve to anchor the dental cement "head cap" in place on the skull, holding the cannulae firmly in place. When screws are in place, lower the cannulae 1 or 2 mm shy of the desired Z coordinate (the microinjector you will later insert in the cannula will extend the extra 1 or 2 mm beyond the cannula into the target region). Repeat this procedure for the other cannula in the opposite hemisphere if bilateral manipulation is intended. When both cannulae are implanted, mix dental cement, and apply it carefully to the skull, making sure to encapsulate both the screws embedded in the skull, as well as the cannulae themselves. You may need to apply the cement in several stages, "building" up the cap to firmly grab both the cannulae as well as the screws—ensure that cement does not dry bonded to the skin or wound. When dried, remove the rat from the stereotaxic instrument, and add a suture behind or in front of the head cap to close the wound if needed. Place steel occluder stylets into the cannulae to close them, preventing foreign material from entering the brain. If rats are housed in groups, do not use occluding stylets with plastic tops, as cage mates will destroy them—occluding stylets made of metal may be a better choice to cover the plastic threads on the cannulae upon which cagemates like to chew. Alternatively, stainless steel wire can be cut to length, with a hook bent at the top to allow it to be removed for microinjections. Apply antibiotic cream to the wound, and treat the rat postoperatively with more cream until healed, pain relievers (e.g., meloxicam), and antibiotics (e.g., cefazolin) as required and approved by local animal care protocols. At least 5 weeks after viral injection surgery, when rats have recovered from cannula surgery (5+ days), behavioral testing and CNO microinjections may commence.

3.1.2 Microinjection Procedures

In preparation for microinjections, attach a gas or heat-sterilized intracranial microinjector onto the end of ~1 m of PE20 tubing, and fill the tubing/microinjector with sterile water. Ensure there are no bubbles in the tubing, then secure the free end onto a 1 μL Hamilton syringe located on a syringe pump. Wipe the microinjector (cut to extend 1 or 2 mm beyond the end of the cannulae into the targeted region) with an alcohol pad, and pull back the

Hamilton plunger ~0.05 μL to generate a small air bubble. Then, place the dry microinjector tip into the CNO (1 mM) or vehicle solution (0.5% DMSO in saline), and pull the Hamilton plunger to the 1 μL position. At this point, a small (~1–2 mm) bubble will be visible in the clear tubing directly above the top of the microinjector tip—the movement of this will serve as an indicator that the drug is flowing into the brain. The air bubble also separates the drug from the sterile water in the rest of the tubing. Gently handle the rat, and remove its steel occluder stylets with sharp forceps. Position microinjectors into each guide cannula, and switch on the syringe pump. Flow rates for drug injection should be tested with a pilot cohort of animals, but we generally use a 1 μL/100 s rate. Switch off the syringe pump upon reaching the appropriate injection volume according to the Hamilton syringe marking, followed by a 1 min waiting period during which microinjector tips remain in place in the brain to allow diffusion. Microinjector tips are then removed, followed by replacement of clean steel occluder stylets. Behavioral testing may commence 5 or 10 min after the microinjection finishes. Critically, since tissue damage accrues following repeated microinjection tests as mentioned above, counterbalancing vehicle and CNO doses for all tests is a necessity to limit the possibility of damage-related order effects.

Following all behavioral tests, histological verification of both DREADD vector injection and cannulation site are critical final steps for pathway-specific DREADD experiments. We perform immunofluorescent staining on every sixth 40 μm coronal section (sliced on a cryostat) to confirm the localization of the DREADD tag (e.g., mCherry, green fluorescent protein, hemagglutinin), and quantify the extent to which expression is specific to the brain region of interest. One may also examine known projection targets of these neurons, to verify axonally expressed DREADDs in pathways of interest. It is advisable to conduct immunohistochemistry for the tag to amplify its visibility, which is facilitated by cardiac perfusion of the animal with cold saline, then cold 4% paraformaldehyde, followed by cryoprotection in 20–30% sucrose with 0.5% sodium azide to preserve tissue prior to slicing (*see* Chap. 15). We then compare DREADD tag expression patterns to a rat brain atlas, to determine its localization relative to clear anatomical landmarks [61]. When possible, use a co-stain that provides objective borders for anatomical localization (in our case, substance P defines ventral pallidum borders, and tyrosine hydroxylase helps visualize VTA). For histological verification of cannula placement, we perform a Nissl stain on every third section of tissue containing cannula damage, comparing the tip location to landmarks in a rat brain atlas [61]. Careful observation of histological data from both viral expression and cannula sites is very important for understanding behavioral effects of DREADD manipulations, so we strongly suggest meticulous mapping of viral expression and microinjection sites in each behaviorally tested animal.

4 Conclusion

In sum, DREADDs afford excellent, repeatable, and precise control over targeted neural pathways, greatly facilitating investigations of the neural circuit substrates of behavior. Though all techniques have their disadvantages, including DREADDs, microinjecting a DREADD agonist drug in the vicinity of DREADD-expressing axons in vivo has great power for advancing our understanding of the neural circuit substrates of behavior across species.

References

1. Rajasethupathy P, Ferenczi E, Deisseroth K (2016) Targeting neural circuits. Cell 165(3): 524–534
2. Roth BL (2016) DREADDs for neuroscientists. Neuron 89(4):683–694
3. Guettier J-M et al (2009) A chemical-genetic approach to study G protein regulation of β cell function in vivo. Proc Natl Acad Sci 106(45): 19197–19202
4. Coward P et al (1998) Controlling signaling with a specifically designed Gi-coupled receptor. Proc Natl Acad Sci 95(1):352–357
5. Magnus CJ et al (2011) Chemical and genetic engineering of selective ion channel–ligand interactions. Science 333(6047):1292–1296
6. Vardy E et al (2015) A new DREADD facilitates the multiplexed chemogenetic interrogation of behavior. Neuron 86(4):936–946
7. Sternson SM, Roth BL (2014) Chemogenetic tools to interrogate brain functions. Annu Rev Neurosci 37(1):387–407
8. Campbell EJ, Marchant NJ (2018) The use of chemogenetics in behavioural neuroscience: receptor variants, targeting approaches and caveats. Br J Pharmacol 175(7):994–1003
9. Chen R et al (2021) Deep brain optogenetics without intracranial surgery. Nat Biotechnol 39(2):161–164
10. Kim T-I et al (2013) Injectable, cellular-scale optoelectronics with applications for wireless optogenetics. Science 340(6129):211–216
11. Gong X et al (2020) An ultra-sensitive step-function opsin for minimally invasive optogenetic stimulation in mice and macaques. Neuron 107(1):38–51. e8
12. Smith KS et al (2016) DREADDS: use and application in behavioral neuroscience. Behav Neurosci 130(2):137
13. Zhang F et al (2010) Optogenetic interrogation of neural circuits: technology for probing mammalian brain structures. Nat Protoc 5(3): 439–456
14. Nagai Y et al (2020) Deschloroclozapine, a potent and selective chemogenetic actuator enables rapid neuronal and behavioral modulations in mice and monkeys. Nat Neurosci 23(9):1157–1167
15. Eldridge MA et al (2016) Chemogenetic disconnection of monkey orbitofrontal and rhinal cortex reversibly disrupts reward value. Nat Neurosci 19(1):37–39
16. Grayson DS et al (2016) The rhesus monkey connectome predicts disrupted functional networks resulting from pharmacogenetic inactivation of the amygdala. Neuron 91(2): 453–466
17. Nagai Y et al (2016) PET imaging-guided chemogenetic silencing reveals a critical role of primate rostromedial caudate in reward evaluation. Nat Commun 7(1):1–8
18. Bonaventura J et al (2019) High-potency ligands for DREADD imaging and activation in rodents and monkeys. Nat Commun 10(1): 1–12
19. Upright NA et al (2018) Behavioral effect of chemogenetic inhibition is directly related to receptor transduction levels in rhesus monkeys. J Neurosci 38(37):7969–7975
20. Raper J et al (2019) Chemogenetic inhibition of the amygdala modulates emotional behavior expression in infant rhesus monkeys. Eneuro 6(5)
21. Galvan A, Caiola MJ, Albaugh DL (2018) Advances in optogenetic and chemogenetic methods to study brain circuits in non-human primates. J Neural Transm 125(3):547–563
22. Wess J, Nakajima K, Jain S (2013) Novel designer receptors to probe GPCR signaling and physiology. Trends Pharmacol Sci 34(7): 385–392

23. Dong S et al (2010) A chemical-genetic approach for precise spatio-temporal control of cellular signaling. Mol BioSyst 6(8): 1376–1380
24. Farrell MS, Roth BL (2013) Pharmacosynthetics: reimagining the pharmacogenetic approach. Brain Res 1511:6–20
25. Szablowski JO et al (2018) Acoustically targeted chemogenetics for the non-invasive control of neural circuits. Nat Biomed Eng 2(7): 475–484
26. Chan KY et al (2017) Engineered AAVs for efficient noninvasive gene delivery to the central and peripheral nervous systems. Nat Neurosci 20(8):1172–1179
27. Song J et al (2021) Chemogenetics as a neuromodulatory approach to treating neuropsychiatric diseases and disorders. Mol Ther
28. Mahler SV et al (2014) Designer receptors show role for ventral pallidum input to ventral tegmental area in cocaine seeking. Nat Neurosci 17(4):577–585
29. Stachniak TJ, Ghosh A, Sternson SM (2014) Chemogenetic synaptic silencing of neural circuits localizes a hypothalamus→ midbrain pathway for feeding behavior. Neuron 82(4): 797–808
30. Mahler SV et al (2019) Chemogenetic manipulations of ventral tegmental area dopamine neurons reveal multifaceted roles in cocaine abuse. J Neurosci 39(3):503–518
31. Halbout B et al (2019) Mesolimbic dopamine projections mediate cue-motivated reward seeking but not reward retrieval in rats. elife 8:e43551
32. Jayachandran M et al (2019) Prefrontal pathways provide top-down control of memory for sequences of events. Cell Rep 28(3): 640–654. e6
33. Mahler SV, Aston-Jones G (2018) CNO evil? considerations for the use of DREADDs in behavioral neuroscience. Neuropsychopharmacology 43(5):934–936
34. Brodnik ZD et al (2020) Chemogenetic manipulation of dopamine neurons dictates cocaine potency at distal dopamine transporters. J Neurosci 40(45):8767–8779
35. Mahler SV, Berridge KC (2009) Which cue to "want?" central amygdala opioid activation enhances and focuses incentive salience on a prepotent reward cue. J Neurosci 29(20): 6500–6513
36. Mahler SV, Berridge KC (2012) What and when to "want"? amygdala-based focusing of incentive salience upon sugar and sex. Psychopharmacology 221(3):407–426
37. Mahler SV, Smith KS, Berridge KC (2007) Endocannabinoid hedonic hotspot for sensory pleasure: anandamide in nucleus accumbens shell enhances 'liking' of a sweet reward. Neuropsychopharmacology 32(11):2267–2278
38. Mahler SV, Smith RJ, Aston-Jones G (2013) Interactions between VTA orexin and glutamate in cue-induced reinstatement of cocaine seeking in rats. Psychopharmacology 226(4): 687–698
39. Smith KS et al (2010) Hedonic hotspots: generating sensory pleasure in the brain. In: Pleasures of the brain. Oxford University Press, New York, NY, pp 27–49
40. Pecina S, Berridge KC (2000) Opioid site in nucleus accumbens shell mediates eating and hedonic 'liking' for food: map based on microinjection Fos plumes. Brain Res 863(1–2): 71–86
41. Pribiag H et al (2021) Ventral pallidum DRD3 potentiates a pallido-habenular circuit driving accumbal dopamine release and cocaine seeking. Neuron 109:2165
42. Lichtenberg NT et al (2017) Basolateral amygdala to orbitofrontal cortex projections enable cue-triggered reward expectations. J Neurosci 37(35):8374–8384
43. Ge F et al (2017) Glutamatergic projections from the entorhinal cortex to dorsal dentate gyrus mediate context-induced reinstatement of heroin seeking. Neuropsychopharmacology 42(9):1860–1870
44. McGlinchey EM, Aston-Jones G (2018) Dorsal hippocampus drives context-induced cocaine seeking via inputs to lateral septum. Neuropsychopharmacology 43(5):987–1000
45. Pardo-Garcia TR et al (2019) Ventral pallidum is the primary target for accumbens D1 projections driving cocaine seeking. J Neurosci 39(11):2041–2051
46. Hupalo S et al (2019) Prefrontal corticotropin-releasing factor (CRF) neurons act locally to modulate frontostriatal cognition and circuit function. J Neurosci 39(11):2080–2090
47. Venniro M et al (2017) The anterior insular cortex→ central amygdala glutamatergic pathway is critical to relapse after contingency management. Neuron 96(2):414–427. e8
48. Gomez JL et al (2017) Chemogenetics revealed: DREADD occupancy and activation via converted clozapine. Science 357(6350): 503–507
49. Jann M, Lam Y, Chang W (1994) Rapid formation of clozapine in guinea-pigs and man following clozapine-N-oxide administration. Arch Int Pharmacodyn Ther 328(2):243–250

50. MacLaren DA et al (2016) Clozapine N-oxide administration produces behavioral effects in Long–Evans rats: implications for designing DREADD experiments. eneuro 3(5)
51. Manvich DF et al (2018) The DREADD agonist clozapine N-oxide (CNO) is reverse-metabolized to clozapine and produces clozapine-like interoceptive stimulus effects in rats and mice. Sci Rep 8(1):3840
52. Weston M et al (2019) Olanzapine: a potent agonist at the hM4D (Gi) DREADD amenable to clinical translation of chemogenetics. Sci Adv 5(4):eaaw1567
53. Thompson KJ et al (2018) DREADD agonist 21 is an effective agonist for muscarinic-based DREADDs in vitro and in vivo. ACS Pharmacol Trans Sci 1(1):61–72
54. Chen X et al (2015) The first structure–activity relationship studies for designer receptors exclusively activated by designer drugs. ACS Chem Neurosci 6(3):476–484
55. Oyama K et al (2021) Chemogenetic dissection of the primate prefronto-subcortical pathways for working memory and decision-making. Sci Adv 7(26):eabg4246
56. Upright NA, Baxter MG (2020) Effect of chemogenetic actuator drugs on prefrontal cortex-dependent working memory in nonhuman primates. Neuropsychopharmacology 45(11): 1793–1798
57. McGlinchey EM et al (2016) Prelimbic to accumbens core pathway is recruited in a dopamine-dependent manner to drive cued reinstatement of cocaine seeking. J Neurosci 36(33):8700–8711
58. Doucette WT et al (2020) Chronic chemogenetic manipulation of ventral pallidum targeted neurons in rats fed an obesogenic diet. bioRxiv
59. Chang SE et al (2017) Optogenetic inhibition of ventral pallidum neurons impairs context-driven salt seeking. J Neurosci 37(23): 5670–5680
60. Rayaprolu V et al (2013) Comparative analysis of adeno-associated virus capsid stability and dynamics. J Virol 87(24):13150–13160
61. Paxinos G, Watson C (2006) The rat brain in stereotaxic coordinates: hard cover edition. Elsevier

Part III

Vector Delivery and Verification of Expression

Chapter 12

Convection Enhanced Delivery of Viral Vectors

Devon Griggs, Tiphaine Belloir, Jasmine Zhou, and Azadeh Yazdan-Shahmorad

Abstract

Convection-enhanced delivery (CED) is a pressure-based approach of delivering agents, such as viral vectors, throughout large regions of tissue. CED is chiefly characterized by injection volumes and flow rates that far exceed traditional diffusion-based approaches, which improves the feasibility of large-scale, targeted experimental and clinical research such as optogenetics and gene therapy. We present methods for MRI-guided CED of optogenetic viral vectors in cortical and deep brain structures of nonhuman primates as a case study of CED application in large animal models. We also include a discussion of important experimental materials and parameters, such as cannula designs and infusion flow rates, and a description of bench-side surgical planning approaches for CED.

Key words Convection enhanced delivery, CED, Optogenetics, Viral vector delivery, Gene therapy, Nonhuman primates, MRI

1 Introduction

Delivering therapeutic compounds into the brain has been an ongoing challenge for clinicians and scientists for many years. Traditional delivery methods such as intravenous administration are confronted with the blood-brain barrier (BBB) that restricts the passage of many substances, including viruses, from the systemic circulation to the brain parenchyma [1, 2].

A typical method to deliver viral vectors to the central nervous system (CNS) consists of mechanically breaching the BBB by direct injection of the vector into the region that contains the cells targeted for transduction. The number of transduced cells is limited by the titer of the vector, its efficiency of infection, and the volume of vector solution that can be delivered (for example, only 1–5 μL can be introduced into the mouse parenchyma by direct

Devon Griggs, Tiphaine Belloir and Jasmine Zhou contributed equally.

Mark A. G. Eldridge and Adriana Galvan (eds.), *Vectorology for Optogenetics and Chemogenetics*, Neuromethods, vol. 195, https://doi.org/10.1007/978-1-0716-2918-5_12,

injection [3]). The subsequent diffusion of the vector is usually limited to cells within a few millimeters surrounding the injection site [4]. To overcome this limited spread of viral vector solutions, convection-enhanced delivery (CED) has been widely used in recent years in different animal models and clinical trials [5] to deliver a variety of therapeutic compounds to the CNS. Originally developed in the early 1990s, this technique involves injecting therapeutic agents directly into the brain interstitial spaces (brain parenchyma) by creating a pressure gradient at the tip of a catheter. Unlike diffusion that operates along a concentration gradient requiring long time scales and high concentrations for injected therapeutic agents to drive their flow, the increased flow rate with CED results in a pressurized extracellular bulk flow which creates a more homogeneous distribution of agents, larger infusion volumes, and reduced infusion time [6].

In practice, a cannula is inserted into the interstitial spaces of the brain using stereotaxic guidance (or visual identification of anatomical structures), and a pump is used to create the pressure gradient that will drive the flow and infuse the desired agent. The technique comes with some physical limitations such as backflow [5]. Backflow, also known as reflux, occurs when the injected fluid travels back up the path of the cannula rather than penetrating the targeted tissue [7]. To limit this problem, Krauze et al. developed a reflux-resistant step cannula design [8]. This can effectively reduce reflux by placing fused silica tubing within the injection needle to create a perpendicular step that reduces the backflow of fluid. The optimization of more efficient cannula designs has also been reported elsewhere [9], and the choice of the appropriate cannula design for specific experimental needs will be further discussed in the Notes section of this chapter (*see* Subheading 4.2), along with other experimental considerations including flow rates and safety measures (*see* Subheading 4.8).

Another difficulty researchers face while implementing CED is the accuracy of cannula placement for delivery. While CED can be performed without real time monitoring such as magnetic resonance imaging (MRI), live MRI during CED is usually desired both in humans and large animal models to make the delivery methods more efficient by preventing off-target injections, and thus increasing the likelihood of surgical success [10–12]. Viral vectors can be administered along with an MRI tracer such as gadoteridol or gadopentetic acid (Gd-DTPA) to monitor optimal placement of the injection. This also has the advantages of identifying if any reflux is occurring (especially at higher flow rates) or if the injected sample is leaking, and to allow for immediate corrections of flow rate or needle placement if necessary. In nonhuman primates (NHPs) in particular, it has been shown that there is a direct correlation between gadoteridol distribution and virus transduction [13], suggesting that MR contrast agents of an appropriately calibrated concentration can approximate viral vector spread. CED was originally investigated for the treatment of tumors like gliomas because it offers several advantages including targeted delivery, a homogeneous distribution (as opposed to the Gaussian distribution observed in diffusion-based methods), and a high concentration of delivered molecules with lower undesired systemic absorption [5, 14, 15]. Since the distribution of the delivered agent is affected by hydrostatic pressure rather than diffusion, the molecular weight of the molecule is not usually a limiting factor, making CED applicable to a wide range of compounds including viral vectors for chemogenetics, optogenetics, and gene therapy in general.

In the early years, CED was used in gene therapy research to increase the distribution of adeno associated virus (AAV) in the brain of non-human primates [16]. Since then, CED has been used to deliver a variety of viral vectors including lentiviruses [17] and AAVs of different serotypes such as AAV2 [18]. In addition to large animal models such as NHPs and pigs [19, 20], CED has also been used in smaller animal models including rats [16, 21] and mice [22]. Specifically, Carty et al. demonstrated successful use of CED for increased transduction of the mouse CNS with different serotypes of recombinant AAV vectors. CED, including CED of viral vectors, has also been conducted in human brains experimentally [5]. However, results have been mixed (e.g., [5]) and further refinements are required before CED becomes standard clinical practice.

Most of the time, the injection directly targets the cells to be transduced, but in some cases, cells located in other regions distanced from the injection site can also be transduced by using anterograde or retrograde viral vectors. CED's broad coverage can facilitate transduction of downstream or upstream neurons based on anterograde or retrograde transport, depending on existing neuronal projections. For example, Kells et al. have previously shown that CED infusion of AAV2 vector into NHP thalamus results in efficient transduction of thalamocortical neurons, leading to widespread expression of the transgene in the cortex through axonal transport [23, 24].

This chapter describes in detail the methods for MRI-guided convection-enhanced delivery of optogenetic viral vectors into NHP brains through injections in either cortical [25] or deep structures such as the thalamus [24]. The protocol is divided into three major steps: first, the acquisition of presurgical MR images as a baseline reference to help plan for the injection location and trajectory, second, the implantation of a cannula array providing support and guidance to the cannula during the injection, and third, the injection of the viral vector to the targeted region. Key points of histological examination relevant to these methods are also discussed.

2 Materials

In this section, the materials needed to perform a CED injection procedure to achieve large-scale expression in NHP are listed. For our cortical CED procedure, two adult male rhesus macaques were involved: monkey G (8 yom, 17.5 kg) and monkey J (7 yom, 16.5 kg). Two female and one male adult rhesus macaques were involved in our thalamic CED procedure: monkey A (9 yof, 7.5 kg), monkey B (11 yof, 6.5 kg) and monkey H (8 yom, 8.7 kg).

2.1 Baseline Imaging

To allow for precise targeting and online monitoring, CED infusions are performed under MR guidance. Presurgical anatomical MR images are required as baseline reference to determine the optimal injection location and trajectory. Materials needed for the baseline MR imaging are the following: MR-compatible volatile anesthesia device (isoflurane), MR-compatible stereotaxic frame (KOPF, 1430 MRI), MR imaging software (Pixmeo, OsiriX MD 10.0), macaque brain atlas [26].

2.2 Cannula Array Implantation

The infusion system consists of a reflux-resistant cannula inserted into the brain with the help of a cannula array. This section lists the materials needed for the implantation of the cannula array, including materials needed to fabricate a custom-made cannula array and a reflux-resistant cannula.

2.2.1 Reflux-Resistant Cannula

The choice of the cannula type and design depends on the experimental needs and targeted brain region and is further discussed in the Notes section (*see* Subheading 4.2). For the cortical CED procedure, we fabricated a custom 1 mm stepped-tip reflux-resistant cannula using: 30 cm long silica tubing with 0.32 mm ID and 0.43 mm OD, 7.5 mm long silica tubing with 0.45 mm ID and 0.67 mm OD (Polymicro Technologies), cyanoacrylate, and a razor blade. For the thalamic CED procedure, a ceramic/fused silica reflux-resistant cannula with a 3 mm stepped tip can either be purchased (MRI Interventions, Irvine, CA) or custom-made using similar materials as the cortical version.

2.2.2 MR-Compatible Cannula Array and Chamber

The cannula array is used to provide guidance and stability for the cannula during the injection procedure. The cannula array design is to be adapted based on the cannula design itself and on the brain region targeted. For our cortical delivery procedure, we fabricated a custom MR-compatible cannula array designed to sit in a custom MR-compatible cranial chamber. Both the cannula array and the cranial chamber were made with the following materials: 3D design software (Inventor), a 3D printer (uPrinter SE 3D printer Stratasys, MN, USA) and acrylonitrile butadiene styrene (ABS) plastic filament (Stratasys, MN, USA). The cannula array also includes a nylon cannula injection grid that can be machined according to the steps described in Subheading 3.2.2.

2.2.3 Cannula Array Implantation

Materials needed for the cannula array implantation are as follows: a device for volatile anesthesia (isoflurane), MR-compatible stereotaxic frame (KOPF, 1430 MRI), MR imaging software (Pixmeo, OsiriX MD 10.0), macaque brain atlas [26], heating pad, clippers, cleaning solutions, saline solution (0.9% NaCl), set of surgical tools (forceps, needle drivers, scalpel, scissors), a surgical elevator (Langenbeck Elevator, Wide Tip,VWR International, LLC.), trephine (GerMedUSA Inc, SKU:GV70-42), suture, custom-made CED cannula array, plastic screws (Plastics 1 0-80 × 1/8 N), titanium screws, dental acrylic (Henry Schein, Inc. 1013117).

2.3 Viral Vector Delivery

A surgical procedure using standard aseptic technique is required for the MR-guided CED infusion delivering the viral vector to the targeted brain region.

Materials needed for this step are the following: Saline (0.9% NaCl), wet sterile absorbable gelatin sponge ("Gelfoam", Pfizer Inc.), sterile antimicrobial incise drape (3 M), vitamin E capsule (Pure Encapsulations, LLC.), MR-compatible volatile anesthesia device (isoflurane), MR-compatible stereotaxic frame (KOPF, 1430 MRI), MR imaging software (Pixmeo, OsiriX MD 10.0), contrast agent gadoteridol 2 mM Gd-DTPA (Prohance, Bracco Diagnostics, Princeton, NJ), 0.2 and 0.5 mL High-Pressure IV Tubing (Smiths Medical Inc., Dublin, OH, USA), MR-compatible pump (Harvard Apparatus, Holliston, MA, USA), MR-compatible 3 cc syringe (Harvard Apparatus, Holliston, MA, USA), Perifix Clamp Style Catheter Connector (B-Braun, Bethlehem, PA, USA), reflux-resistant cannula (*see* Subheading 3.2.1). Viral vector (*see* Subheading 4.4): for our cortical delivery procedure, we used AAV vectors with a CamKIIa promoter driving expression of C1V1 fused to EYFP, titered at 2.5×10^{12} virus molecules/mL (AAV2.5-CamKII-C1V1-EYFP, Penn Vector Core, University of Pennsylvania). For our thalamic delivery procedure, we used AAV vectors with CamKIIa promoter and ChR2 opsin (AAV2.2-CamKII-ChR2(H134R)-YFP), titer: 1.02×10^{13} virus molecules/mL and (AAV2.9-CamKII-ChR2(H134R)-YFP), titer: 5.26×10^{12} virus molecules/mL.

2.4 Verification of Expression with Immunochemistry

Materials needed for harvesting and sectioning the brain are the following: regular and heparinized phosphate-buffered saline (PBS), 4% paraformaldehyde in PBS, 30% sucrose solution, cryostat or microtome and cryoprotectant. For immunostaining: 3% hydrogen peroxide in PBS, 50% ethanol, normal donkey serum and Triton X-100 required for blocking, primary rabbit polyclonal anti-YFP antibody (Abcam, RRID: AB_303395 or similar), biotinylated donkey anti-rabbit antibody (Jackson Immunoresearch, RRID: AB_2340593 or similar), ExtrAvidin (Sigma-Aldrich), and diaminobenzidine chromogenic reagent (Sigma-Aldrich). For mounting the stained sections: gelatin-coated slides, xylene, coverslip, and D.P.X. mounting media (Sigma-Aldrich).

3 Methods

In this section, we are presenting the practical steps to perform an MRI-guided CED delivery of optogenetic viral vectors into the NHP brain. The technique can be used for injecting into both surface and deep regions as we have demonstrated in our previous work by successfully injecting into the cortex [25] and the thalamus [24] of rhesus macaques. The steps described here are based on these two examples: they present many similarities; however, we include information about their differences and how to adapt the technique to the experimenter's needs. We recommend that the experimenters test CED with relevant parameters and cannula designs in a gel model prior to surgical operations (*see* Subheading 4.3).

3.1 Baseline Imaging

Baseline anatomical MR images are needed to determine the optimal placement of the cannula array implant and to define the stereotaxic coordinates to guide the injection of the viral vector to the target site. The method to acquire baseline images is detailed hereafter.

- The animal is sedated and intubated and general anesthesia is maintained under isoflurane (concentration changed between 0% to 5% as needed, depending on vital signs such as respiratory rate and heart rate). The animal's temperature, heart rate, oxygen saturation, electrocardiographic responses, and end-tidal partial pressure of CO_2 are monitored throughout the procedure.
- The animal is placed in the MR-compatible stereotaxic frame, connected to the portable MR-compatible isoflurane machine and transported to the MRI scanner (3 T).
- Standard T1 (flip angle = 9°, repetition time/echo time = 668.6, matrix size = 192 × 192 × 80, slice thickness = 1 mm) and T2 (flip angle = 130°, repetition time/echo time = 52.5, matrix size = 256 × 256 × 45, slice thickness = 1 mm) anatomical MR images are acquired as baseline reference and for surgical planning.

The acquired T1 and T2 images are used to determine the placement of the craniotomy for the implantation of the cannula array. The location of the craniotomy can be precisely determined by matching the area of interest from MRI with the macaque brain atlas [26]. In addition, these baseline images can be used to define the trajectories to target the brain region for viral delivery in stereotaxic coordinates using, for example, Osirix software.

- After imaging, the animal is recovered from anesthesia and transported to the animal housing area.

3.2 Cannula Array Implantation

The viral vector is injected via a reflux-resistant cannula. (The steps needed to fabricate the cannula itself (Fig. 1a, b) are described in Subheading 3.2.1.). An MR-compatible cannula array providing access to the brain is needed to support the cannula during injection (Fig. 1). The following section describes the implantation of the cannula array (Fig. 1g, h) and the fabrication steps of its different components including the injection grid (Fig. 1c, d) and the cylinder (Fig. 1e, f) supporting the injection grid in the cranial chamber (Fig. 1e, f).

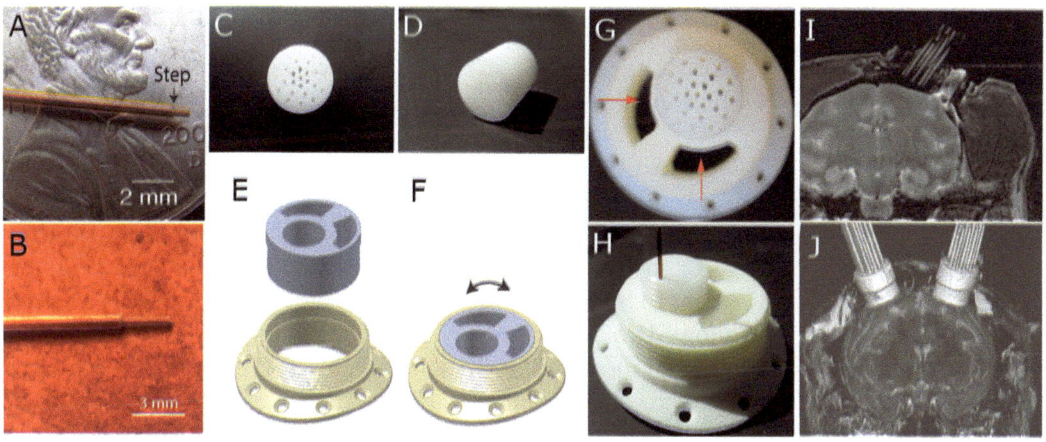

Fig. 1 Reflux-resistant cannula, MR-compatible cannula array and MR imaging. (**a**) Photograph of the custom-made 1 mm tipped reflux-resistant injection cannula used for our cortical delivery. (**b**) Photograph of the 3 mm tipped reflux-resistant cannula used for our thalamic delivery procedure (**c**, **d**) Custom-designed nylon injection grid. (**e**, **f**) Schematics of the MR-compatible cylinder that can be rotated within the cranial chamber for more flexible positioning of the cannula. (**g**) MR-compatible cannula array with fixed cylinder position. The arrows point to the cavities designed to be filled with wet sterile absorbable gelatin keeping the surface of the brain moist during the injection. (**h**) Cannula inserted in the injection grid. (**i**) MR image of the brain taken after cannula array implantation for the cortical injection. (**j**) MR image of the brain showing the implanted cannula arrays for a thalamic injection. (Modified from refs. [24, 25] with permission)

3.2.1 Reflux-Resistant Cannula Fabrication

The choice of the cannula type and design depends on the experimental needs and targeted brain region (*see* Subheading 4.2). For our thalamic delivery experiments, we used a ceramic/fused silica reflux-resistant cannula with a 3 mm stepped tip (MRI Interventions, Irvine, CA; Fig. 1b). For the cortical delivery, we designed and fabricated a 1 mm stepped-tip cannula (Fig. 1a) following the steps below:

- A 30 cm long silica tubing with 0.32 mm ID and 0.43 mm OD is cut for the inner silica tubing using a razor blade.
- Two pieces of different lengths of silica tubing with 0.45 mm ID and 0.67 mm OD are cut: one 7.5 cm long and the other 5 cm long. The 7.5 cm outer tubing is adhered to the 30 cm inner tubing with cyanoacrylate such that the inner tube extends 1 mm beyond the outer tube, making the reflux-resistant step. To ensure that the cyanoacrylate does not enter the inside of the inner tubing, the cyanoacrylate is placed on the outside of the inner tubing far from the cannula tip and the inner tubing is then slid inside the outer tubing. In practice, the glue may fuse the two tubes together with a stepped tip of inappropriate length—resolve this by trimming the tip to the appropriate length with a razor blade.
- The 5 cm long silica tubing is glued to the other end of the inner tubing for attachment to the clamp style catheter connector.

Test the cannula in a gel model before surgical use (*see* Subheading 4.3).

3.2.2 MR-Compatible Cannula Array and Cranial Chamber Fabrication

MR-compatible cannula arrays are needed to provide support to the cannula during the MR-guided CED infusion. For cortical delivery, we designed and fabricated a custom cannula array (Fig. 1g, h) which consisted of a cannula injection grid (Fig. 1c, d) meant to be inserted into a cylinder. The cylinder was either fixed or allowed to rotate within the MR-compatible cranial chamber (Fig. 1e, f) mounted on the animal's skull. The cranial chamber is designed to fit the skull curvature of the animal and 3D printed out of acrylonitrile butadiene styrene (ABSO plastic). The fabrication steps of the cortical CED-specific cannula array are described hereafter.

- The nylon cannula injection grid (Fig. 1c, d) is machined according to the following dimensions: the length and diameter of the injection grid are 15.00 and 12.00 mm, respectively. The injection grid pattern consists of 0.8 mm diameter holes forming three concentric circles (diameter to hole center: 1.6, 3.2, and 4.8 mm) around the center of the grid.

- The MR-compatible cylinder is 3D printed out of acrylonitrile butadiene styrene (ABS0 plastic) and allows for rotation of the cylinder, thus extending the range of the injection region (Fig. 1e, f).
- The cannula injection grid is threaded and the corresponding cavity of the cylinder is tapped such that both pieces exhibit the same threading. The nylon injection grid is inserted by screwing it into the tapped hole of the MR-compatible cylinder.

3.2.3 Cannula Array Implantation

The cannula array implantation location depends on the targeted injection sites. In our case, for cortical CED, the regions of interest were the motor cortex and the somatosensory cortex (M1 and S1). For the thalamic delivery, we centered the implant on the pre-planned trajectories defined in the baseline imaging section. The steps to implant the array are otherwise similar:

- The animal is sedated and intubated and general anesthesia is maintained with standard anesthetic monitoring as described in Subheading 3.1. The animal is then placed in the MR-compatible stereotaxic frame and connected to a portable MR-compatible isoflurane machine.
- A sagittal incision is made with a scalpel, approximately 2 cm from the midline with a length of about 5 cm. The underlying soft tissue is retracted from the skull using elevators.
- A circular craniotomy is created using a trephine, to provide access to the pre-planned trajectories for injections. The centering point of the trephine is lowered past the cutting edge of the trephine. An indentation is created at the center of the planned craniotomy sufficiently deep in the skull to anchor the trephine using the adjustable centering point. Caution should be exercised to avoid completely penetrating through the depth of the skull as this could cause damage to the underlying neural tissue. The area is periodically flushed with saline as needed to maintain tissue moisture and provide cooling throughout the craniotomy procedure. Once the center is made, the trephine is lowered onto the skull and rotated clockwise and counterclockwise while applying downward pressure until the bone cap can be removed with forceps. Caution should be taken to avoid damaging the underlying tissue with the trephine. The size of the craniotomy depends on the size of the array to be implanted. In our case of cortical CED, the craniotomy has a 25 mm diameter; for the thalamic CED, 15 mm.
- A fine suture (size 6-0) is threaded (surgical loupes can be used for more precision) through the dura in the center of the craniotomy and the dura is lifted by gently pulling the suture from the surface of the brain creating a tent in the center of the

craniotomy. The dura is punctured close to the center of the tent using fine ophthalmic scissors to avoid damaging the brain. Then the dura is cut from the center to the edge of the craniotomy and along the edge with fine ophthalmic scissors.

- The custom-made, MR-compatible cranial chamber is mounted to the skull on top of the craniotomy to provide cannula support during CED infusion. The chamber is secured to the skull such that the curvature of the chamber flange aligns well with the curvature of the skull. Plastic screws and dental acrylic or a few titanium screws are used to secure the chamber to the skull.
- The cannula array is then inserted into the cranial chamber.

3.3 Viral Vector Delivery

3.3.1 Preinfusion Preparation

Before starting the CED infusion process, anatomical MR images are obtained to guide the delivery of viral vectors to the target site (*see* Subheading 3.1 for details of MR sequence specifications)

- The cannula array implantation and viral vector delivery can be performed in either one or two surgeries. If the viral infusion surgery is scheduled after the animal fully recovers from the implantation, the animal's skin around the implant needs to be prepared using standard aseptic techniques prior to the following procedures.
- The animal needs to be kept in the MR-compatible stereotaxic frame throughout the viral vector delivery procedure. The implanted grid is filled with sterile saline (0.9% NaCl) for visualizing the injection trajectories with T2 MR images similar to that in Fig. 1i, j. Meanwhile, the chamber cavities are filled with a sterile gelatin solution to maintain the moisture in the brain.
- To maintain the sterility of the MR-compatible cylinder during animal transport from the preparation suite to the MRI scanner and during MR infusions, the cylinder and exposed skin need to be covered with a sterile drape. Then, the animal is connected to a portable MR-compatible isoflurane machine and transported to the MRI scanner.
- To measure the distance from the cannula injection grid to the surface of the brain during imaging, the top of the grid is marked with a vitamin E capsule which is visible in T1 images (not shown). Then, T1 and T2 MR images are acquired, which are used to determine the depth of injection from the top of the grid and the optimal cannula guides for reaching the target infusion site, respectively.

3.3.2 MR-Guided Infusion

Both cortical and thalamic CED infusions are accompanied by real-time MR imaging to guide the delivery and monitor the spread of viral vectors. For infusions without live MRI monitoring, *see* Subheading 4.5.

- The thawed viral vector is mixed with the MR contrast agent gadoteridol in a 250:1 ratio using a sterile pipette. The mixture is then vortexed and loaded into a 0.2 mL high-pressure IV tubing using the pipette.
- A sterile field is created outside the MR scanner, and a long extension line (3–5 m high-pressure IV tubing primed with saline) is connected to an MR-compatible 3 cc syringe. This syringe-tubing assembly is then connected to the 0.2 mL IV tubing previously loaded with the virus. Finally, the reflux-resistant cannula is connected to the distal end of this assembly with a clamp style catheter connector. Note that to account for the possible loss of viral vector mixture in the IV tubing and connectors (dead volume), we prepared twice the amount of virus solution needed in this MR-compatible setup.
- The syringe should be placed into an MR-compatible pump while the pump controller (not MR-compatible) is left in the scanner control room. The baseline anatomical MR images obtained previously can be used to select the cannula injection grid location and insertion depth needed to reach the target infusion site. (For comments on planning infusions deep in the brain, *see* Subheading 4.6). Specifically, T1 images can be used to calculate the distance between the top of the injection grid (marked earlier by the vitamin E capsule) and the cortical surface. Then, the target insertion depth is marked on the cannula using sterile tape.
- As the cannula is manually inserted (*see* Subheading 4.7 on cannula insertion and removal speed) through the cannula-guide array to the target depth labeled by the tape, a slow flow (0.5 μL/min) should be maintained in the infusion line so that the penetrated tissue will be less likely to clog the cannula needle during the insertion process. Once the insertion is completed and infusion begins, the flow rate should start at 1 μL/min and be gradually increased to 5 μL/min with 1 μL/min steps for both cortical and thalamic delivery (*see* Subheading 4.8 for comments on flow rates and safety).
- Once the infusion starts, flash T1 weighted images are taken every 2 min to achieve online monitoring of the viral vector infusion. Around 10 μL of viral vectors should be infused first, after which MR images are used to verify the correct cannula placement indicated by the spread of the contrast agent. The cannula should be adjusted within the grid when necessary to reach the correct infusion site and depth.
- For cortical delivery, the infusion rate should be reduced by 1 μL/min steps after infusing ~40 μL of the vector and stopped after ~50 μL are injected (*see* Fig. 2a). Multiple injections can be done at the same time if the placement of the cannula array

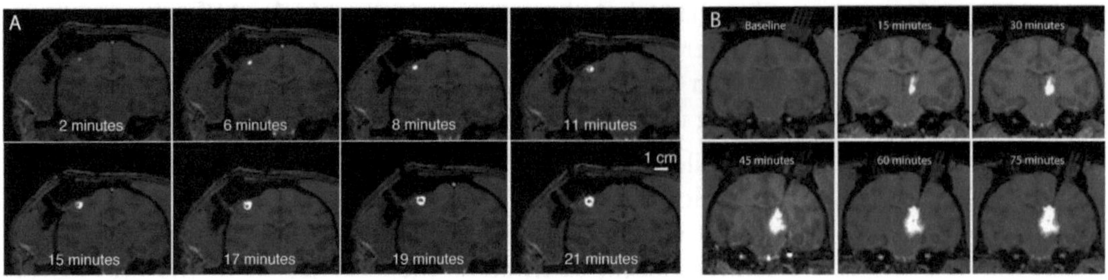

Fig. 2 MR imaging during infusion to monitor the distribution of viral vectors. (**a**) Temporal monitoring of the spread of 50 μL viral vectors during cortical infusion in S1 of monkey G. (**b**) Temporal monitoring of the spread of 246 μL viral vector during medial thalamic infusion in monkey H. (Modified from refs. [24, 25] with permission)

allows. For two of our animals that received thalamic CED, we infused the viral vector at the anterior and posterior thalami simultaneously. The infusion was monitored through online MR images and stopped when the entire thalamic area was covered (*see* Fig. 2b). Based on our animals, the infusion volume to the thalamus varied between 80 and 250 μL per site depending on the number of infusion locations and the size of the thalamus. The MR images should be continuously monitored for signs of reflux during the infusion process (*see* Subheading 4.5).

- After the infusion, the cannula is left in place for 10 min to prevent the backflow of viral vectors and then removed from the brain (*see* Subheading 4.7). The cylinder should be covered with a sterile drape at the end of injections before the animal is transported back to the operating room. Depending on the next experimental step, the cannula array can be left in place, explanted, or replaced with other implantable devices such as a titanium chamber and artificial dura for optogenetic experiments [25].

3.4 Verification of Expression with Immunohistochemistry

If the construct delivered via viral vectors includes a reporter gene (e.g., YFP, HA, etc.), histological analysis is performed through immunohistochemistry staining to confirm the intensity and location of reporter expression in the brain tissues.

- We allowed between 4 and 12 weeks for the virus to express in the brain tissue before validating optogenetic expression. For histological validation of viral vector expression at the end of all experiments, the animal is deeply anesthetized with pentobarbital or euthanasia solution, and then transcardially perfused with heparinized phosphate-buffered saline (PBS) and 4% paraformaldehyde (PFA) in PBS. The brain is then harvested and post-fixed by immersion in 4% PFA/PBS at 4 °C for 24 h. It is then sliced into 6 mm thick coronal blocks using a custom matrix and transferred into 30% sucrose for incubation. For sectioning, each brain block needs to be frozen and then cut into 50 μm thick sections on a cryostat to be stored at −20 °C in cryoprotectant or at 4 °C in PBS.

- For immunoperoxidase staining, the 50 μm sections are washed in PBS and incubated in 3% hydrogen peroxide/PBS for 10 min to quench endogenous peroxidase activity. Then, the sections are rinsed in 50% ethanol once and washed with PBS, after which they are blocked in 5% normal donkey serum in PBS for at least 1 h. After blocking, the sections are incubated in the primary rabbit polyclonal anti-YFP antibody mixture for 48 h at 4 °C. They are then rinsed with PBS and incubated in the biotinylated donkey anti-rabbit antibody at 4 °C for 12 h. Finally, the sections are rinsed and incubated in ExtrAvidin at room temperature for 6 h, and a diaminobenzidine (DAB) chromogenic reagent is added to detect the presence of peroxidase and thus mark the location of YFP expression. Sections are incubated with DAB for at least 10 min until desired stain intensity develops. The sections are then mounted onto gelatin-coated slides and imaged using a wide-field light microscope (Fig. 3c, g).
- For cortical CED, we saw that the YFP expression in S1 (Fig. 3c, e) matched the viral vector distribution observed from MR images (*see* Fig. 3b for an example). For thalamic CED, in addition to the strong YFP expression in the thalamus, layer-specific YFP-positive cells were observed at distinct cortical regions corresponding to their thalamocortical projections (Fig. 3f, g).

4 Notes

4.1 CED for Smaller Brains

A similar detailed protocol for implementing CED in rodent brains can be found elsewhere [27]. Both deep [28] and cortical [21] CED have been demonstrated in rat brains. Typically, these surgical procedures are carried out without MRI guidance and the cannula is directly attached to the syringe. Additionally, the implantation of a cannula array may not be necessary. For deep injection, flow rates up to 3 μL/min with a stepped-tip cannula were shown [28], which is similar to successful CED in large brains [24]. However, the cortical injection was limited to a flow rate of 0.6 μL/min, even with a stepped-tip cannula [21]. Of note, flow rates in the range of 0.4–1.0 μL/min tested in agarose gel produced less reflux when preceded by a flow rate of 0.25 μL/min, a technique that was used in vivo [21]. These rates still exhibit advantages in speed and volume of delivery over standard diffusion-based methods although they are up to roughly an order of magnitude lower than those which have been achieved in large brains.

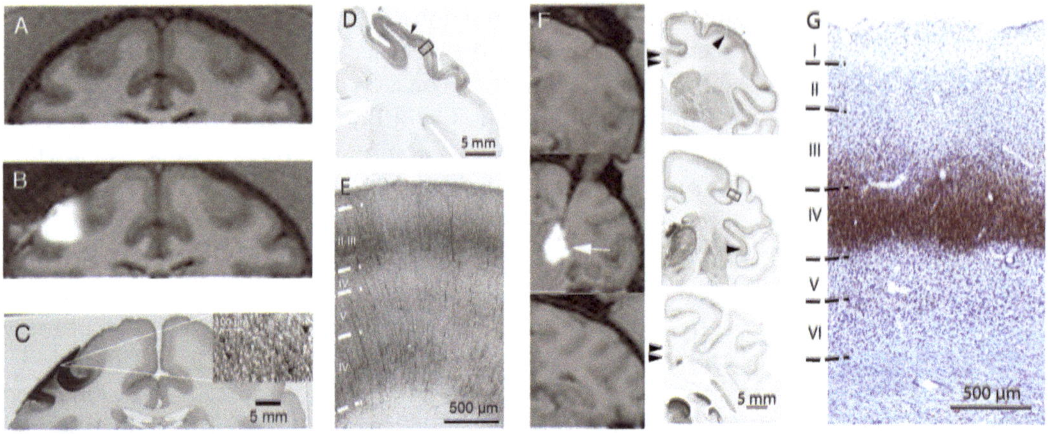

Fig. 3 Histological sections to verify cortical and thalamic CED expression. (**a**) Baseline coronal MR image of monkey G. (**b**) Spread of contrast agent after cortical infusion for the MR coronal slice in A. (**c**) A coronal tissue section stained with peroxidase reflecting expression of the EYFP-reporter, from approximately the same site in A and B. The inset shows stained tissue in higher magnification. (**d**) A coronal section stained with anti-YFP antibody showing the mediolateral aspect of YFP expression in S1 of monkey J. The black arrowhead indicates the location of the cannula track. The black frame is microscopically enlarged in E to show the laminar distribution of the YFP-positive cells. (**e**) YFP-positive cells are located predominantly in layers II–III and V–VI after cortical infusion. (**f**) Coronal sections to verify expression after thalamic infusion in monkey H. Row 1: single arrowheads point to prefrontal cortex; double arrowheads point to YFP in the cingulate cortex; Row 2: single arrowheads show YFP in the insular cortex. The black frame is microscopically enlarged in G to show the laminar distribution of the YFP-positive cells (including corticothalamic cell bodies and thalamocortical axon terminals); Row 3: double arrowheads point to YFP expression in retrosplenial and posterior cingulate cortex. While thalamic YFP expression matches the MRI distribution of viral vector, the distinct YFP expression in cortical areas produced by retrograde and anterograde transmission of the virus also corresponds to the known distribution of thalamocortical and corticothalamic projections. (**g**) Nissl and Immunoperoxidase co-staining that shows the layer-specific distribution of YFP-positive axons and cell bodies across layers in S1 of monkey H. (Modified from refs. [24, 25] with permission)

4.2 Cannula Designs

Experimenters should carefully consider which cannula design would be most appropriate for their specific experimental needs. The basic cannula design is the open port cannula, which is essentially a cylinder and is susceptible to reflux. To improve reflux resistance a number of designs have been proposed. We and others use a cannula with a stepped tip which is a good balance between cannula design complexity and reflux resistance [8, 21, 24, 25, 28, 29]. Another design is the porous hollow fiber cannula which has millions of nanoscale openings. Some special cases may call for a balloon-tipped catheter, where the catheter tip may be inflated to fill a resection cavity. An additional cannula variety is a multi-port cannula which has multiple openings along the cannula, but we recommend caution in selecting this version because it is common for fluid to favor the proximal openings of the cannula [30]. The designs mentioned above are discussed in depth elsewhere [9]. More recently, a recessed-step catheter has been proposed

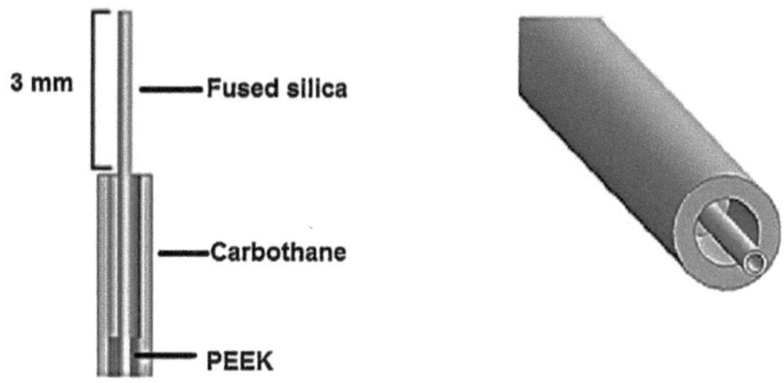

Fig. 4 Recessed-step catheter. A catheter designed to reduce reflux by incorporating a "recessed" step. (Modified from ref. [31] with permission)

(Fig. 4) [31] which shows promising improvements over the stepped-tip design, however, the technology has not been widely adopted [29].

4.3 Gel Models

We suggest that CED be tested bench-side in advance of surgery. Agarose, a clear hydrogel that approximates the fluidic and mechanical properties of the brain, is commonly used for modeling bench-side CED tests [7, 8, 21, 28, 29, 31–36]. Gel concentrations vary by publication, but we suggest 0.6% (w/v) [29, 32, 37]. For the dry component of the mixture, we differ from our peers by using agar powder (Benchmark Scientific, A1700, Sayreville, NJ, USA) and locust bean gum powder (Modernist Pantry LLC, Eliot, ME, USA) in a 4:1 ratio [29, 37] because we have found that injections into 0.6% gel produced from this mixture are more replicable than a gel made from agar powder alone (data not shown). To make the agarose, we mix the dry powder mixture into phosphate-buffered saline. Note that locust bean gum tends to clump and should be chopped finely before it is mixed into the phosphate-buffered saline. Then we heat the solution in a microwave oven until it begins to boil. At this point, we pause to swirl the solution for a few seconds, and then place it back in the microwave to repeat the process until the cumulative microwaving time has reached 2 min. After cooling briefly to produce minimal warping of our plastic 3D-printed molds, we pour the solution into the molds and allow them to set at room temperature for about 2 h. Further details on our agarose production method are published elsewhere [37]. Mold choices may differ depending on experimental aims. When infusion test runs with a specific location of the brain of a specific subject are needed, use a mold of the brain produced with our previously published method [37]. When CED parameters (e.g., flow rate, injection volume, and cannula design) require investigation, use molds which produce gel blocks with flat sides for imaging (e.g., [8, 29]). In some cases, a flat side can be added to a brain mold so as to capitalize on both strategies simultaneously

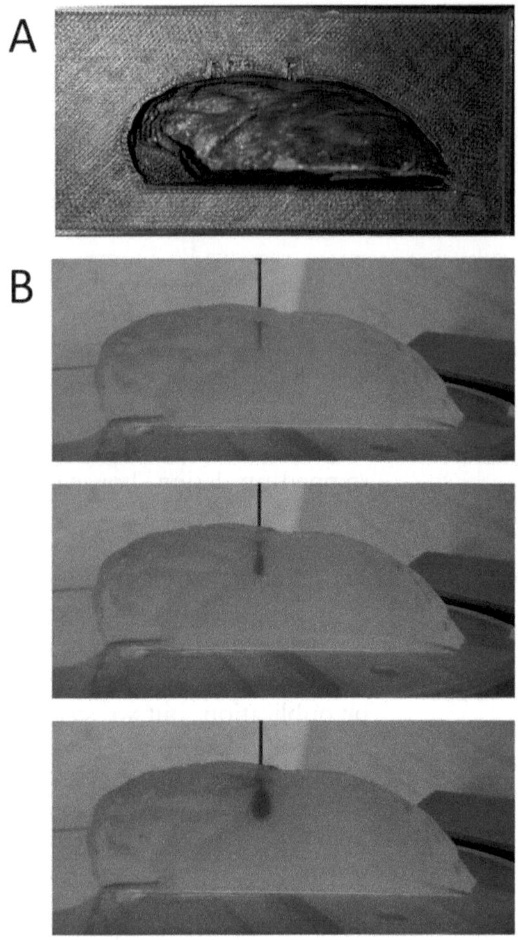

Fig. 5 Gel modeling. (**a**) 3D printed plastic mold of a partial NHP hemisphere. (**b**) CED infusion of yellow food coloring into gel at three different points in time. (Modified from ref. [37] with permission)

(Fig. 5). Trypan blue dye is commonly used to visualize the spread of injection through the gel [8, 28, 31, 34], although we use yellow food coloring as a non-toxic alternative [29, 37]. Note that injected dye may diffuse throughout the gel over time, so measurements of dye spread should be made during injection or shortly afterwards. This is of particular importance in cases where multiple injections in close proximity are needed (e.g., [25]): we still encourage experimentation with gel models but we caution that the first infusion may diffuse during the second infusion, making measurement of the injected region more difficult.

4.4 Viral Vector Selection

Selecting the most appropriate viral vector construct for a given experiment is a critical question that other chapters of this textbook discuss deeply (e.g., *see* Chapters 6 and 16). In this note, we chiefly aim to make the reader more acutely aware of two experimental factors which become more pertinent with CED. The first factor is the region of viral delivery. Because CED can spread viral vectors throughout large volumes of neural tissue, experimenters should carefully plan injection procedures so as to avoid off-target effects. In some cases, existing publications may contain methods for targeting specific brain regions in specific species (e.g., [28]). In other cases, experimenters may need to develop their own infusion protocols. Regardless, we recommend experimenting with gel-based infusion models (*see* Subheading 4.3) and, where appropriate, in vivo dye infusions to refine surgical techniques and thus reduce experimental risk during viral vector infusions. The second factor is more nuanced. Viral vectors injected into one region of the brain can, in some cases, transport antero- or retrogradely to distant regions of the brain and produce off-target effects. To illustrate this, consider adeno-associated viruses (AAVs), a family of viruses whose behavior varies by serotype. For example, in primates AAV6 exhibits predominantly retrograde transport [38] while AAV2 exhibits predominantly anterograde transport [23]. These phenomena make AAV serotypes powerful tools to investigate neural connections in the brain. In conjunction with CED, these properties enable experimenters to spread optogenetic expression across wide volumes of the brain. For example, we have shown that CED of optogenetic viral vectors in the thalamus of macaques can generate expression throughout large regions of the cortex [24]. However, with this experimental flexibility comes risk of error: researchers should plan experiments and surgical procedures carefully to meet their desired outcomes and avoid off-target effects.

4.5 Infusion Without Live MRI Monitoring

Live MRI monitoring of infusion is not always logistically available or feasible. In such cases, surgical materials and methods are largely similar to those described above, albeit somewhat simplified. Firstly, equipment does not need to be MRI-compatible. Additionally, one can eliminate the cannula array and some of the tubing because the animal does not need to be moved into and out of the MRI machine. Depending on the cannula design, the experimenters may be able to connect the cannula directly to the syringe mounted to the injection pump.

Lack of MRI monitoring does present the challenge of assessing the success of delivery; thus, we encourage the experimenters to closely watch the cortex near the cannula for any sign of reflux during the cannula insertion, infusion, and extraction process. If reflux is detected, we recommend terminating the infusion and, if it is experimentally feasible to do so, make a second infusion attempt at a nearby location. As an additional measure, tubing between the injection pump and the cannula can be partially filled with dark oil (located far from the viral vector being delivered) to track the progress of the fluid through the line. Alternatively, others have used food coloring to aid in tracking fluid flow [39]. This will help assess that the line is not clogged and that fluid is moving through the line during surgery. A simple calculation based on IV line inner diameter will help experimenters confirm the volume and flow rate of delivery. Additionally, experimenters may co-infuse the viral agent with a manganese-based MRI contrast agent which has a longer half-life than gadolinium-based contrast agents used in live MRI monitoring and is effective for post-operative imaging [29, 40] (*see* Chapter 13).

4.6 Targeting Deep Brain Structures

Baseline MRI data should be used for surgical planning of CED in large animal models. This is especially true when targeting deep structures to ensure that the cannula does not pass through ventricles or other brain structures of interest. For example, we inserted a cannula at an angle in the macaques to reach the thalamus without passing through any ventricles [24].

4.7 Cannula Insertion and Removal Speed

Insertions of needles with 0.7176 mm diameter were tested at different insertion speeds, 1.8, and 0.2 mm/s, in multiple tests. The data indicate that faster insertions reduce tissue damage and backflow [7]. We reason that tissue with less damage seals around the cannula better, and thus reduces backflow. We suggest that, as a general rule, insertions be performed quickly and smoothly. In contrast with cannula insertion speed, the optimal speed of cannula removal has not been researched to our knowledge. In our earlier work we removed the cannula slowly [41], however, we have tried quick and smooth manual withdrawals in a few in vivo cases and we have not observed issues.

4.8 Flow Rates and Safety

While CED can accommodate flow rates much higher than standard diffusion-based techniques, there remain practical limitations. The risk of reflux generally increases in likelihood and severity with increased flow rates [21], as illustrated by Fig. 6. One study found that with a stepped-tip cannula, a flow rate of up to 10 μL/min can be used [8]. Similarly, in our work, we have effectively used rates up to 5 μL/min with stepped-tip cannulas in NHPs [25]. In our cases, we increased the flow rate gradually from 1 to 5 μL/min in increments of 1 μL/min, in order to mitigate the potential negative effects of a sudden flow rate change. An additional parameter to consider is cannula diameter: in general, CED can be performed at higher flow rates without reflux when smaller cannula diameters are used [8]. Decisions about the flow rate for a particular experiment should be made in light of other practical considerations, such as the amount of time an animal can be anesthetized, the experimental tolerance for backflow, and data available from injection trials with gel models (*see* Subheading 4.3) or other in vivo experiments. Flow rates for smaller brains (e.g., rats) are discussed in *see* Subheading 4.1.

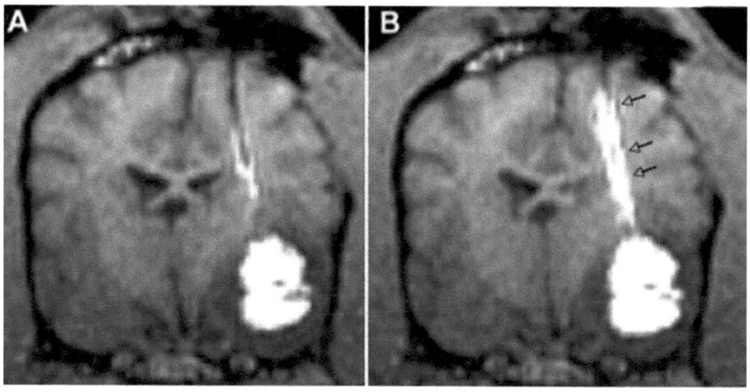

Fig. 6 MRI of CED with and without significant reflux. Low reflux along the path of the cannula at 5 μL/min (**a**) and high reflux at 8 μL/min (**b**) as indicated by arrows. (Modified from ref. [53] with permission)

5 Conclusion

In this chapter, we have discussed methods of MRI-guided CED in cortical and deep structures of the NHP brain as representative methods of CED in large animal models. CED is an efficient method for widespread and uniform viral vector transduction, which makes it a powerful tool for neuroscientific pursuits, especially when complemented with large-scale neuroengineering techniques. For example, we have combined CED of optogenetic viral vectors with large-scale imaging, electrophysiological recording, and optogenetic modulation [25, 42–44] for the purpose of studying stimulation-induced plasticity across large regions of the cortex [45–47]. Considering our recent developments of imaging techniques [48] and transparent neural interface technology in NHPs [49–51], along with a growing database of NHP optogenetic experiments [52], the prospective demand for large-scale viral infusion methods in translational research is high. More generally, CED equips researchers to study viral vector delivery-based gene therapies at large scales with greater efficiency. While CED is not yet widely adopted in the clinic, recent work suggests that CED is promising for future use in humans [5], especially when considered together with therapies that would benefit from large-scale infusions, e.g., treatment of brain tumors [9]. We expect that continued refinements in CED methods, such as improvements in cannula designs and utilization of MRI-guided infusions, will continue to increase the adoption of CED in neuroscience research and as a therapeutic tool.

Acknowledgments

This chapter was supported by the Eunice Kennedy Shriver National Institute of Child Health & Human Development of the National Institutes of Health (K12HD073945, A Y), the University of Washington Royalty Research Fund (A Y), the Washington National Primate Research Center (WaNPRC, P51 OD010425), the Center for Neurotechnology (CNT, a National Science Foundation Engineering Research Center, EEC-1028725, D G), the National Institute of Neurological Disorders and Stroke of the National Institute of Health (1R01NS116464-01, A Y, T B), and the Weill Neurohub (J Z).

References

1. Chen Y, Liu L (2012) Modern methods for delivery of drugs across the blood-brain barrier. Adv Drug Deliv Rev 64:640–665. https://doi.org/10.1016/j.addr.2011.11.010
2. Gosselet F, Loiola RA, Roig A et al (2021) Central nervous system delivery of molecules across the blood-brain barrier. Neurochem Int 144:104952. https://doi.org/10.1016/j.neuint.2020.104952
3. Davidson BL, Breakefield XO (2003) Viral vectors for gene delivery to the nervous system. Nat Rev Neurosci 4:353–364. https://doi.org/10.1038/nrn1104
4. Stanimirovic DB, Sandhu JK, Costain WJ (2018) Emerging technologies for delivery of biotherapeutics and gene therapy across the blood–brain barrier. BioDrugs 32:547–559. https://doi.org/10.1007/s40259-018-0309-y
5. Mehta AM, Sonabend AM, Bruce JN (2017) Convection-enhanced delivery. Neurotherapeutics 14:358–371. https://doi.org/10.1007/s13311-017-0520-4
6. Lonser RR, Sarntinoranont M, Morrison PF, Oldfield EH (2015) Convection-enhanced delivery to the central nervous system. J Neurosurg 122:697–706. https://doi.org/10.3171/2014.10.JNS14229
7. Casanova F, Carney PR, Sarntinoranont M (2012) Influence of needle insertion speed on backflow for convection-enhanced delivery. J Biomech Eng 134:1–8. https://doi.org/10.1115/1.4006404
8. Krauze MT, Saito R, Noble C et al (2005) Reflux-free cannula for convection-enhanced high-speed delivery of therapeutic agents. J Neurosurg 103:923–929. https://doi.org/10.3171/jns.2005.103.5.0923
9. Debinski W, Tatter SB (2009) Convection-enhanced delivery for the treatment of brain tumors. Expert Rev Neurother 9:1519–1527. https://doi.org/10.1586/ern.09.99
10. Mardor Y, Rahav O, Zauberman Y et al (2005) Convection-enhanced drug delivery: increased efficacy and magnetic resonance image monitoring. Cancer Res 65:6858–6863. https://doi.org/10.1158/0008-5472.CAN-05-0161
11. Salegio EA, Samaranch L, Kells AP et al (2012) Guided delivery of adeno-associated viral vectors into the primate brain. Adv Drug Deliv Rev 64:598–604. https://doi.org/10.1016/j.addr.2011.10.005
12. San Sebastian W, Richardson RM, Kells AP et al (2012) Safety and tolerability of magnetic resonance imaging-guided convection-enhanced delivery of AAV2-hAADC with a novel delivery platform in nonhuman primate striatum. Hum Gene Ther 23:210–217. https://doi.org/10.1089/hum.2011.162
13. Su X, Kells AP, Aguilar Salegio EA et al (2010) Real-time MR imaging with gadoteridol predicts distribution of transgenes after convection-enhanced delivery of AAV2 vectors. Mol Ther 18:1490–1495. https://doi.org/10.1038/mt.2010.114
14. Saito R, Tominaga T (2012) Convection-enhanced delivery: from mechanisms to clinical drug delivery for diseases of the central nervous system. Neurol Med Chir (Tokyo) 52:531–538. https://doi.org/10.2176/nmc.52.531
15. Hunt Bobo R, Laske DW, Akbasak A et al (1994) Convection-enhanced delivery of macromolecules in the brain. Proc Natl Acad Sci U S A 91:2076–2080. https://doi.org/10.1073/pnas.91.6.2076

16. Cunningham J, Oiwa Y, Nagy D et al (2000) Distribution of AAV-TK following intracranial convection-enhanced delivery into rats. Cell Transplant 9:585–594. https://doi.org/10.1177/096368970000900504
17. White E, Bienemann A, Megraw L et al (2012) Distribution properties of lentiviral vectors administered into the striatum by convection-enhanced delivery. Hum Gene Ther 23:115–127. https://doi.org/10.1089/hum.2010.185
18. Hadaczek P, Kohutnicka M, Krauze MT et al (2006) Convection-enhanced delivery of adeno-associated virus type 2 (AAV2) into the striatum and transport of AAV2 within monkey brain. Hum Gene Ther 17:291–302. https://doi.org/10.1089/hum.2006.17.291
19. Sonabend AM, Stuart RM, Yun J et al (2011) Prolonged intracerebral convection-enhanced delivery of topotecan with a subcutaneously implantable infusion pump. Neuro-Oncology 13:886–893. https://doi.org/10.1093/neuonc/nor051
20. Pleticha J, Maus TP, Christner JA et al (2014) Minimally invasive convection-enhanced delivery of biologics into dorsal root ganglia: validation in the pig model and prospective modeling in humans: technical note. J Neurosurg 121:851–858. https://doi.org/10.3171/2014.6.JNS132364
21. Yu Z, Nurmikko A, Ozden I (2017) Wide-spread functional opsin transduction in the rat cortex via convection-enhanced delivery optimized for horizontal spread. J Neurosci Methods 291:69–82. https://doi.org/10.1016/j.jneumeth.2017.08.008
22. Carty N, Lee D, Dickey C et al (2010) Convection-enhanced delivery and systemic mannitol increase gene product distribution of AAV vectors 5, 8, and 9 and increase gene product in the adult mouse brain. J Neurosci Methods 194:144–153. https://doi.org/10.1016/j.jneumeth.2010.10.010
23. Kells AP, Hadaczek P, Yin D et al (2009) Efficient gene therapy-based method for the delivery of therapeutics to primate cortex. Proc Natl Acad Sci U S A 106:2407–2411. https://doi.org/10.1073/pnas.0810682106
24. Yazdan-Shahmorad A, Tian N, Kharazia V et al (2018) Widespread optogenetic expression in macaque cortex obtained with MR-guided, convection enhanced delivery (CED) of AAV vector to the thalamus. J Neurosci Methods 293:347–358. https://doi.org/10.1016/j.jneumeth.2017.10.009
25. Yazdan Shahmorad A, Diaz-Botia C, Hanson TL et al (2016) A large-scale Interface for Optogenetic stimulation and recording in nonhuman primates. Neuron 89:927–939. https://doi.org/10.1016/j.neuron.2016.01.013
26. Paxinos G, Huang X, Petrides M et al (2008) The rhesus monkey brain in stereotaxic coordinates, 2nd edn. Academic Press, San Diego
27. Nash KR, Gordon MN (2016) Convection enhanced delivery of recombinant adeno-associated virus into the mouse brain. Methods Mol Biol 1382:285–295. https://doi.org/10.1007/978-1-4939-3271-9_21
28. Yin D, Forsayeth J, Bankiewicz KS (2010) Optimized cannula design and placement for convection-enhanced delivery in rat striatum. J Neurosci Methods 187:46–51. https://doi.org/10.1016/j.jneumeth.2009.12.008
29. Griggs DJ, Garcia AD, Au WY, Ojemann WKS, Johnson AG, Ting JT, Buffalo EA, Yazdan-Shahmorad A (2022) Improving the efficacy and accessibility of intracranial viral vector delivery in non-human primates. Pharmaceutics 14(7) 1435. https://doi.org/10.3390/pharmaceutics14071435
30. Raghavan R, Brady ML, Rodríguez-Ponce MI et al (2006) Convection-enhanced delivery of therapeutics for brain disease, and its optimization. Neurosurg Focus 20:E12. https://doi.org/10.3171/foc.2006.20.4.7
31. Gill T, Barua NU, Woolley M et al (2013) In vitro and in vivo testing of a novel recessed-step catheter for reflux-free convection-enhanced drug delivery to the brain. J Neurosci Methods 219:1–9. https://doi.org/10.1016/j.jneumeth.2013.06.008
32. Pervin F, Chen WW (2011) Mechanically similar gel simulants for brain tissues. In: Proulx T (ed) Dynamic behavior of materials, Conference Proceedings of the Society for Experimental Mechanics Series, vol 1. Springer, New York, pp 9–13
33. Chen ZJ, Gillies GT, Broaddus WC et al (2004) A realistic brain tissue phantom for intraparenchymal infusion studies. J Neurosurg 101:314–322. https://doi.org/10.3171/jns.2004.101.2.0314
34. Linninger AA, Somayaji MR, Mekarski M, Zhang L (2008) Prediction of convection-enhanced drug delivery to the human brain. J Theor Biol 250:125–138. https://doi.org/10.1016/j.jtbi.2007.09.009
35. Orozco GA, Córdoba G, Urrea F et al (2021) Finite element model to reproduce the effect of pre-stress and needle insertion velocity during infusions into brain phantom gel. Irbm 42:180–188. https://doi.org/10.1016/j.irbm.2020.04.005

36. Urrea FA, Casanova F, Orozco GA, García JJ (2016) Evaluation of the friction coefficient, the radial stress, and the damage work during needle insertions into agarose gels. J Mech Behav Biomed Mater 56:98–105. https://doi.org/10.1016/j.jmbbm.2015.11.024
37. Ojemann WKS, Griggs DJ, Ip Z et al (2020) A mri-based toolbox for neurosurgical planning in nonhuman primates. J Vis Exp 2020:1–16. https://doi.org/10.3791/61098
38. San Sebastian W, Samaranch L, Heller G et al (2013) Adeno-associated virus type 6 is retrogradely transported in the non-human primate brain. Gene Ther 20:1178–1183. https://doi.org/10.1038/gt.2013.48
39. Diester I, Kaufman MT, Mogri M et al (2011) An optogenetic toolbox designed for primates. Nat Neurosci 14:387–397. https://doi.org/10.1038/nn.2749
40. Fredericks JM, Dash KE, Jaskot EM et al (2020) Methods for mechanical delivery of viral vectors into rhesus monkey brain. J Neurosci Methods 339:108730. https://doi.org/10.1016/j.jneumeth.2020.108730
41. Khateeb K, Griggs DJ, Sabes PN, Yazdan-Shahmorad A (2019) Convection enhanced delivery of optogenetic adeno-associated viral vector to the cortex of rhesus macaque under guidance of online MRI images. J Vis Exp 147: 1–8. https://doi.org/10.3791/59232
42. Yazdan-Shahmorad A, Diaz-Botia C, Hanson T et al (2015) Demonstration of a setup for chronic optogenetic stimulation and recording across cortical areas in non-human primates. SPIE BiOS 9305:93052K. https://doi.org/10.1117/12.2080405
43. Griggs DJ, Belloir T, Yazdan-Shahmorad A (2021) Large-scale neural interfaces for optogenetic actuators and sensors in non-human primates. SPIE BiOS 1166305:17. https://doi.org/10.1117/12.2579431
44. Ledochowitsch P, Yazdan-Shahmorad A, Bouchard KE et al (2015) Strategies for optical control and simultaneous electrical readout of extended cortical circuits. J Neurosci Methods 256:220–231. https://doi.org/10.1016/j.jneumeth.2015.07.028
45. Yazdan-Shahmorad A, Silversmith DB, Kharazia V, Sabes PN (2018) Targeted cortical reorganization using optogenetics in non-human primates. Elife 7:1–21. https://doi.org/10.7554/eLife.31034
46. Bloch JA, Khateeb K, Silversmith DB et al (2019) Cortical stimulation induces network-wide coherence change in non-human primate somatosensory cortex*. Annu Int Conf IEEE Eng Med Biol Soc 2019:6446–6449. https://doi.org/10.1109/EMBC.2019.8856633
47. Bloch J, Greaves-Tunnell A, Shea-Brown E, Harchaoui Z, Shojaie A, Yazdan-Shahmorad A (2022) Network structure mediates functional reorganization induced by optogenetic stimulation of non-human primate sensorimotor cortex. iScience 25(5) 104285. https://doi.org/10.1016/j.isci.2022.104285
48. Macknik SL, Alexander RG, Caballero O et al (2019) Advanced circuit and cellular imaging methods in nonhuman primates. J Neurosci 39:8267–8274. https://doi.org/10.1523/JNEUROSCI.1168-19.2019
49. Griggs DJ, Khateeb K, Philips S et al (2019) Optimized large-scale optogenetic interface for non-human primates. SPIE BiOS 1086605:3. https://doi.org/10.1117/12.2511317
50. Griggs DJ, Khateeb K, Zhou J et al (2021) Multi-modal artificial dura for simultaneous large-scale optical access and large-scale electrophysiology in non-human primate cortex. J Neural Eng 18:055006. https://doi.org/10.1101/2021.02.03.429596
51. Griggs DJ, Bloch J, Fisher S, Ojemann WKS, Coubrough KM, Khateeb K, Chu M, Yazdan-Shahmorad A (2022) Demonstration of an optimized large-scale optogenetic cortical interface for non-human primates. 44th Annual International Conference of the IEEE Engineering in Medicine & Biology Society (EMBC) 3081–3084. https://doi.org/10.1109/EMBC48229.2022.9871332
52. Tremblay S, Acker L, Afraz A et al (2020) An open resource for non-human primate optogenetics. Neuron 108:1–16. https://doi.org/10.1016/j.neuron.2020.09.027
53. Fiandaca MS, Forsayeth JR, Dickinson PJ, Bankiewicz KS (2008) Image-guided convection-enhanced delivery platform in the treatment of neurological diseases. Neurotherapeutics 5:123–127. https://doi.org/10.1016/j.nurt.2007.10.064

Chapter 13

Multichannel Microinjector Arrays for Efficient Viral Vector Delivery into Rhesus Monkey Brain

Nanami L. Miyazaki, Bing Li, and Mark A. G. Eldridge

Abstract

Effective delivery of viral vectors to target brain regions is critical for using genetically encoded molecular tools such as chemogenetics, optogenetics, and fluorescent indicators. Compared to rodent models, applications of genetically encoded tools in nonhuman primates have progressed slowly due to challenges in injecting larger volumes of brain tissue and the limited availability of transgenic lines. Stereotaxic and hand-held injections have been used to infuse viral constructs into the nonhuman primate brain, but at the cost of speed and dense expression, respectively. Here, we present a novel injection method using custom multichannel microinjector arrays to deliver viral vectors efficiently and with dense expression into cortical regions of rhesus macaques. We describe in detail the methods for building and implementing three types of multichannel microinjector arrays. In addition, we describe the methods for co-infusing manganese for postoperative validation of injection targeting using MRI.

Key words Viral vector delivery, Chemogenetics, Optogenetics, Nonhuman primates

1 Introduction

Many applications of genetically encoded molecular tools—chemogenetics, optogenetics, fluorescent indicators—depend on the precise delivery of viral constructs to target brain regions. Methods for intracranial delivery of viral vectors and drugs through stereotaxic injections have been well-described in rodent models [1–3]. The availability of a wide variety of transgenic lines, particularly in mice, have further enhanced the applications of genetically encoded molecular tools [4–6].

In comparison, the use of optogenetics, chemogenetics, and genetically encoded indicators has expanded more slowly in nonhuman primate (NHP) models ([7–11] and for comprehensive review [12]). Methods for stereotaxic injection have proven successful when the target region is relatively small [13, 14], but the larger brain size of many NHPs poses challenges in expressing viral

Mark A. G. Eldridge and Adriana Galvan (eds.), *Vectorology for Optogenetics and Chemogenetics*,
Neuromethods, vol. 195, https://doi.org/10.1007/978-1-0716-2918-5_13,

constructs densely and uniformly enough to produce behavioral effects. The difficulty in generating transgenic NHPs has limited cell-type specific expression [15], although recent developments of selective enhancers show promise for addressing this issue [16–19].

Currently, there are two main strategies for delivering viral vectors into the NHP brain: (1) stereotaxic injections [13, 14, 20] and (2) hand-held injections [21, 22]. Stereotaxic injections allow for precise delivery but are slow, which causes difficulty when targeting a large volume of tissue. The use of convection-enhanced delivery (*see* Chap. 12) can increase the speed of stereotaxic injections, but it often comes at the cost of reduced localization, as infusate under high pressure can travel back up the needle track, resulting in the transduction of tissue overlying the target site [23, 24], *see* Chap. 12 for approaches to reduce back reflux]. In addition, injections into deep ventral regions from the dorsal surface can produce unwanted damage, and obstruction by the temporal bone can limit access to temporal regions. Hand-held injections, which rely on referencing visual landmarks on the brain surface, have been used to deliver chemogenetic constructs to ventral regions, but this approach can result in patchy expression [21].

Recently, we reported a novel approach to delivering viral vectors into cortical areas of the rhesus monkey using a suite of multichannel microinjector arrays, allowing for efficient and precise viral vector delivery [25]. In this chapter, we describe in detail the methods for (1) building previously reported and newly designed multichannel microinjector arrays, (2) implementing these new devices for delivery of viral vectors into the rhesus brain and (3) co-infusing manganese for postoperative validation of targeting accuracy using MRI. The three injector arrays described are the ventral 2 × 2 array, linear 1 × 4 array, and linear 3 × 3 array. The purpose of the ventral 2 × 2 array is to allow stable injections of large volumes of the virus at low speed into visualized cortex on the lateral and ventral aspects of the brain that cannot be reached with a standard arm-mounted injection device without deriving stereotaxic coordinates (Fig. 1a). The linear 1 × 4 array allows for efficient injections into sulcal regions (Fig. 1b). The linear 3 × 3 array allows for a dense coverage of dorsal aspects of the brain.

2 Multichannel Microinjector Array Assembly

2.1 Ventral 2 × 2 Array Assembly

1. Cut the 31G needles to 2.5 mm length using the Dremel rotary tool with a cutting wheel attachment or the KLS 246 laser. Clear any potential blockages in the needles after cutting using a Hamilton cleaning wire.
2. Insert hypodermic tubing into each channel of 3D-printed manifold (Fig. 2b).

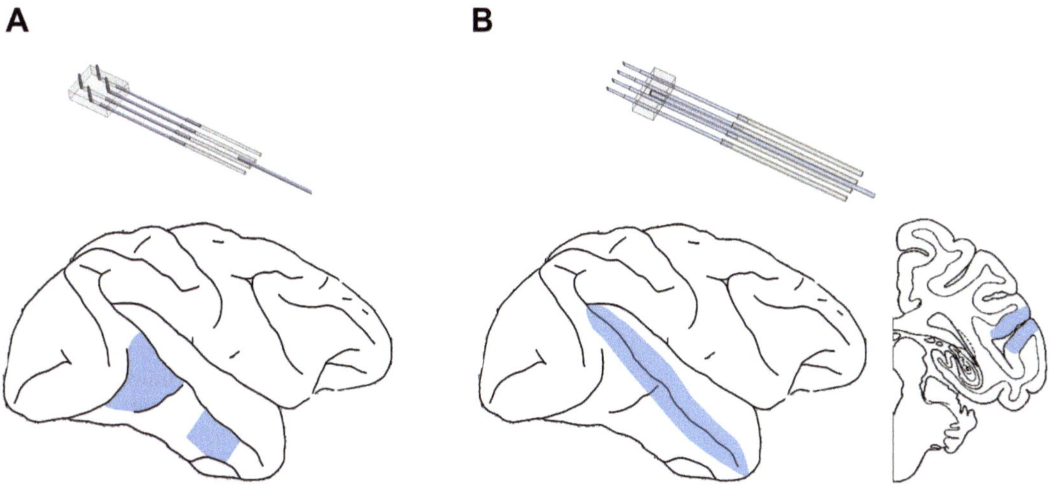

Fig. 1 Target regions accessible by multichannel microinjector arrays. (**a**) 2 × 2 ventral array enables access to ventral areas such as area TEO and TE. (**b**) 1 × 4 linear array enables access to sulcal regions at varying depths

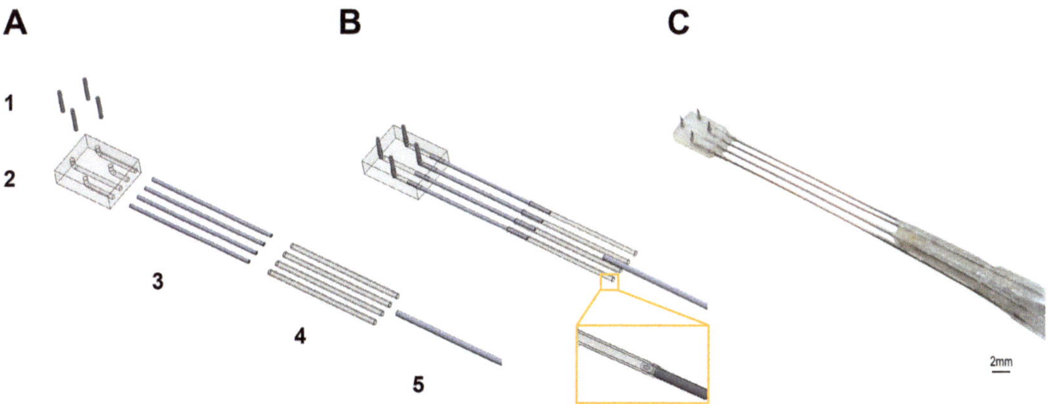

Fig. 2 Ventral 2 × 2 microinjector array. (**a**) Components of ventral 2 × 2 array: (1) 31G needles, (2) 3D-printed manifold, (3) hypodermic tubing, (4) silicone tubing, (5) 23G wire. (**b**) CAD model of assembled array. Orange inset indicates attachment of 26G needle to silicone tubing. (**c**) Photograph of assembled array

3. Secure silicone tubing to each hypodermic tube (Fig. 2b). On the other end of the silicone tubing, insert the beveled end of the 26G NanoFil needle (Fig. 2b inset). The blunt end of the 26G NanoFil needle will attach to a NanoFil syringe.
4. Check flow through each channel:
    1. Draw up deionized water (Milli-Q water) into the NanoFil syringe.
    2. Secure the NanoFil syringe to the 26G NanoFil needle at the end of each silicone tubing.

Table 1
Components for ventral 2 × 2 microinjector array assembly. Assembly under a microscope is recommended

Label	Item	Quantity	Company	Catalog #	Comments
(1)	31G needles	4	BD	328468	Cut to 2.5 mm length
(2)	3D-printed manifold	1	n/a	n/a	3D-printed https://3dprint.nih.gov/discover/3dpx-016894
(3)	30G hypodermic tubing	4	Component Supply	HTX-30R	0.14 mm inner diameter 0.305 mm outer diameter Cut to 35 mm length
(4)	Tygon ND100–80 microbore silicone tubing	4	Component Supply	TND80–010	0.01 mm inner diameter 0.03 mm outer diameter Cut to 15 cm length
(5)	23G wire	1	Beebeecraft	CWIR-BC0006-02A-S	Cut to 20 cm length
(8)	26G NanoFil needles	4	World Precision Instruments	NF26BV-2	
(6)	100 μL NanoFil syringe	1	World Precision Instruments	NANOFIL-100	
	Krazy Glue Maximum Bond	As needed	Elmer's Products		85 cps viscosity at room temperature
	Hamilton cleaning wire	As needed	Hamilton	18,300	
	Dremel 4000 High Performance Variable Speed Rotary Tool		Robert Bosch Tool Corporation	F0134000AW	Used to cut 31G needles
	409 15/16″ cutting wheel		Robert Bosch Tool Corporation	2615000420	Attachment for Dremel rotary tool
	KLS 246 pulsed Nd: YAG laser		LASAG		Alternative tool to cut 31G needles
	ProJet 6000 HD 3D printer		3D Systems		3D printer used to print multichannel microinjector manifolds
	Accura ClearVue		3D Systems		Material used for 3D printer

3. Slowly infuse deionized water through each channel to confirm flow. Infusion can be done by hand or using the pump control unit.
4. Once the flow has been confirmed, remove and empty the NanoFil syringe. Flush out each channel with air to expel any remaining deionized water.
5. Secure interface between hypodermic tubing and 3D-printed manifold using Krazy glue.
6. Insert 31G needles into the dorsal surface of the 3D-printed manifold. Ensure that the bevels are facing the same orientation and all needles protrude 2 mm from the dorsal surface of the 3D-printed manifold (Fig. 2b).
7. Secure 31G needles in place using Krazy glue, ensuring no glue flows into the needle itself.
8. Check flow through each channel (repeat **Step 4**).
9. Secure 23G wire to the hypodermic tubing using Krazy glue, approximately 20 mm from the 3D-printed manifold (Fig. 2b). The wire will be used to secure the multichannel microinjector array to the injector pump apparatus during surgery.
10. Optional: Apply Krazy glue across the 4 hypodermic tubes where the 23G wire is attached for additional stability.

2.2 Linear 1 × 4 Array Assembly

1. Insert 33G NanoFil needles through each channel of the 3D-printed manifold. Ensure that the bevels are facing the same orientation and all needles are the same length with respect to the 3D-printed manifold (Fig. 3b).
2. Secure silicone tubing to the blunt end of each NanoFil needle (Fig. 3b). On the other end of the silicone tubing, insert the beveled end of the 26G NanoFil needle (Fig. 2b inset). The blunt end of the 26G NanoFil needle will attach to a NanoFil syringe.
3. Check flow through each channel (*see* Subheading 2.1, **step 4** for details).
4. Secure the interface between the NanoFil needles and the 3D-printed manifold using Krazy glue.
5. Secure 23G wire into the center slot of the 3D-printed manifold using Krazy glue (Fig. 3b). The wire will be used to secure the multichannel microinjector array to the injector pump apparatus during surgery.

2.3 Linear 3 × 3 Array Assembly

1. Insert 33G NanoFil needles through each channel of the 3D-printed manifold. Ensure that the bevels are facing the same orientation and all needles are the same length with respect to the 3D-printed manifold (Fig. 4b).

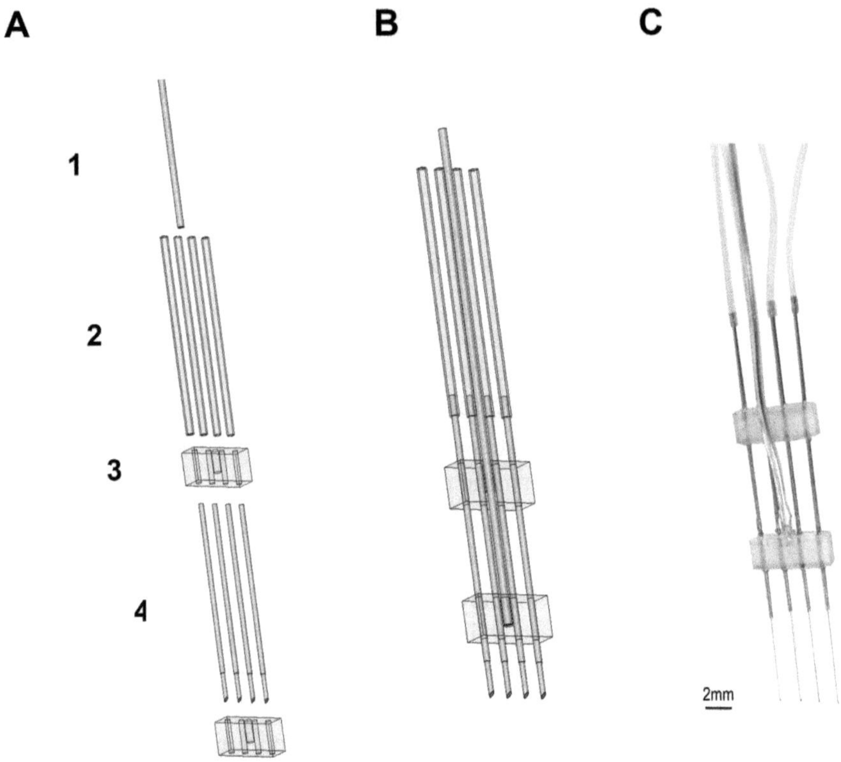

Fig. 3 Linear 1 × 4 microinjector array. (**a**) Components of linear 1 × 4 array: (1) 23G wire, (2) silicone tubing, (3) 3D-printed manifold, (4) 33G needles. (**b**) CAD model of assembled array. (**c**) Photograph of assembled array

Table 2
Components for linear 1 × 4 microinjector array assembly. Assembly under a microscope is recommended

Label	Item	Quantity	Company	Catalog #	Comments
(1)	23G wire	1	Beebeecraft	CWIR-BC0006-02A-S	Cut to 20 cm length
(2)	Tygon ND100–80 microbore silicone tubing	4	Component Supply	TND80–010	0.01 mm inner diameter 0.03 mm outer diameter Cut to 15 cm length
(3)	3D-printed manifold	2	n/a	n/a	3D-printed https://3dprint.nih.gov/discover/3dpx-016896
(4)	33G NanoFil needles	4	World Precision Instruments	NF33BV-2	
(8)	26G NanoFil needles	4	World Precision Instruments	NF26BV-2	

(continued)

Table 2 (continued)

Label	Item	Quantity	Company	Catalog #	Comments
(6)	100 μL NanoFil syringe	4	World Precision Instruments	NANOFIL-100	
	Krazy Glue Maximum Bond		Elmer's Products		85 cps viscosity at room temperature
	ProJet 6000 HD 3D printer		3D Systems		3D printer used to print multichannel microinjector manifolds
	Accura ClearVue		3D Systems		Material used for 3D printer

2. Attach silicone tubing to the blunt end of each NanoFil needle (Fig. 4b). On the other end of the silicone tubing, insert the beveled end of the 26G NanoFil needle (Fig. 2b inset). The blunt end of the 26G NanoFil needle will attach to a NanoFil syringe.
3. Check flow through each channel (*see* Subheading 2.1, **step 4** for details).
4. Secure the interface between the NanoFil needles and the 3D-printed manifold using Krazy glue.
5. Secure 23G wire into the outside slot of the 3D-printed manifold using Krazy glue (Fig. 4b). The wire will be used to secure the multichannel microinjector array to the injector pump during surgery.

3 Multichannel Microinjector Array Use

3.1 Array Use Protocol

#	Item	Quantity	Company	Catalog #	Comments
	Unsterilized items:				
(1)	Pump 11 Elite Nanomite Infusion/Withdrawal Programmable Syringe Pump	1	Harvard Apparatus	70–4507	
	Manipulator stand	1	Kopf Instruments	1749	
	Micromanipulator	1	Kopf Instruments	1760–61 or 1760	

(continued)

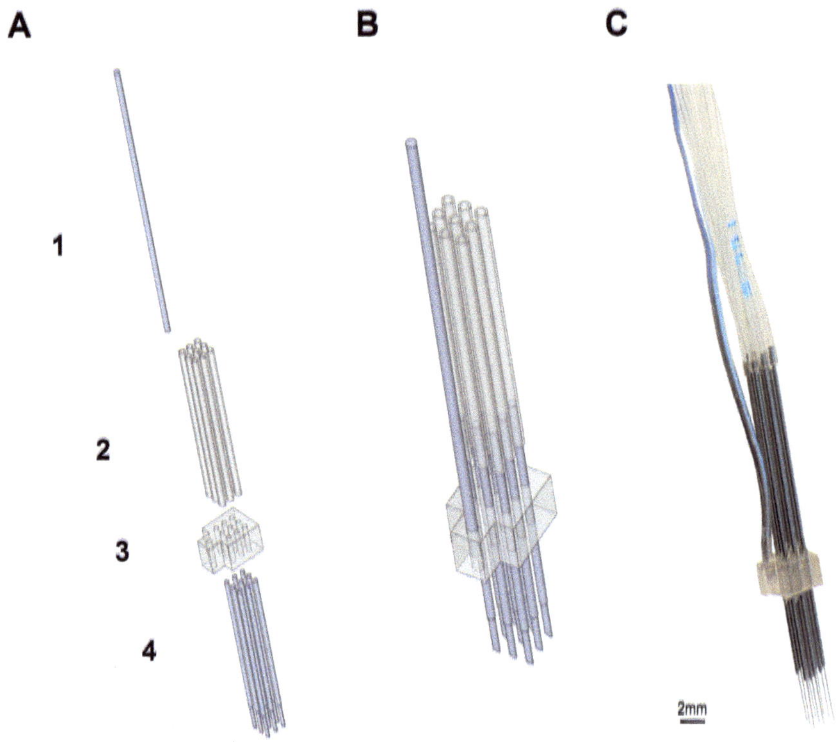

Fig. 4 Linear 3 × 3 microinjector array. (**a**) Components of linear 3 × 3 array: (1) 23G wire, (2) silicone tubing, (3) 3D-printed manifold, (4) 33G needles. (**b**) CAD model of assembled array. (**c**) Photograph of assembled array

Table 3
Components for 3 × 3 microinjector array assembly. Assembly under a microscope is recommended

# Item	Quantity	Company	Catalog #	Comments
(1) 23G wire	1	Beebeecraft	CWIR-BC0006-02A-S	Cut to 20 cm length
(2) Tygon ND100–80 microbore silicone tubing	9	Component Supply	TND80–010	0.01 mm inner diameter 0.03 mm outer diameter Cut to 15 cm length
(3) 3D-printed manifold	1	n/a	n/a	3D-printed https://3dprint.nih.gov/discover/3dpx-016895
(4) 33G NanoFil needles	9	World Precision Instruments	NF33BV-2	
(8) 26G NanoFil needles	9	World Precision Instruments	NF26BV-2	

(continued)

Table 3
(continued)

#	Item	Quantity	Company	Catalog #	Comments
(6)	100 μL NanoFil syringe	9	World Precision Instruments	NANOFIL-100	
	Krazy Glue Maximum Bond		Elmer's Products		85 cps viscosity at room temperature
	ProJet 6000 HD 3D printer		3D Systems		3D printer used to print multichannel microinjector manifolds
	Accura ClearVue		3D Systems		Material used for 3D printer

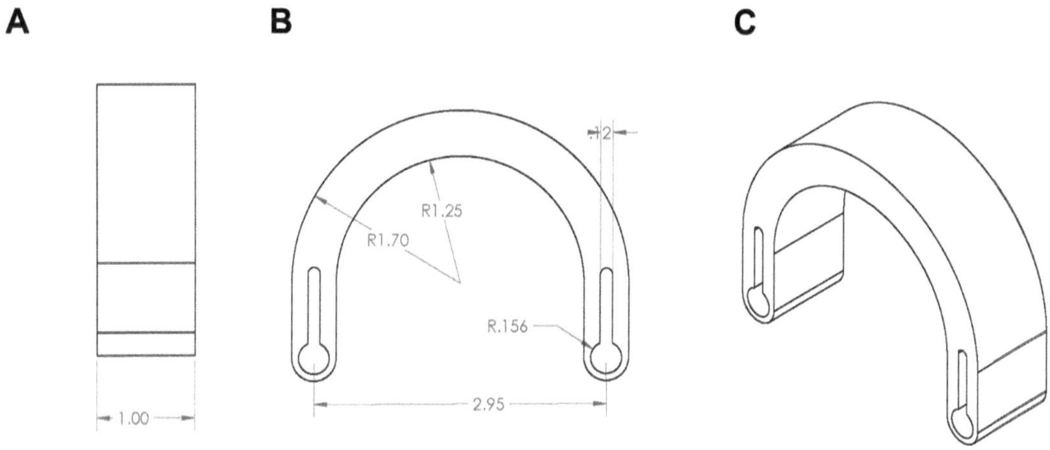

Fig. 5 C-arm dimensions (in inches). (**a**) Side view. (**b**) Top view. (**c**) 3/4 view

#	Item	Quantity	Company	Catalog #	Comments
	Sterilized items:				
(2)	C-arm	1	n/a	n/a	Dimensions in Fig. 5 Machined in-house
(3)	Metal dowel	1	n/a	n/a	5/16″ (7 mm) diameter ~9″ (23 cm) length
(4)	6–32 nylon screws	6–12	McMaster-Carr	94323A581	6 screws for 4-syringe holder 12 screws for 9-syringe holder
(5)	Syringe holder	2	n/a	n/a	3D-printed

(continued)

#	Item	Quantity	Company	Catalog #	Comments
					1 for holding syringes 1 for holding vials containing virus https://3dprint.nih.gov/discover/3dpx-016897 (for 4 syringes) https://3dprint.nih.gov/discover/3dpx-016898 (for 9 syringes)
(6)	100 uL NanoFil syringe		World Precision Instruments		Quantity should match number of channels on array
(7)	T-depressor	1	n/a	n/a	3D-printed https://3dprint.nih.gov/discover/3dpx-016899 (for 4 syringes) https://3dprint.nih.gov/discover/3dpx-016900 (for 9 syringes)
(8)	26G NanoFil needles	>2	World Precision Instruments	NF35BV-2	Used to load mineral oil and virus into syringes. If the vial containing virus is larger than a 1.5 mL Eppendorf tube, use a longer needle to accommodate for a larger vial.
	Mineral oil				
	Plastic petri dish				Used to hold sterile water to test multichannel array flow
	Multichannel microinjector array	1	n/a	n/a	Custom-made (refer to *2. Multichannel Microinjector Array Assembly*)
	Hamilton cleaning wire	1	Hamilton	18300	Used for troubleshooting
	Scissors	1			Used for troubleshooting

(continued)

#	Item	Quantity	Company	Catalog #	Comments
	Allen wrenches (various sizes)				Used to adjust micromanipulator-infusion pump interface
	Phillips-head screwdriver	1			

Before surgery day

Before gas (ethylene oxide) sterilizing all items, check the micromanipulator-infusion pump setup. The orientation of the C-arm can change the orientation of the syringes which can then impact infusion. While it is possible to make adjustments during surgery, testing the setup in advance can minimize time loss during surgery.

Surgery day

1. Store the virus on dry ice until ready for use. If using manganese for postoperative MR imaging, prepare the virus according to the protocol outlined in Subheading 3.4.
2. Set up a separate station near the surgical table to prepare the infusion pump and microinjector array. Attach infusion pump unit to the micromanipulator (Figs. 6a and 7a).
3. Don sterile PPE. Place a sterile drape over the setup area.
4. Attach the C-arm and metal dowel component to the micromanipulator (Figs. 6c and 7b).
5. Secure syringe holder to the bottom of the metal dowel. Secure the T-depressor into the plunger block/capture of the infusion pump injection unit (Figs. 6d and 7c).
6. Load NanoFil syringes with 30 μL of mineral oil per syringe using a 26G NanoFil needle, avoiding the introduction of air bubbles.
7. Secure the second syringe holder to the top of the metal dowel. The vial containing the virus can be placed in this syringe holder (Fig. 6d).
8. Don a second pair of gloves to handle nonsterile items. Thaw virus and place vial into the second syringe holder (Fig. 6d).
9. Discard the second pair of gloves to expose sterile undergloves. Attach a 26G NanoFil needle to one preloaded syringe and expel mineral oil to 20 μL in the syringe. Keeping the NanoFil needle attached, load the desired volume of virus, avoiding the introduction of air bubbles. Repeat for the remaining syringes (Fig. 6d, inset).
10. Place the virus-loaded syringes into the syringe holder, ensuring that the syringe flange aligns with the barrel flange capture

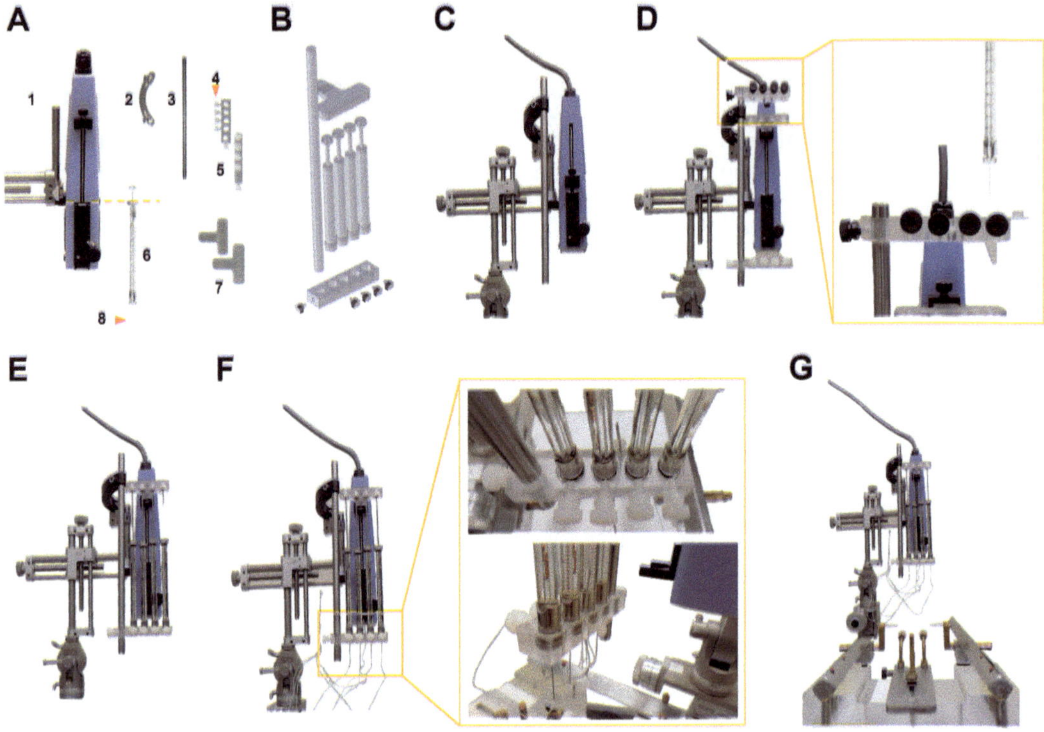

Fig. 6 Multichannel microinjector array apparatus for ventral 2 × 2 array. (**a**) Components of multichannel microinjector array apparatus: (1) Pump 11 Elite Nanomite Infusion/Withdrawal Programmable Syringe Pump, (2) C-arm, (3) metal dowel, (4) nylon screws, (5) syringe holder, (6) 100 μL NanoFil syringe, (7) T depressor, (8) 26G NanoFil needle. The orange dotted line indicates proper alignment of 100 μL NanoFil syringe and barrel flange capture on the syringe pump. (**b**) Schematic of apparatus assembly. (**c**) Attachment of C-arm. (**d**) Attachment of T depressor, syringe holder and additional syringe holder for virus withdrawal. (**e**) Placement of syringes. (**f**) Attachment of multichannel microinjector array. The orange arrowhead indicates the screw which holds the 23G wire in place. (**g**) Assembled micromanipulator-injector pump apparatus mounted onto a stereotaxic frame

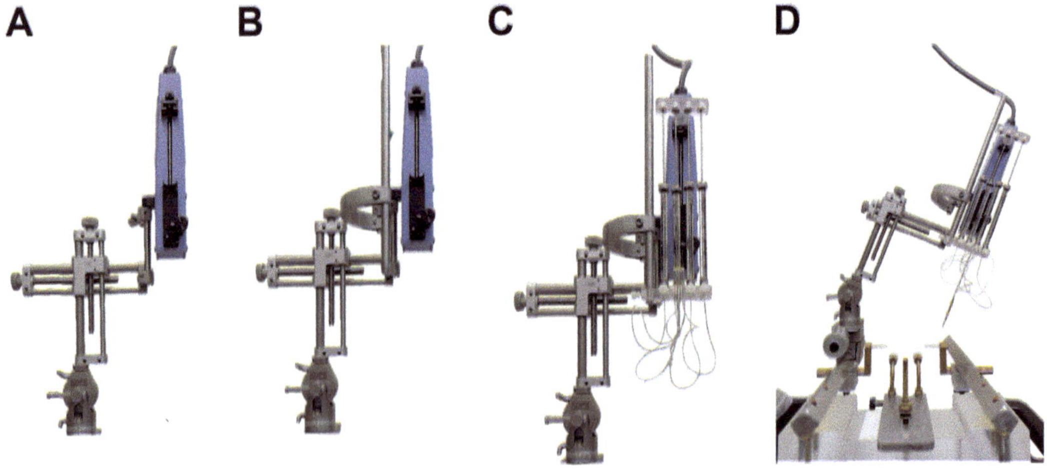

Fig. 7 Multichannel microinjector array apparatus for linear arrays. (**a**) Attachment of injector pump to a micromanipulator. (**b**) Attachment of C-arm. (**c**) Assembly of microarray injector array components. (**d**) Assembled micromanipulator-injector pump apparatus mounted onto a stereotaxic frame. Rotation along the coronal plane allows access to lateral areas

on the infusion pump injector unit (indicated by the orange dotted line in Fig. 6a, e).

11. Using the control unit, lower the T-depressor until the syringe plungers fit snugly into the recessed wells of the T-depressor. Ensure that syringes are straight and parallel from all directions to ensure even infusion (Fig. 6e).
12. Fill a spare Nanofil syringe with sterile water. Attach the needle end of one channel of the multichannel microinjector array to the spare Nanofil syringe. By hand, slowly infuse sterile water through the channel and check flow. Repeat for the remaining channels. Once the flow is confirmed, flush each channel with air.
13. Secure the flexible 23G wire on the multichannel microinjector array to the syringe holder (Fig. 6f right panel). Connect the needles of the multichannel microinjector array, connected to the end of the tubing, to each syringe (Figs. 6f and 7c).
14. Move the micromanipulator-injector pump unit to the stereotaxic frame on the operating table contralateral to the injection site (Figs. 6g and 7d).
15. Check initial flow through all channels of the multichannel microinjector array using an infusion rate of 1–3 μL/min. Once the proper flow is confirmed and air bubbles are flushed out, reduce the infusion rate to 0.05 μL/min. Do not stop the pump.
16. Position the multichannel microinjector array into the area of interest. In addition to moving the array using the dials on the micromanipulator, the flexible 23G wire on the multichannel microinjector array can be used to adjust the angle and positioning. For injections using the linear arrays, position the multichannel microinjector array parallel to the metal dowel holding the C-arm (Fig. 7d). This ensures that the angle measurements taken from the micromanipulator dial should closely reflect the angle of the needles on the multichannel microinjector array. Refer to Fig. 8a–b for an example of the ventral 2 × 2 microarray placement on area TEO.
17. Begin infusion at 0.5 μL/min for the desired volume (typically 10–20 μL per channel).
18. Once the infusion is finished, reduce the infusion rate to 0.05 μL/min and wait 10 min before withdrawing the multichannel microinjector array. If performing multiple injections within a single tract using the linear array, wait 1 min before moving between depths and 10 min before withdrawing completely from the tract.
19. Once the injections are completed, flush each channel of the array using alcohol, deionized water, and air three times each.

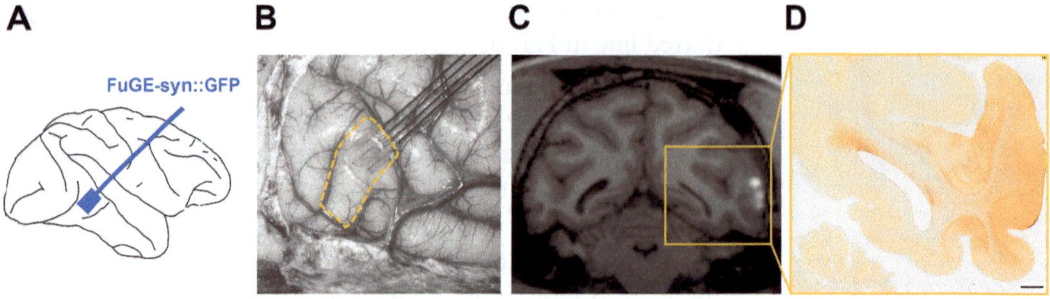

Fig. 8 Ventral 2 × 2 microinjector array efficiently delivers viral vector to area TEO. (**a**) Schematic of the injection site. (**b**) Intraoperative photograph showing multichannel microinjector array placement (orange dashed area indicates area TEO). (**c**) Injection site visualization with postoperative MR imaging of manganese contrast agent (white hyper signal indicates injection site). (**d**) DAB staining of GFP expression. The scale bar is 2 mm

3.2 Array Use Additional Notes

- Leaks can occur from the interface between the array needles and tubing. Usually, this can be resolved by cutting the end of the tubing and reattaching to the array needles.
- Blockages can occur within the array needles. Use the Hamilton cleaning wire to clear out any debris or blockage within the needle. Maintaining positive pressure by running the pump at a slow rate can facilitate this process.
- Although this protocol states that a final volume of 20 μL of mineral oil should be used in each syringe, the volume can be altered to accommodate more or less virus. The mineral oil helps to expel the small volume of virus remaining in the array tubing. At least 5 μL of mineral oil in each syringe is recommended to account for dead volume in the array tubing.
- Ensure that the tubes containing the virus can fit into the syringe holder prior to surgery. The syringe holder can accommodate 0.2 mL PCR tubes, 1.5 mL Eppendorf tubes and snugly fits 100 μL virus tubes. If the tubes containing virus do not fit, use an alternative method or transfer to a smaller tube.
- The C-arm dimensions (Fig. 5) provided are the standard for all multichannel arrays. The diameter of the C-arm can, however, be altered if needed as seen in Fig. 6.

3.3 Manganese Contrast Visualization

In vivo methods to visualize injections, other than histological analyses, are critical to verify injection accuracy and ensure that the intended area was targeted prior to the commencement of behavioral and/or electrophysiological testing or histology. They also offer an opportunity for touch-up injections. PET imaging using radioligands that bind to DREADD receptors, opsins, or other proteins of interest offer visualization of expression in target regions [25; *see* Chap. 14]. PET imaging is the gold standard for visualizing and quantifying expression. Co-infusion of manganese contrast reagent with the viral vector, followed by a postoperative MRI scan, enables rapid verification of the injection targeting [25] and has been validated in in vitro studies [24; *see* Chap. 1]. We demonstrate the utility of this method with lentiviral injections into area TEO expressing FuGE-syn::GFP (20 μL/site at 0.5 μL/min, 8 sites, 160 μL total, 2×10^9 IU/mL) using the ventral 2 × 2 array followed by a postoperative MRI scan. Comparison between the manganese contrast signal and the histological analyses confirm that the signals align (Fig. 8c, d).

Here, we outline the methods for preparing the manganese contrast reagent and mixing it with the virus. The manganese-virus solution is compatible with the microarrays described above (refer to *3. Multichannel Microarray Array Use*).

3.4 Manganese-Virus Co-infusion Protocol

Item	Quantity	Company	Catalog #	Comments
Manganese(II)chloride tetrahydrate		Sigma	M8054	
Molecular grade water		Thermo fisher	BP2819100	
0.22 μm filter	1	Corning	430513	
500 mL sterile receiver bottle	1	Corning	430282	

1. Make a 100 mM solution of manganese(II)chloride in molecular grade water under a sterile hood.
2. Filter solution through a 0.22 μm filter into a sterile receiver bottle.
3. Dilute 100 mM stock solution to 10 mM working manganese solution.
4. Add 10 mM manganese solution to virus to produce a final concentration of 0.1–0.5 mM manganese (e.g., 1–5 μL of stock solution in a 100 μL aliquot). Ensure the virus is at room temperature prior to mixing to prevent precipitation of manganese. Mix thoroughly by gently pipetting the full volume a few times.
5. Use the virus-manganese solution for the desired injection method.

Funding

This work was supported by the Intramural Research Program, National Institute of Mental Health, National Institutes of Health, Department of Health and Human Services (annual report number ZIAMH002032).

References

1. Cetin A, Komai S, Eliava M et al (2006) Stereotaxic gene delivery in the rodent brain. Nat Protoc 1:3166–3173. https://doi.org/10.1038/nprot.2006.450
2. Lowery RL, Majewska AK (2010) Intracranial injection of adeno-associated viral vectors. J Vis Exp. https://doi.org/10.3791/2140
3. Gonzalez-Perez O, Guerrero-Cazares H, Quiñones-Hinojosa A (2010) Targeting of deep brain structures with microinjections for delivery of drugs, viral vectors, or cell transplants. J Vis Exp. https://doi.org/10.3791/2082
4. Zhu H, Roth BL (2015) DREADD: a chemogenetic GPCR signaling platform. Int J Neuropsychopharmacol 18:1–6. https://doi.org/10.1093/ijnp/pyu007
5. Zeng H, Madisen L (2012) Mouse transgenic approaches in optogenetics. In: Progress in brain research, pp 193–213
6. Gong S, Doughty M, Harbaugh CR et al (2007) Targeting cre recombinase to specific neuron populations with bacterial artificial chromosome constructs. J Neurosci 27:9817–9823. https://doi.org/10.1523/JNEUROSCI.2707-07.2007
7. Han X, Qian X, Bernstein JG et al (2009) Millisecond-timescale optical control of neural dynamics in the nonhuman primate brain. Neuron 62:191–198. https://doi.org/10.1016/j.neuron.2009.03.011
8. Heider B, Nathanson JL, Isacoff EY et al (2010) Two-photon imaging of calcium in virally transfected striate cortical neurons of behaving monkey. PLoS One 5:e13829. https://doi.org/10.1371/journal.pone.0013829
9. Diester I, Kaufman MT, Mogri M et al (2011) An optogenetic toolbox designed for primates. Nat Neurosci 14:387–397. https://doi.org/10.1038/nn.2749
10. Galvan A, Caiola MJ, Albaugh DL (2018) Advances in optogenetic and chemogenetic methods to study brain circuits in non-human primates. J Neural Transm 125:547–563. https://doi.org/10.1007/s00702-017-1697-8
11. Tremblay S, Acker L, Afraz A et al (2020) An open resource for non-human primate optogenetics. Neuron 108:1075–1090. https://doi.org/10.1016/j.neuron.2020.09.027
12. Raper J, Galvan A (2022) Applications of chemogenetics in non-human primates. Curr Opin Pharmacol 64:102204. https://doi.org/10.1016/j.coph.2022.102204
13. Stauffer WR, Lak A, Yang A et al (2016) Dopamine neuron-specific optogenetic stimulation in rhesus macaques. Cell 166:1564–1571. https://doi.org/10.1016/j.cell.2016.08.024
14. Grayson DS, Bliss-Moreau E, Machado CJ et al (2016) The rhesus monkey connectome predicts disrupted functional networks resulting from pharmacogenetic inactivation of the amygdala. Neuron 91:453–466. https://doi.org/10.1016/j.neuron.2016.06.005
15. Izpisua Belmonte JC, Callaway EM, Caddick SJ et al (2015) Brains, genes, and primates. Neuron 86:617–631. https://doi.org/10.1016/j.neuron.2015.03.021
16. Dimidschstein J, Chen Q, Tremblay R et al (2016) A viral strategy for targeting and manipulating interneurons across vertebrate species. Nat Neurosci 19:1743–1749. https://doi.org/10.1038/nn.4430
17. Vormstein-Schneider D, Lin JD, Pelkey KA et al (2020) Viral manipulation of functionally distinct interneurons in mice, non-human primates and humans. Nat Neurosci 23:1629–1636. https://doi.org/10.1038/s41593-020-0692-9
18. Martel A-C, Elseedy H, Lavigne M et al (2020) Targeted transgene expression in cholinergic interneurons in the monkey striatum using canine adenovirus serotype 2 vectors. Front Mol Neurosci 13. https://doi.org/10.3389/fnmol.2020.00076
19. Nair RR, Blankvoort S, Lagartos MJ, Kentros C (2020) Enhancer-Driven Gene Expression (EDGE) enables the generation of viral vectors specific to neuronal subtypes. iScience 23: 10.1016/j.isci.2020.100888
20. Saunders RC, Aigner TG, Frank JA (1990) Magnetic resonance imaging of the rhesus

monkey brain: use for stereotactic neurosurgery. Exp Brain Res 81:443–446. https://doi.org/10.1007/BF00228139

21. Eldridge MAG, Lerchner W, Saunders RC et al (2016) Chemogenetic disconnection of monkey orbitofrontal and rhinal cortex reversibly disrupts reward value. Nat Neurosci 19:37–39. https://doi.org/10.1038/nn.4192
22. Upright NA, Brookshire SW, Schnebelen W et al (2018) Behavioral effect of chemogenetic inhibition is directly related to receptor transduction levels in rhesus monkeys. bioRxiv 331694. https://doi.org/10.1101/331694
23. San Sebastian W, Samaranch L, Heller G et al (2013) Adeno-associated virus type 6 is retrogradely transported in the non-human primate brain. Gene Ther 20:1178–1183. https://doi.org/10.1038/gt.2013.48
24. Varenika V, Dickinson P, Bringas J et al (2008) Detection of infusate leakage in the brain using real-time imaging of convection-enhanced delivery. J Neurosurg 109:874–880. https://doi.org/10.3171/JNS/2008/109/11/0874
25. Fredericks JM, Dash KE, Jaskot EM et al (2020) Methods for mechanical delivery of viral vectors into rhesus monkey brain. J Neurosci Methods 339:108730. https://doi.org/10.1016/j.jneumeth.2020.108730
26. Nagai Y, Kikuchi E, Lerchner W et al (2016) PET imaging-guided chemogenetic silencing reveals a critical role of primate rostromedial caudate in reward evaluation. Nat Commun 7. https://doi.org/10.1038/ncomms13605

Chapter 14

Methods to Verify Expression and Function of DREADDs Using PET

Yuji Nagai and Takafumi Minamimoto

Abstract

The chemogenetic technology, designer receptors exclusively activated by designer drugs (DREADDs), offers reversible and remote control of neuronal activity and behavior in rodents and monkeys. In vivo verification of DREADD receptors and their function is valuable for long-term studies in nonhuman primates and future clinical application. Here we describe the protocols for positron emission tomography (PET) imaging to verify the expression and function of DREADDs in living monkeys.

Key words DREADDs, Chemogenetics, Positron emission tomography, Monkey, Nonhuman primates, Reporter imaging, Functional imaging

1 Introduction

Chemogenetics, such as designer receptors exclusively activated by designer drugs (DREADDs), are a suite of genetic tools used to manipulate neuronal/non-neuronal cell activity through receptors or channels that are genetically engineered to bind specific exogenous ligands, usually biologically inert molecules. Among chemogenetic tools, muscarinic receptor-based DREADDs are the most popular and widely used [1, 2]. Activation of a modified human M3 muscarinic receptor (hM3Dq) enhances neuronal activity, and a modified M4 receptor (hM4Di) silences it. DREADDs have been widely used to modify neuronal activity and behavior in rodents. They have also proven to be effective in nonhuman primates, allowing reversible manipulation of activity across a specific neuronal population in large and discontinuous brain regions that are beyond the reach of pharmacological, electrical, or optogenetic manipulation strategies [3–8]. In addition, they allow for pathway and cell-type specific manipulation in nonhuman primates [9–12]. These represent the therapeutic potential of DREADDs for

Mark A. G. Eldridge and Adriana Galvan (eds.), *Vectorology for Optogenetics and Chemogenetics*, Neuromethods, vol. 195, https://doi.org/10.1007/978-1-0716-2918-5_14,

human brain disorders [13, 14]. Yet, successful chemogenetic behavioral manipulation is still challenging in nonhuman primates' studies. Arguably, efficient transduction of the DREADD gene into target neuronal populations and establishing sufficient and stable receptor expression are key factors for successful chemogenetic manipulation. Fluorescent light systems are available to detect fluorescent markers that can be used for verifying DREADD expression in vivo. However, such verification is not easily attained from deep brain structures and does not provide direct information about the level and extent of transgene expression. Another challenge is to readout chemogenetic effects on neuronal systems in nonhuman primates, where in vitro assays such as slice preparations are generally impractical.

Positron emission tomography (PET) is an in vivo molecular imaging technique that allows for the three-dimensional and quantitative mapping of positron-emitting radiotracers typically labeled with ^{11}C and ^{18}F. Using a radiolabeled compound that selectively binds to DREADDs as a radiotracer, PET enables the visualization of the location, area, and level of DREADD expression in vivo. We have visualized DREADD expression in mice [15] and monkeys [4] using ^{11}C-labeled clozapine as a PET probe, enabling 'imaging-guided chemogenetics' (Fig. 1). Currently, several PET probes are available with a superior selectivity for muscarinic DREADDs over clozapine, including [^{18}F]JHU37107 [16], [^{18}F]7b [17], and [^{11}C]deschroloclozapine (DCZ) [3]. Moreover, PET also enables the study of neuronal function using [^{18}F]fluorodeoxyglucose (FDG) as a metabolic marker. When combined with DREADDs, FDG-PET provides metabolic mapping reflecting chemogenetic induced neuronal activity [3, 18, 19] (*see* **Note 1**: other functional imaging). These two imaging methods are useful for verifying

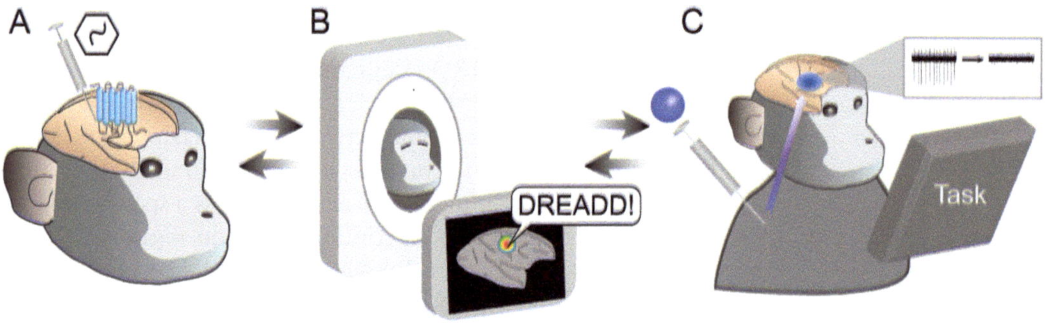

Fig. 1 Concept of PET-imaging guided chemogenetics. PET imaging provides a means to visualize the location of, and measure the stability of, DREADD expression in vivo. After viral vector injection (**a**), it provides critical information on whether the correct region has been affected and when the expression level has stabilized (**b**). It is a useful screening procedure prior to beginning a series of behavioral experiments (**c**). When in vivo expression is found to be insufficient or not targeted to the expected tissue, additional vector injections can be made to achieve ideal expression (**b** → **a**)

DREADD function in neuroscience research from mice to nonhuman primates, and will probably be a valuable tool for future clinical applications.

Here we describe detailed protocols for PET imaging of DREADD expression using [^{11}C]DCZ and DREADD-induced metabolic changes using [^{18}F]FDG, specifically in nonhuman primates.

2 Materials

2.1 Animal Preparation

This chapter focuses on Macaque monkeys injected with viral vectors expressing muscarinic DREADDs (hM3Dq or hM4Di). Since it takes several weeks for the expression level of DREADDs to stabilize [4], PET measurement is recommended to be performed at least 30 days (4 weeks), ideally 45 days or later (>6 weeks), after vector injection. Structural magnetic resonance (MR) images can be acquired beforehand to create an overlay image to confirm the location of DREADD expression with reference to brain structures. The animal needs to be fasted overnight before the PET scan.

2.2 PET Scanner

A microPET Focus220 scanner (Siemens Medical Solutions USA, Malvern, PA), which yields a 258 mm diameter × 76 mm axial field of view (FOV) and a spatial resolution of 1.3 mm full width at half maximum at the center of FOV, is used.

2.3 PET Tracer and Drug Preparation

We have developed PET imaging methods to visualize the expression of muscarinic DREADDs using [^{11}C]DCZ which is applicable for mice and monkeys [3, 10]. [^{11}C]DCZ is synthesized in the laboratory as described previously [3].

[^{18}F]FDG is used to visualize DREADD-induced metabolic changes in neuronal activity. [^{18}F]FDG is an established tracer for glucose metabolism and is commercially available from radiopharmaceutical suppliers. It can be synthesized in-house using methods described previously [20, 21].

To activate the excitatory muscarinic DREADDs, DCZ (HY-42110, MedChemExpress) is dissolved in 1–2% dimethyl sulfoxide (DMSO) in saline to a final volume of 0.1 mL/kg.

3 Methods

3.1 DREADD Expression Imaging by PET with [^{11}C]DCZ

[^{11}C]DCZ-PET is used for visualizing muscarinic DREADDs (hM4Di and hM3Dq) in vivo to evaluate the location, area, and level of expression (Fig. 2a) (*see* **Note 2**: occupancy).

3.1.1 PET Scan

1. Anesthetize the monkey. Initially, sedation is achieved with intramuscular injection of ketamine (5–10 mg/kg) and

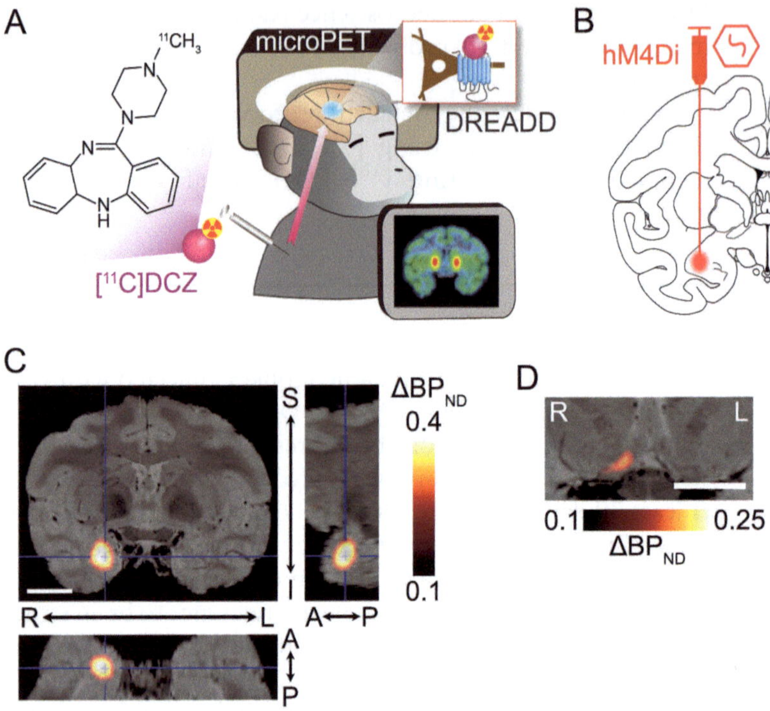

Fig. 2 Visualization of hM4Di expression in the amygdala using PET with [^{11}C]DCZ. (**a**) Illustration of PET imaging with [^{11}C]DCZ in monkeys. (**b**) Illustration of the location of viral vector injections. hM4Di expressing vector is injected into the right amygdala. (**c**, **d**) Subtracted parametric images of [^{11}C]DCZ specific binding (BP_{ND}) superimposed on to an individual MR image. hM4Di expression in the right amygdala is clearly visible (**c**). Also, hM4Di expression in the projection area, ventral striatum, is visible in the subtracted image (**d**). Scale bars are 10 mm. R, right; L, left; S, superior; I, inferior; A, anterior; P, posterior

xylazine (0.2–0.5 mg/kg) and maintained with isoflurane (1–3%) through an intubated tracheal tube. The subject is then placed in a supine position, and the head is secured.

2. Vital information (body temperature, heart rate, SpO_2, end-tidal CO_2, etc.) is monitored throughout the PET procedures.
3. Transmission scan: to correct the scatter and attenuation, transmission scan is acquired for about 20 min with a ^{68}Ge source.
4. Emission scan: emission scan is acquired for 90 min after intravenous bolus injection of [^{11}C]DCZ (185–370 MBq).
5. Reconstruction: PET images are reconstructed with filtered back-projection using a Hanning filter cut-off at a Nyquist frequency (0.5 mm^{-1}) with attenuation and scatter corrections. Time frames consist of 27 frames (frames × min: 5 × 1, 5 × 2, 5 × 3, and 12 × 5).

3.1.2 Data Analysis

To assess the DREADD expression level, the specific binding of [^{11}C]DCZ is calculated as the regional binding potential relative to nondisplaceable radioligand (BP_{ND}). BP_{ND} is calculated by using an original multilinear reference tissue model (MRTMo) [3, 22, 23] (*see* **Note 3**: other models, **Note 4**: signal underestimate). The cerebellum is used for a reference region where displaceable binding is negligible (*see* **Note 5**: reference region). The analysis is performed as follows using PMOD software (PMOD Technologies Ltd, Zurich, Switzerland):

1. Average the PET images over a period from 30 to 90 min.
2. Perform Rigid matching function of PMOD Fusion tool to anatomically match the averaged PET images to individual MR images (*see* **Note 6**: extracted brain image). Then, apply the same parameters to PET images from individual time points.
3. Draw volumes of interest (VOIs) on the target region(s) and cerebellum by referring to the MR images.
4. Obtain a time-activity curve for each VOI.
5. Calculate the VOI-based BP_{ND} using MRTMo with the cerebellum as a reference region and an initial equilibration time (t^*) is 15 min.
6. Create the BP_{ND} images from the dynamic-series PET image using MRTMo with the cerebellum as a reference region and $t^* = 15$ min.
7. Perform the above procedure for the data from each scan.

3.1.3 Consideration of Off-Target Binding

The BP_{ND} value of [^{11}C]DCZ consists of target binding to muscarinic DREADDs and off-target binding to endogenous binding sites [23]. Therefore, to estimate expression level, it is important to consider off-target binding in the target region. The simplest procedure is to calculate a ΔBP_{ND} image by subtracting the BP_{ND} image of the pre-vector scan from that of the post-vector scan.

1. Perform a Rigid matching function to match the BP_{ND} images to individual MR images using the parameter obtained from Subheading 3.1.2, **step 2**.
2. Perform subtraction between the pre- and the post-vector images using the statistical parametric mapping software (SPM12, The Wellcome Centre for Human Neuroimaging, UCL Queen Square Institute of Neurology, London, UK; www.fil.ion.ucl.ac.uk).

When PET data before viral vector injection is not available and the target region is confined to one hemisphere without expressing DREADD in the other hemisphere (such as axon terminal), the following procedure can be used to construct subtraction images using BP_{ND} images with left-right flipping and anatomical standardization.

1. Perform a Rigid matching function to match BP_{ND} images to individual MR images using the parameters obtained in Subheading 3.1.2, **step 2**.
2. Duplicate the individual MR and matched BP_{ND} images, and flip them right and left.
3. Original and flipped MR images are anatomically standardized to standard MR images [24] by Elastic deformation function of PMOD Fusion tool. The deformable matching parameters are followings: sampling rate = 1.0 mm, calculate inverse transformation, nonlinear warping (with iterations = 16, frequency cutoff = 25, and regularization = 1.0), and affine transformation before nonlinear warping. The original and flipped BP_{ND} images are also anatomically standardized using the parameters obtained from the original and flipped MR images, respectively (*see* **Note 7**: standardization).
4. Calculate the subtraction image between the standardized original and the flipped BP_{ND} images using SPM12.
5. Return the subtracted image to the individual MR space using inverse standardization parameters to verify the DREADD-positive locations and areas.

The subtraction BP_{ND} image allows visualization of the location and area of DREADD expression (Fig. 2c). It can also detect DREADD expression at axon terminals in dense projections (Fig. 2d) (*see* **Note 8**: detect projection). If a high expression level is predicted (BP_{ND} ~0.5 from baseline), [^{11}C]DCZ PET can detect the DREADD expression without subtraction (Fig. 3c). We routinely use [^{11}C]DCZ PET to monitor the development and stabilization of DREADD expression for long-term behavioral experiments (~3 years).

3.2 DREADD Functional Imaging by PET with [^{18}F]FDG

3.2.1 PET Imaging

PET with [^{18}F]FDG can assess changes in neuronal activity induced by DREADD manipulation [3, 18] because changes in regional [^{18}F]FDG uptake reflect changes in local neuronal activity [25, 26]. We routinely evaluate DREADD-induced metabolic changes by comparing the difference of [^{18}F]FDG uptake between DREADD agonist and vehicle treatment conditions (Fig. 3a). We also estimate agonist dose-dependent metabolic changes [3]. Note that there is a risk of excitotoxicity and/or epileptic seizures when excitatory DREADDs are strongly activated. We recommend starting with a low dose of DCZ (1–10 μg/kg) for hM3Dq and 100 μg/kg or lower for hM4Di [3].

1. Fasting for at least 6 h to prevent the increase in blood glucose level.
2. Anesthetize the monkey (sedated with ketamine and xylazine and maintained with isoflurane), place it in a supine position, and immobilize its head (*see* **Note 9**: awake condition).

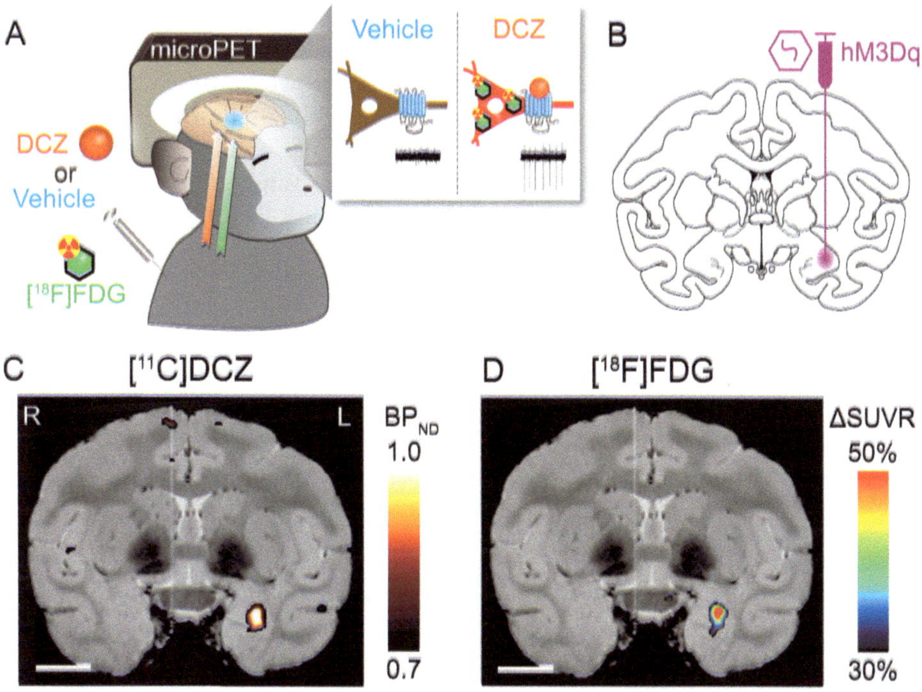

Fig. 3 Functional metabolic mapping of hM3Dq activation in a monkey using PET with [^{18}F]FDG. (**a**) Illustration of hM3Dq activation imaging with [^{18}F]FDG. After DCZ or vehicle administration, [^{18}F]FDG is injected and a PET scan is immediately started to assess DREADD-induced metabolic changes. (**b**) Illustration of viral vector injection expressing hM3Dq into one side of the amygdala. (**c**) A coronal section of parametric [^{11}C]DCZ image of specific binding (BP_{ND}) overlaying the corresponding MR image. R, right; L, left. (**d**) A subtraction image of [^{18}F]FDG SUVR (DCZ minus Vehicle conditions) in the same position as shown in (**c**). Scale bars are 10 mm

3. Monitor vital information throughout the PET procedures.
4. Measure plasma glucose level before the scan because the glucose level may affect the distribution of [^{18}F]FDG in the brain [27, 28]. If the glucose level is not within a normal range [67 ± 16 mg/dL (mean ± twice of standard deviation), male rhesus monkeys under fasting condition [29]], the PET data should be used carefully.
5. Transmission scan: the transmission scan is acquired for about 20 min with a ^{68}Ge source to correct the scatter and attenuation.
6. Emission scan: the emission scan is acquired for 60 min immediately after intravenous bolus injection of [^{18}F]FDG (111–222 MBq). Pretreatment of DCZ or vehicle is 1 min before [^{18}F]FDG injection (*see* **Note 10**: pretreatment timing).
7. Reconstruct the images in the same way as [^{11}C]DCZ-PET. Time frames consist of 12 frames × 5 min.

3.2.2 Data Analysis

The [^{18}F]FDG uptake is usually evaluated by using standardized uptake value [SUV, regional radioactivity (Bq/mL) × body weight (g)/injected radioactivity (Bq)]. However, the SUV can be affected by body conditions (i.e., the depth of anesthesia, body temperature, and plasma glucose level). We regularly use the SUV ratio (SUVR) normalized by the mean value of the whole brain. The analysis is performed as follows using PMOD software.

1. PET images are motion-corrected and converted to SUV images.
2. SUV images are created as an average of 30–60 min. The duration can be determined independently for each experiment since it depends on the agonist used and the timing of pretreatment. The scan duration also needs to be adjusted accordingly.
3. Perform Rigid matching function to match the averaged images with individual MR images.
4. Draw volumes of interest (VOIs) on the target region(s) and the whole brain with reference to individual MR images.
5. Calculate the SUV for each VOI.
6. Calculate the SUVR; dividing the SUV of the target region (s) by the whole brain SUV. Construct the SUVR image by dividing the SUV image by the whole brain SUV.
7. Calculate the difference in SUVRs between DCZ and vehicle pretreatment conditions. This difference reflects the metabolic change induced by DREADD manipulation.

The subtraction SUVR images are also useful for mapping metabolic change (Fig. 3d). The subtraction image is constructed as follows:

1. Perform Rigid matching function to match the SUVR images to the individual MR images using the same parameters as obtained in the previous **step 3**.
2. Create the subtraction images between DCZ and vehicle pretreatment conditions using SPM12 software.

The network-level statistical analysis can also be performed based on multi-subject or multi-scan data [18, 19].

Validation

The parametric [^{11}C]DCZ binding map reflects the distribution of DREADD expression. This can be verified via histological assessment. We routinely perform immunohistochemical staining using anti-M3 and -M4 antibodies [3]. Because endogenous M4 receptors are abundant in the brain, we also visualize coexpressed protein tags (e.g., GFP and HA) [4].

4 Notes

1. Functional MR imaging [30, 31] has also been used for assessing DREADD-induced changes in neuronal activity.
2. PET measurement with [^{11}C]DCZ is used to evaluate the occupancy of the DREADD agonist. An occupancy study can estimate the optimal dose range of agonists [3].
3. BP_{ND} may be obtained using other kinetic models, such as the simplified reference tissue model [32] and Logan noninvasive model [33].
4. The reference model was validated to provide useful information about the DREADD expression, although it underestimates BP_{ND} value of [^{11}C]DCZ at DREADD-positive brain regions, compared to the 'gold-standard' kinetic model using the arterial input function [23].
5. When the cerebellum is the target region, VOI can be delineated except for the DREADD expressed area if the area is smaller than one-third of the cerebellum. However, if the expression area is larger, the arterial input function is needed.
6. When Rigid matching between PET and MR images, it is recommended to extract brain images from MR images to obtain precise matching.
7. Since left and right hemispheres are not perfectly symmetrical, anatomical standardization is essential. Anatomical standardization can be performed by other softwares, such as SPM, FMRIB software library (FSL, https://fsl.fmrib.ox.ac.uk/fsl/fslwiki), Advanced normalization tools (ANTs, http://stnava.github.io/ANTs/), etc.
8. We visualized hM4Di expression at the axon terminal in the substantia nigra (terminals of the putamen neurons) [3] and in the caudate nucleus and mediodorsal nucleus of the thalamus (those of the dorsolateral prefrontal cortex neurons) [9]. Mapping of axon terminal hM4Di expression is useful for synaptic silencing via local agonist infusion [9].
9. We successfully assessed the changes in neuronal activity induced by hM3Dq excitation regardless of anesthetized states. However, assessing activity change due to inhibition via hM4Di can be relatively difficult under an anesthetized state, in which neuronal activity is largely decreased. In such cases, we recommend to consider [^{18}F]FDG PET measurement under conscious, head-fixed conditions [34–36].
10. For [^{18}F]FDG functional imaging, it is necessary that the agonist starts activating DREADDs prior to the onset of [^{18}F]FDG uptake. Therefore, when using a poor brain penetrable agonist like CNO, pretreatment should be done at least 15 min before [^{18}F]FDG injection.

Acknowledgments

This work was supported by JSPS KAKENHI Grant Number JP19K08138 (to YN) and AMED Grant Number JP21dm0107146 (to TM).

References

1. Armbruster BN, Li X, Pausch MH et al (2007) Evolving the lock to fit the key to create a family of G protein-coupled receptors potently activated by an inert ligand. Proc Natl Acad Sci U S A 104:5163–5168
2. Roth BL (2016) DREADDs for neuroscientists. Neuron 89:683–694
3. Nagai Y, Miyakawa N, Takuwa H et al (2020) Deschloroclozapine, a potent and selective chemogenetic actuator enables rapid neuronal and behavioral modulations in mice and monkeys. Nat Neurosci 23:1157–1167
4. Nagai Y, Kikuchi E, Lerchner W et al (2016) PET imaging-guided chemogenetic silencing reveals a critical role of primate rostromedial caudate in reward evaluation. Nat Commun 7:13605
5. Upright NA, Brookshire SW, Schnebelen W et al (2018) Behavioral effect of chemogenetic inhibition is directly related to receptor transduction levels in rhesus monkeys. J Neurosci 38:7969–7975
6. Raper J, Murphy L, Richardson R et al (2019) Chemogenetic inhibition of the amygdala modulates emotional behavior expression in infant rhesus monkeys. eNeuro 6: ENEURO.0360-19.2019. https://doi.org/10.1523/ENEURO.0360-19.2019
7. Hayashi T, Akikawa R, Kawasaki K et al (2020) Macaques exhibit implicit gaze bias anticipating others' false-belief-driven actions via medial prefrontal cortex. Cell Rep 30:4433–4444.e5
8. Hori Y, Mimura K, Nagai Y et al (2021) Single caudate neurons encode temporally discounted value for formulating motivation for action. Elife 10:e61248. https://doi.org/10.7554/eLife.61248
9. Oyama K, Hori Y, Nagai Y et al (2021) Chemogenetic dissection of the primate prefronto-subcortical pathways for working memory and decision-making. Sci Adv 7:eabg4246
10. Mimura K, Nagai Y, Inoue K-I et al (2021) Chemogenetic activation of nigrostriatal dopamine neurons in freely moving common marmosets. iScience 24:103066
11. Oguchi M, Tanaka S, Pan X et al (2021) Chemogenetic inactivation reveals the inhibitory control function of the prefronto-striatal pathway in the macaque brain. Commun Biol 4:1088
12. Perez P, Chavret-Reculon E, Ravassard P, Bouret S (2022) Using inhibitory DREADDs to silence LC neurons in monkeys. Brain Sci 12:206
13. Desloovere J, Boon P, Larsen LE et al (2019) Long-term chemogenetic suppression of spontaneous seizures in a mouse model for temporal lobe epilepsy. Epilepsia 60:2314–2324
14. Walker MC, Kullmann DM (2020) Optogenetic and chemogenetic therapies for epilepsy. Neuropharmacology 168:107751
15. Ji B, Kaneko H, Minamimoto T et al (2016) Multimodal imaging for DREADD-expressing neurons in living brain and their application to implantation of iPSC-derived neural progenitors. J Neurosci 36:11544–11558
16. Bonaventura J, Eldridge MAG, Hu F et al (2019) High-potency ligands for DREADD imaging and activation in rodents and monkeys. Nat Commun 10:4627
17. Hu F, Morris PJ, Bonaventura J et al (2021) ^{18}F-labeled radiotracers for in vivo imaging of DREADD with positron emission tomography. Eur J Med Chem 213:113047
18. Michaelides M, Anderson SAR, Ananth M et al (2013) Whole-brain circuit dissection in free-moving animals reveals cell-specific mesocorticolimbic networks. J Clin Invest 123:5342–5350
19. Michaelides M, Hurd YL (2015) DREAMM: a biobehavioral imaging methodology for dynamic in vivo whole-brain mapping of cell type-specific functional networks. Neuropsychopharmacology 40:239–240
20. Ido T, Wan CN, Fowler JS, Wolf AP (1977) Fluorination with molecular fluorine. A convenient synthesis of 2-deoxy-2-fluoro-D-glucose. J Org Chem 42:2341–2342
21. Hamacher K, Coenen HH, Stöcklin G (1986) Efficient stereospecific synthesis of no-carrier-added 2-[^{18}F]-fluoro-2-deoxy-D-glucose using aminopolyether supported nucleophilic substitution. J Nucl Med 27:235–238

22. Ichise M, Ballinger JR, Golan H et al (1996) Noninvasive quantification of dopamine D2 receptors with iodine-123-IBF SPECT. J Nucl Med 37:513–520
23. Yan X, Telu S, Dick RM et al (2021) [^{11}C] deschloroclozapine is an improved PET radioligand for quantifying a human muscarinic DREADD expressed in monkey brain. J Cereb Blood Flow Metab 41:2571–2582
24. McLaren DG, Kosmatka KJ, Oakes TR et al (2009) A population-average MRI-based atlas collection of the rhesus macaque. Neuroimage 45:52–59
25. Phelps ME, Huang SC, Hoffman EJ et al (1979) Tomographic measurement of local cerebral glucose metabolic rate in humans with (F-18)2-fluoro-2-deoxy-D-glucose: validation of method. Ann Neurol 6:371–388
26. Poremba A, Malloy M, Saunders RC et al (2004) Species-specific calls evoke asymmetric activity in the monkey's temporal poles. Nature 427:448–451
27. Kawasaki K, Ishii K, Saito Y et al (2008) Influence of mild hyperglycemia on cerebral FDG distribution patterns calculated by statistical parametric mapping. Ann Nucl Med 22:191–200
28. Burns CM, Chen K, Kaszniak AW et al (2013) Higher serum glucose levels are associated with cerebral hypometabolism in Alzheimer regions. Neurology 80:1557–1564
29. Magden ER, Mansfield KG, Simmons JH, Abee CR (2015) Nonhuman primates. In: Fox JG, Anderson LC, Otto G et al (eds) Laboratory animal medicine, 3rd edn. Academic Press, pp 771–930
30. Grayson DS, Bliss-Moreau E, Machado CJ et al (2016) The rhesus monkey connectome predicts disrupted functional networks resulting from pharmacogenetic inactivation of the amygdala. Neuron 91:453–466
31. Hirabayashi T, Nagai Y, Hori Y et al (2021) Chemogenetic sensory fMRI reveals behaviorally relevant bidirectional changes in primate somatosensory network. Neuron 109:3312–3322.e5. https://doi.org/10.1016/j.neuron.2021.08.032
32. Lammertsma AA, Hume SP (1996) Simplified reference tissue model for PET receptor studies. NeuroImage 4:153–158
33. Logan J, Fowler JS, Volkow ND et al (1996) Distribution volume ratios without blood sampling from graphical analysis of PET data. J Cereb Blood Flow Metab 16:834–840
34. Noda A, Ohba H, Kakiuchi T et al (2002) Age-related changes in cerebral blood flow and glucose metabolism in conscious rhesus monkeys. Brain Res 936:76–81
35. Itoh T, Wakahara S, Nakano T et al (2005) Effects of anesthesia upon 18F-FDG uptake in rhesus monkey brains. Ann Nucl Med 19: 373–377
36. Eberling JL, Roberts JA, De Manincor DJ et al (1995) PET studies of cerebral glucose metabolism in conscious rhesus macaques. Neurobiol Aging 16:825–832

Chapter 15

Reporter Selection and Postmortem Methods to Verify Transgene Expression

Kate S. Heffernan, Yoland Smith, and Adriana Galvan

Abstract

The accurate localization of transgene expression after viral vector delivery is essential to the interpretation of experiments based on genetic-based approaches, such as chemo- or optogenetics. Postmortem histological analysis can be used to examine the injection target, the extent of the virus transduction, the types of cells expressing the transgene, and the subcellular localization of the protein. In this chapter, we will provide a general description of methods to identify transgene expression, immunocytochemistry protocols, and examples of specific protocols. We close the chapter with an example of an application of electron microscopy to identify the localization of transgene expression.

Key words Immunohistochemistry, Fluorescence, Immunoperoxidase, Immunogold light microscope, Electron microscope, Reporter protein

1 Introduction

Genetic-based approaches to manipulate or monitor neuronal activity have become routine tools in neuroscience research. Most of these studies rely on the use of viral vectors to deliver the genetic material of interest to neurons or other brain cell types. The accurate and sufficient expression of the transgene is the most important component of these experiments. In vivo methods (such as imaging or fiber photometry) have been used by some experimenters to assess transgene expression [1, 2]. Evidently, there is value in these methods since they may allow researchers to correct insufficient expression or mistargeted injections prior to extensive behavioral testing. However, the currently available in vivo approaches to analyze transgene expression have limited spatial resolution and do not provide information about cellular localization. To date, postmortem histological analysis remains the gold standard for accurate and detailed assessment of transgene expression. Postmortem examination can help to address questions about the precision of

Mark A. G. Eldridge and Adriana Galvan (eds.), *Vectorology for Optogenetics and Chemogenetics*,
Neuromethods, vol. 195, https://doi.org/10.1007/978-1-0716-2918-5_15,

the injection target, the extent of the virus transduction, the types of cells expressing the transgene, and the subcellular localization. Thus, the information gained with histological examination is integral to the interpretation of functional data garnered during studies based on viral vector administration and can be used to refine future studies.

Target of Postmortem Verification: Transgene of Interest or Tag Protein

Histological assessment is based on brain tissue localization of the proteins (or the corresponding mRNA, see below) for which genetic sequences were delivered by the viral vector. In a few cases, the protein of interest (POI, e.g., the opsin or chemogenetic receptor) will be detected, but most often, a reporter (label or "tag" protein) is used to label the transgene in the tissue. Since the pioneering optogenetic studies that used the fusion of the channelrhodopsin (ChR2) with a green fluorescent protein to visualize ChR2-expressing neurons with fluorescent and confocal microscopes [3], the fusion of the opsin to tag proteins has become the most commonly used approach to identify transduced proteins in in vivo or ex vivo experiments and postmortem tissue.

However, fluorescent proteins are not needed in many applications, and overexpression could result in their aggregation or retention in the cytoplasm [4]. An alternative approach is to use epitope tags, that is, short antigenic peptide sequences that can be identified with immunohistochemistry [5]. A recently developed group of tag proteins ("spaghetti monster" fluorescent proteins) combine GFP-like fluorescent proteins and various tag epitopes [6].

The tag proteins are either expressed as fusions to the protein of interest, or as separate entities. In the former case, they are termed "fused" proteins, while in the latter case, their expression is regulated by inclusion in the viral vector of an internal ribosome entry site (IRES) or a P2A sequence, that directs the viral genetic material to express the two proteins separately. Although this can be a convenient approach, the expression of the tag protein does not reliably indicate the expression of the POI (since the two proteins may be expressed in different subcellular compartments). In addition, some reports indicate that the addition of an IRES sequence can reduce the expression of the transgene of interest [7].

Methods for Postmortem Verification

While other methods can provide information about levels of expression of a transgene (e.g., Western blots), histological methods provide information about its pattern of regional, cellular, and subcellular localization. The efficient use of these approaches depends on the integrity of the brain structures under study and the resolution scale, which expands from nanometer (electron microscopy studies using very small blocks of brain) to centimeter (light or confocal microscopy analysis using whole brain sections), depending on the type of microscopy being used. At any level, innovative approaches such as serial block-face scanning electron microscopy or clarification of brain tissue, may provide three-dimensional information and allow digital reconstruction of the neuronal structures.

Histologic methods can be divided in two types: those based on the visualization of proteins (the POI itself or the tag protein associated with it), and those based on identification of nucleic acid sequences coding for these proteins. The latter type uses complementary DNA or RNA probes to localize the corresponding sequence in the tissue sections by means of in situ hybridization, FISH, and RNAscope. These approaches will not be discussed here, the reader is referred to [8, 9] for more details. In this chapter, we will describe histologic methods to identify the POIs using (in most cases) antibodies directed against the POI or the tag proteins (immunohistochemistry). We will first comment on the advantages and limitations of these methods and will then provide a general protocol to use these approaches at the light and electron microscope level.

The simplest approach to visualize viral-mediated expression of endogenous proteins is through the endogenous fluorescence signal of the fluorescent tag proteins. This method requires only a few preparatory steps for observations under the microscope, allowing for rapid discrimination of transduced tissue. However, if the endogenous fluorescence of the proteins is weak, it may not be detectable. In such cases, an immunohistochemical approach could be advantageous. One could use a primary antibody against the fluorescent protein, followed by a secondary antibody that provides a fluorescent signal, thus enhancing the endogenous fluorescence (Fig. 1a). A major advantage of the fluorescence method is that it allows for the use of multiple fluorophores that have emission signals at different wavelengths and serve to discriminate more than one protein. The method is commonly used, for example, to determine if the POI is expressed in a specific phenotypically-characterized cell type. A drawback of fluorescent preparations (endogenous or antibody-enhanced) is that the fluorescence fades with time.

Regular light microscopy is commonly used to visualize brain sections that have been treated with antibodies and revealed with peroxidase or a similar technique (see below). Immunoperoxidase is a highly sensitive technique that allows the identification of even very low levels of antigens. Tissue prepared with this method can be counterstained to facilitate the identification of brain regions or

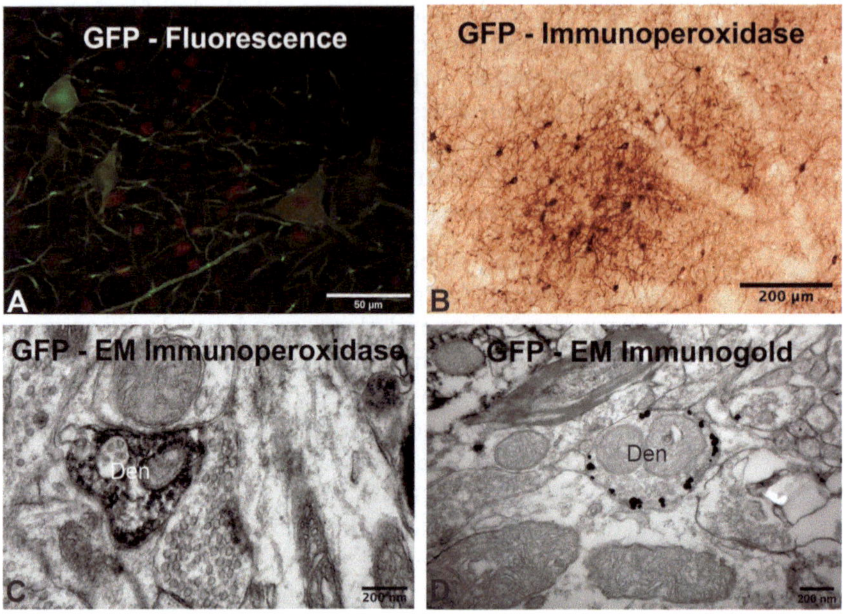

Fig. 1 Examples of visualization of GFP in macaque brain tissue using different immunohistochemistry methods. Neurons in various brain regions of macaque express GFP after AAV transduction. (**a**) GFP-expressing neurons in the motor cortex. GFP was revealed with primary antibodies against the protein and secondary antibodies conjugated to a green fluorophore. The nuclei of neurons were identified with NeuN and a red fluorophore, to verify the neuronal expression of GFP. (**b**) GFP-expressing neurons in the putamen. GFP was revealed using immunoperoxidase. (**c**, **d**) Electron microscope micrographs of GFP-expressing dendrites (Den) in ventrolateral (**c**) and pulvinar (**d**) nuclei of the thalamus. In (**c**), GFP was visualized using the immunoperoxidase method. Note the dark amorphous product of the peroxidase reaction. In (**d**), gold particles were used to identify GFP. Note that the gold particles are located at the plasma membrane. Panel A from Galvan et al., 2016 and panel C from Galvan et al., 2012. (Both reproduced under Creative Commons Attribution 4.0 International License)

cellular profiles during the light microscope observations. Light microscopy observations are used to assess regional and subregional patterns of expression of POIs (Fig. 1b).

Electron microscopy (EM) is the gold standard when one aims to characterize the subcellular and subsynaptic localization of POIs. The successful detection of antibodies at the EM level results from a balance of technical conditions that preserve the antigenic sites and the tissue ultrastructure. The antibodies bound to the POIs must be labeled using electron dense markers to allow EM visualization (such as peroxidase reaction product or gold particles (Fig. 1c, d, see below).

These various visualization methods offer complementary information on protein localization, and thus a comprehensive assessment of transgene distribution at the cellular, subcellular and subsynaptic levels can be achieved through a combination of these techniques.

In this chapter, we will provide a general description of immunohistochemistry protocols, followed by examples of specific protocols. We close the chapter with an example of an application of EM methods to identify the localization of transgenes. **

General Protocol Description

Note: Timing for all incubation steps can be found in Subheading 10.

2 Tissue Preparation

We use rhesus macaques in our laboratory, but mice and rats may also be used with minor adjustments in the procedure. Monkeys are deeply anesthetized with pentobarbital (100 mg/kg, i.v.), and transcardially perfused with an oxygenated Ringer's solution followed by a fixative solution of 4% paraformaldehyde and 0.1% glutaraldehyde. Paraformaldehyde is commonly used and appropriate for most antigens. In some cases, formaldehyde can mask epitopes, but this is a consequence of all fixation procedures. The addition of glutaraldehyde to the perfusion fixative solution is necessary for electron microscopy methods [10, 11]. Compared to formaldehyde, glutaraldehyde yields higher cross-linking, making it an excellent choice when ultrastructural preservation is needed. However, glutaraldehyde does have disadvantages, such as the addition of free aldehyde groups which must be removed or blocked-in later steps.

Following perfusion, the brain is extracted from the skull and post-fixed in 4% paraformaldehyde for 24 h. Successful perfusion will result in white tissue. Pink tissue suggests that blood is still present in the tissue, causing increased background staining and poor structural preservation. Poor perfusion can be mitigated to a degree by extending the post-fixation period.

Sections are usually cut in 60 μm-thick coronal sections using a vibrating microtome for electron microscopy and immunoperoxidase studies, and a freezing microtome for fluorescent studies. If sectioning on a freezing microtome, tissue is first cryoprotected in a phosphate buffer (PB) solution containing 30% sucrose to prevent the formation of ice crystals.

3 Storage of Tissue

Following cutting, sections can be stored in an antifreeze solution containing 3 parts ethylene glycol, 3 parts glycerol, 3 parts distilled water, and 1 part sodium dihydrogen orthophosphate buffer and stored at −20°C.

4 Preparation of Tissue for Light Microscopy and Fluorescence Microscopy

To remove the anti-freeze solution, rinse the sections with phosphate buffered saline (PBS).

5 Blocking Aldehydes and Endogenous Signals

5.1 Aldehydes

Following rinses, if glutaraldehyde was used in the fixative solution, incubate the sections in a sodium borohydride solution (1% w/v in phosphate buffered saline). This reduces the aldehydes introduced by glutaraldehyde fixation, a necessary step to allow proper recognition of the epitopes of interest [12].

5.2 Biotin

If a biotinylated secondary antibody is to be used, blocking endogenous biotin may be necessary. Preincubating sections with avidin (0.01% w/v in PBS) will bind to any endogenous biotin, and then subsequently rinsing with PBS and incubating sections with biotin (0.001% w/v in PBS) will occupy any biotin binding sites on the avidin. Egg white and skim milk can be used as sources for avidin and biotin respectively, but rinses with distilled water should replace PBS as PBS will precipitate egg white proteins.

5.3 Peroxidase

In cases where the perfusion was unsuccessful, blocking endogenous peroxidase is necessary. Red blood cells contain peroxidase which will result in high background staining. Preincubation of sections in 10% hydrogen peroxide in distilled water will decrease the unwanted background.

5.4 Lipofuscin

This pigment is common in aging and accumulates in the cytoplasm of most cell types [13]. Lipofuscin can cause high background staining and should be blocked using a cupric sulfate solution or sudan black B [14]. We recommend incubating sections in a solution of cupric sulfate in 50 mM ammonium acetate (pH adjusted to 5.0).

6 Permeabilization and Blocking

Sections must then be permeabilized and nonspecific targets blocked. The degree of permeabilization depends upon the location of the antigen of interest. For antigens in the plasma membrane, permeabilization may destroy the target, and thus incubations without a permeabilization agent should be tried first. For antigens in the cytoplasm, a degree of permeabilization will be needed, and antigens in the nucleus typically require the greatest degree of permeabilization. Common permeabilization detergents include Triton-X 100, Tween-20, and saponin. Triton-X is preferred over Tween-20 and saponin when a greater degree of permeabilization is needed, for example, in the case of a nuclear antigen.

Blocking is necessary to prevent nonspecific binding of the primary antibody and should immediately precede primary antibody incubation. The blocking agent should always be serum of the species in which the secondary antibody is raised (*see* next section). Never use a serum from the same species in which the primary antibody was raised, as this will cause nonspecific binding.

7 Selection and Incubation of the Primary Antibody

The immunohistochemistry technique allows, first and foremost, the detection of specific proteins in biological specimens. The reliability and sensitivity of these techniques are dependent on the use of highly specific antibodies that bind to the POI, without significant cross-reactivity with other unrelated proteins in the tissue under study [15]. Any immunohistochemistry experiment should be supported by adequate controls. The optimal control is to test the antibodies specificity in tissue lacking the target antigen (i.e., in gene knock-out mice). When this is not an option, alternative controls to consider are immunoblotting of brain tissue or transfected cells, pre-adsorption of primary antibodies with their corresponding antigenic peptides, and omission of the antibodies from the tissue incubation [16, 17].

Antibodies have two heavy chains and two light chains. Antibody subtypes vary from species to species. In humans, there are five isotypes: IgM, IgD, IgG, IgA, and IgE. Antibodies have two F (ab) segments and an Fc segment. The F(ab) segments are linked by disulfide bonds and contain two constant domains and two variable domains which are the antigen binding sites. The primary antibody should be raised in a different animal species from the one used to generate the tissue to be studied. When using multiple primary antibodies for co-localization studies in single sections, they should all be raised in different species to avoid cross-reactivity and nonspecific binding. Additionally, primary antibodies can be polyclonal or monoclonal. Polyclonal antibodies recognize multiple epitopes on the antigen of interest and therefore typically provide a greater signal, which can be an advantage if there is low expression of the antigen of interest. However, there is increased susceptibility to cross-reactivity and background staining. Monoclonal antibodies only recognize one epitope on the antigen of interest, and therefore result in less background staining and cross-reactivity but may need to be used at higher concentrations to generate optimal staining.

8 Selection and Incubation of the Secondary Antibody

The choice of the secondary antibodies depends on the method of interest (fluorescent versus immunoperoxidase) and the species in which the primary antibody was raised. The secondary antibody should be raised in the same species as the blocking serum and directed against the species of the primary antibody (i.e., donkey serum used for blocking, rabbit primary antibody, donkey anti-rabbit secondary). For immunoperoxidase methods, a biotinylated secondary antibody is needed. For fluorescent methods, the secondary must be conjugated to a fluorophore. It should be noted that there are primary antibodies that are available already conjugated to a fluorophore (direct method). These types of primary antibodies are usually less sensitive and can also result in a greater degree of nonspecific binding, and thus will only be briefly discussed.

9 Overview of Different Detection Methods

9.1 Indirect Method

In direct methods, the primary antibody is directly conjugated to a reporter. The reporter can be enzymatic, as in the case of horseradish peroxidase (HRP) for chromogenic reactions, fluorescent, or a colloidal gold particle. However, most neuroscience studies rely on indirect methods in which the primary antibody incubation is followed by incubation with a secondary antibody that is conjugated to the reporter (Fig. 2). The indirect method allows for signal amplification since two or more secondary antibodies can bind to one primary antibody. Thus, indirect is preferred when the antigen of interest has low expression, or when there is a greater degree of endogenous signals (i.e., higher background noise) from any of the contributors described in the section above.

9.2 Avidin-Biotin Complex (ABC) Method

The ABC and the peroxidase anti-peroxidase methods (the former described in this section, the latter immediately below) rely on the use of a peroxidase enzyme (usually horseradish peroxidase (HRP)) to reveal the antigenic sites. In the presence of H_2O_2 and substances such as benzidine derivatives, HRP produces an insoluble reaction product, of characteristic color that is easily identified under the light or electron microscope.

The ABC method (Fig. 3) is a chromogenic stain that yields high amplification of the signal, and therefore an excellent noise-to-signal ratio [18]. After incubations with a primary antibody and biotinylated secondary antibody, the sections must be incubated in an avidin-biotin complex (ABC) solution. The ABC solution contains avidin, which has four biotin binding sites, and a biotinylated enzyme (HRP). Since avidin has four biotin-binding sites, multiple biotinylated enzymes can bind to one molecule of avidin, forming a

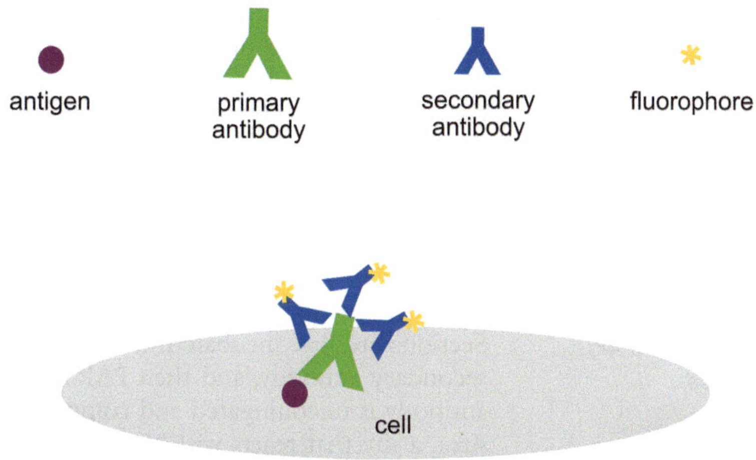

Fig. 2 Indirect method of immunolabeling. The antigen to be located (for the purpose of this chapter, the protein of interest (transgene or tag protein)) is identified by a primary antibody, which is, in turn, bound by a secondary antibody produced against the species the primary antibody was raised in. The secondary antibody is conjugated to a visible marker (in this example, a fluorescent protein)

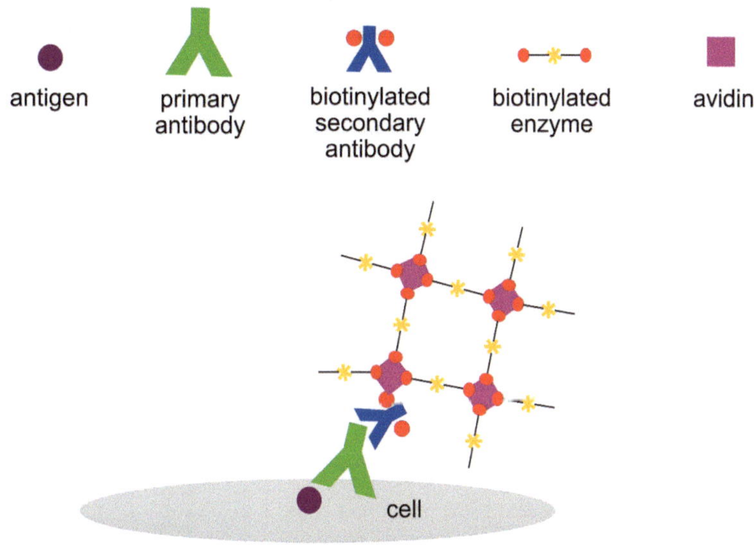

Fig. 3 The avidin-biotin complex (ABC) procedure of immunolabeling. As in Fig. 2, the antigen is bound by a primary antibody, and this by a secondary antibody. The secondary antibody is conjugated to biotin molecules. The peroxidase enzyme is also conjugated to biotin molecules. Avidin, an egg white protein, binds up to four molecules of biotin. The biotinylated enzyme and avidin form a lattice complex of multiple biotinylated enzymes

lattice complex. During incubation, open biotin binding sites are occupied by the biotinylated secondary antibody, linking the ABC to the biotinylated secondary. Following these steps, the application of a solution containing DAB or other chromogen and hydrogen peroxide results in a chromogenic reaction. DAB is oxidized in the presence of peroxide and peroxidase, and since the biotinylated enzyme (horseradish peroxidase) is tethered to the site of the antigen, a brown precipitate forms at the binding sites.

9.3 Peroxidase Anti-Peroxidase (PAP) Method

The PAP method relies on three main steps to increase sensitivity. Sections are first incubated in a primary antibody, followed by a secondary antibody, and then PAP complex [19]. The secondary antibody is unconjugated and comprised of two available binding sites: a site that reacts with the primary antibody, while the other remains available to bind with the PAP complex in the following step. The PAP complex is comprised of two anti-peroxidase antibodies and three peroxidase molecules (Fig. 4). Subsequent steps include a chromogenic reaction with a substrate such as DAB. Since three peroxidase molecules are bound to every secondary antibody, the PAP method results in a greater signal amplification than the indirect method. The PAP method is also useful in tissues with high endogenous biotin because it results in less background staining than the ABC method. Even in tissues with low endogenous biotin, such as the brain, PAP decreases unwanted background signal. PAP does not include avidin, which is positively charged and can bind to nucleic acids as well as other molecules, increasing background staining [18]. This method is only possible for chromogenic stains, and not immunofluorescent approaches since the PAP complex contains peroxidase.

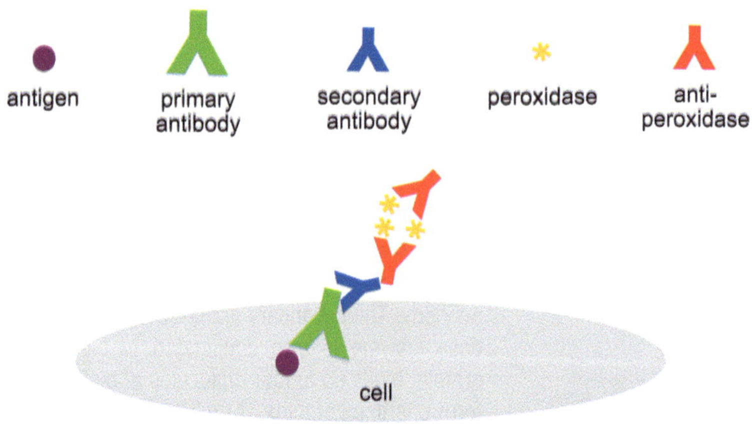

Fig. 4 The peroxidase-antiperoxidase (PAP) procedure of immunolabeling. The antigen is bound by a primary antibody, which is in turn bound by an unconjugated secondary ("linking") antibody. The linking antibody will get bound to an antiperoxidase antibody, generated in the same animal species as the primary antibody which recognized the antigen of interest

Both peroxidase-based approaches can be adapted for EM. The electron-dense, insoluble amorphous reaction product of immunoperoxidase can be easily visualized at the EM. The limitation of this approach is the limited spatial resolution because the diffuse peroxidase reaction product does not allow assessment of the precise localization of the antigen (Fig. 1c). Also, if the antigen is abundant, the amorphous reaction product may conceal important ultrastructural details (e.g., synaptic specializations). It remains, however, an excellent and highly sensitive method to identify in which cellular elements the antigen is expressed.

9.4 Immunogold Method

This method is used, most commonly, for EM preparations. To identify the precise localization of a protein at the subcellular level, it becomes necessary to use nondiffusible markers to identify the antigenic sites. Nondiffusible colloidal gold provides a higher spatial resolution than the diffusible peroxidase reaction product. Thus, the immunogold technique is the method of choice to study the spatial relationship between synapses and receptors or transporter proteins. However, in spite of the improved spatial resolution, the immunogold method is not as sensitive as the immunoperoxidase approach, because it lacks the amplification steps inherent to the immunoperoxidase technique. In the immunogold method, the secondary antibody is coupled to a colloidal gold particle (~1–10 nm in diameter) that shows up as a black dot under the electron microscope. Gold particles can help to discriminate, for example, if the POI is located in the plasma membrane or the cytoplasm (e.g., Fig. 1d). In the case of opsins and chemogenetic receptors, location at the plasma membrane of these transgenes is an essential prerequisite to achieve neuromodulation (see below for examples) [4, 20].

10 Example Protocols

10.1 Immunofluorescence Protocol

1. Take sections out of the antifreeze storage solution and wash with PBS.
2. Pretreat sections with 10% hydrogen peroxide for 20 min (optional—helpful for poor perfusion).
3. If glutaraldehyde was used during fixation, remove excess aldehydes. Prepare 1% sodium borohydride in PBS and incubate sections under chemical hood for 20 min. Rinse sections with PBS at least 4–5× or until all bubbles are gone.
4. Permeabilize and block. Prepare a solution of 1% bovine serum albumin, 5% serum (should be the species in which the secondary antibody was raised), 0.3% TritonX-100, in PBS. Incubate sections on a shaker for 1 h at room temperature.

5. Primary antibody incubation. Prepare a solution of 1% bovine serum albumin, 1% serum (should be the species in which the secondary antibody was raised), 0.3% TritonX-100, and the primary antibody of interest in PBS. Incubate the sections on a shaker overnight at room temperature.
6. Rinse sections 3× for 5 min in PBS.
7. Secondary antibody incubation. Prepare a solution of 1% bovine serum albumin, 1% serum (should be the species in which the secondary antibody was raised), and the secondary antibody of interest in PBS. This is also an appropriate time to add a counterstain, such as DAPI. Incubate sections on a shaker for 1 h at room temperature. Make sure the sections are in a covered container or wrap the container in foil to prevent bleaching.
8. Rinse sections 3× for 5 min in PBS.
9. Quench autofluorescence. Prepare a 10 mM cupric sulfate solution in 50 mM ammonium acetate buffer. Adjust the pH to 5.0. Incubate sections on a shaker for 30 min at room temperature.
10. Rinse sections 3× for 5 min in PBS.
11. Mount the sections on slides and allow them to dry for an hour at room temperature to increase adhesion to the slide. After drying, coverslip using an appropriate mounting medium such as Vectashield (Vector Laboratories) or Fluoromount-G (Thermo Fisher).
12. Seal the outer edges of the coverslip with nail polish.

10.2 Light Microscopy ABC Immunoperoxidase Protocol

1. Take sections out of the antifreeze storage solution and wash with PBS.
2. Pretreat sections with 10% hydrogen peroxide for 20 min (optional—helpful for poor perfusion).
3. If glutaraldehyde was used during fixation, remove excess aldehydes. Prepare 1% sodium borohydride in PBS and incubate sections under chemical hood for 20 min. Rinse sections with PBS at least 4–5× or until all bubbles are gone.
4. Permeabilize and block. Prepare a solution of 1% bovine serum albumin, 1% serum (should be the species in which the secondary antibody was raised), 0.3% TritonX-100, in PBS. Incubate sections on a shaker for 1 h at room temperature.
5. Primary antibody incubation. Prepare a solution of 1% bovine serum albumin, 1% serum (should be the species in which the secondary antibody was raised), 0.3% TritonX-100, and the primary antibody of interest in PBS. Incubate the sections on a shaker overnight at room temperature.

6. Rinse sections 3× for 10 min in PBS.
7. Secondary antibody incubation. Prepare a solution of 1% bovine serum albumin, 1% serum (should be the species in which the secondary antibody was raised), 0.3% TritonX-100, and the secondary antibody of interest in PBS. Incubate sections on a shaker for 90 min at room temperature. During this incubation, prepare avidin-biotin complex (ABC) solution, according to your kit instructions, for step #9.
8. Rinse sections 3× for 10 min in PBS.
9. Incubate sections in ABC solution on a shaker for 90 min at room temperature.
10. Rinse sections 2× for 10 min in PBS. Rinse 1× for 10 min in 0.05 M pH 7.6 Tris–HCl buffer.
11. Prepare DAB solution. Prepare Tris–HCl buffer containing 0.025% DAB, 10 mM imidazole, and 0.05% hydrogen peroxide under a chemical hood on a stir plate. Wait to add the hydrogen peroxide until the sections are ready for incubation. Incubate sections in DAB solution on a shaker for 10 min at room temperature.
12. Rinse sections 4–5× with PBS to stop the peroxidase reaction.
13. Mount sections on slides and allow them to dry out, preferably overnight.
14. Place slides in a slide holder. Dehydrate sections by placing the slide holder briefly in graded dilutions of ethanol (50%, 75%, 95%, 100%).
15. Place the slide holder in xylene. Coverslip slides using a hardening mounting medium.

10.3 Pre-Embedding Electron Microscopy ABC Immunoperoxidase Protocol

1. Take sections out of antifreeze storage solution and wash with PBS.
2. Since glutaraldehyde is highly suggested for sufficient ultrastructural preservation, it should have been used in the perfusion solution. Therefore, one must remove excess aldehydes. Prepare 1% sodium borohydride in PBS and incubate sections under chemical hood for 20 min. Rinse sections with PBS at least 4–5× or until all bubbles are gone.
3. Prepare a cryoprotectant solution of 0.05 M PB, 25% sucrose, and 10% glycerol adjusted to pH 7.4. Place sections in the cryoprotectant solution and freeze at −80°C for 20 min. Thaw sections and move to a graded series of cryoprotectant solution diluted in PBS. For example: place the tissue in 100, 70, 50, and 30% cryoprotectant solution in PBS for 10 min each, followed by 10 min in PBS.
4. Wash sections in PBS 3× for 10 min.

5. Block. Prepare a solution of 1% normal serum (should be the species in which the secondary antibody was raised) and 1% bovine serum albumin in PBS.
6. Primary antibody incubation. Prepare a solution of 1% bovine serum albumin, 1% serum (should be the species in which the secondary antibody was raised), and the primary antibody of interest in PBS. Incubate the sections on a shaker for 48 h at 4°C. Wash sections in PBS 3× for 10 min.
7. Secondary antibody incubation. Prepare a solution of 1% bovine serum albumin, 1% serum (should be the species in which the secondary antibody was raised), and the biotinylated secondary antibody of interest in PBS. Incubate sections on a shaker for 2 h at room temperature. During this incubation, prepare avidin-biotin complex (ABC) solution, according to your kit instructions, for step #9.
8. Rinse sections 3× for 10 min in PBS.
9. Incubate sections in ABC solution on a shaker for 90 min at room temperature.
10. Rinse sections 2× for 10 min in PBS. Rinse 1× for 10 min in 0.05 M pH 7.6 Tris–HCl buffer.
11. Prepare DAB solution. Prepare Tris–HCl buffer containing 0.025% DAB, 10 mM imidazole, and 0.05% hydrogen peroxide under a chemical hood on a stir plate. Wait to add the hydrogen peroxide until the sections are ready for incubation. Incubate sections in DAB solution on a shaker for 10 min at room temperature.
12. Rinse sections 4–5× with PBS to stop the peroxidase reaction.
13. Keep sections in PBS at 4 °C overnight.
14. Desalt. Wash sections 3× for 5 min each in PB (0.1 M pH 7.4).
15. Prepare a 1% osmium tetroxide solution in PB. Make sure the sections are flat (no folding). Remove the PB and add the 1% osmium tetroxide solution dropwise. Let sections incubate for 20 min.
16. Wash sections in PB once for 2 min and 3× for 5 min.
17. Dehydrate. Prepare 1% uranyl acetate in 70% alcohol solution and keep in the dark. The uranyl acetate solution must be stirred for approximately 15 min to dissolve. During this 15-min waiting period, place sections in 50% alcohol. Filter the uranyl acetate solution and then place sections in the solution for 35 min. Then, transfer sections to a 90% alcohol solution for 15 min. Transfer sections to 100% alcohol for 10 min, and then to fresh 100% alcohol for 10 more minutes. Transfer sections to propylene oxide for 10 min, and then to fresh propylene oxide for 10 more minutes.

18. Embed. Prepare resin (our lab uses Durcupan but other resins can be used) as described. Pour resin into disposable aluminum dishes and carefully place sections in the resin, ensuring that the entire section is covered. Let the sections sit for 12 h at room temperature.
19. Mount. Cover a hotplate with aluminum foil and set the temperature to 50 °C. Lightly cover slides and coverslips with mineral oil. Place the aluminum dishes containing the resin and sections from step #18 on the hotplate. Once the resin liquifies, transfer the sections to the oiled slides. Cover the slides with oiled coverslips and spread the resin by carefully pressing on the coverslip to force air bubbles out.
20. Polymerize the resin. Place the slides in an oven at 60 °C for 48 h.
21. Identify areas for block preparation. Using a razor blade, remove the coverslip around the area of interest. Wear safety goggles and protective gloves (such as Kevlar), as the glass will break into small pieces.
22. Once the coverslip over the area of interest has been removed, place the slide on a hot plate set to 50 °C. After a few minutes, remove the slide from the hot plate. Using a scalpel, carefully cut out a square of tissue that contains the area of interest. Remove the square of tissue and superglue to a resin block.
23. Wait 24 h to begin cutting on an ultramicrotome.
24. Cut 60 nm sections on an ultramicrotome and collect on single-slot pioloform-coated copper grids.
25. Prepare lead citrate solution and stain grids for 5 min.

11 Application Example: Use of EM Methods to Identify DREADDs

We have conducted high-resolution EM-immunogold studies to examine the ultrastructural localization of opsins and DREADDs [4, 20]. To be functional, these proteins should be expressed at the cellular plasma membrane where opsins can be activated by light stimulation and DREADDs can be bound by their agonists. Here we discuss our studies to localize DREADDs, hM4Di, and hM3Dq, after transduction with AAVs, in different brain regions of monkeys and mice [4]. We used primary antibodies against mCherry or hemagglutinin (HA) tag, followed by secondary antibodies coupled to nanogold particles.

The first group of animals (mice and monkeys) received intracerebral injections of AAV-hSyn-hM4Di-mCherry. In monkey neurons, the immunogold particles indicating mCherry (hM4Di) were found in the cytoplasm of dendrites, and only a small proportion was bound to the plasma membrane (Fig. 5a). In mice tissue, a

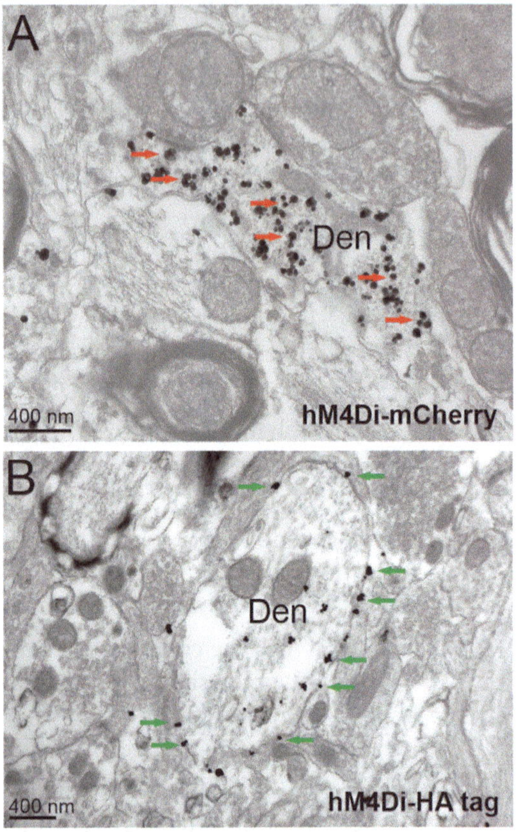

Fig. 5 Electron microscope images showing ultrastructural localization of tag proteins to identify hM4Di in monkeys. The pre-embedding immunogold technique was used to reveal mCherry in the subthalamic nucleus (**a**) or HA tag in the amygdala (**b**). Gold particles were found mostly in the cytoplasm of dendrites (Den, red arrows) when hM4Di was coupled to mCherry, while the fusion of the DREADD to HA tag resulted in expression of the receptor at the plasma membrane (green arrows)

larger proportion of gold particles were found in the plasma membranes of dendrites, but many of the particles were also seen in the cytoplasmic space. The sparse expression of hM4Di at the plasma membrane expression could reduce the functional effects of systemic administration of DREADD agonists, particularly in monkey neurons.

We hypothesized that a different tag fused to the hM4Di might improve the transport of this DREADD to the plasma membrane. The haemagglutinin (HA) tag is a small 9 amino acid epitope of the haemagglutinin of the influenza virus, which was first used as a fusion protein more than 30 years ago [21]. A second group of animals received intracerebral injections of AAVs in which the hM4Di was fused with the HA tag sequence, while the mCherry sequence was omitted. This change resulted in prominent detection of the gold particles at the plasma membranes of dendrites, both in monkey (Fig. 5b) and mouse neurons.

Our results indicate that the tag protein may have an important influence on the localization of the expression of the transgene of interest. In studies where the fluorescence is not needed, the use of HA and other short epitope tags [5] could help improve the transport of the protein of interest to the membrane. These studies underscore the importance of careful and thorough characterization of protein expression to interpret the result of opto- or chemogenetic studies.

References

1. Nagai Y, Kikuchi E, Lerchner W, Inoue KI, Ji B, Eldridge MA, Kaneko H, Kimura Y, Oh-Nishi A, Hori Y, Kato Y, Hirabayashi T, Fujimoto A, Kumata K, Zhang MR, Aoki I, Suhara T, Higuchi M, Takada M, Richmond BJ, Minamimoto T (2016) PET imaging-guided chemogenetic silencing reveals a critical role of primate rostromedial caudate in reward evaluation. Nat Commun 7:13605. https://doi.org/10.1038/ncomms13605
2. Diester I, Kaufman MT, Mogri M, Pashaie R, Goo W, Yizhar O, Ramakrishnan C, Deisseroth K, Shenoy KV (2011) An optogenetic toolbox designed for primates. Nat Neurosci. https://doi.org/10.1038/nn.2749
3. Boyden ES, Zhang F, Bamberg E, Nagel G, Deisseroth K (2005) Millisecond-timescale, genetically targeted optical control of neural activity. Nat Neurosci 8(9):1263–1268. https://doi.org/10.1038/nn1525
4. Galvan A, Raper J, Hu X, Pare JF, Bonaventura J, Richie CT, Michaelides M, Mueller SAL, Roseboom PH, Oler JA, Kalin NH, Hall RA, Smith Y (2019) Ultrastructural localization of DREADDs in monkeys. Eur J Neurosci 50(5):2801–2813. https://doi.org/10.1111/ejn.14429
5. Kimple ME, Brill AL, Pasker RL (2013) Overview of affinity tags for protein purification. Curr Protoc Protein Sci 73:9 9 1-9 9 23. https://doi.org/10.1002/0471140864.ps0909s73
6. Viswanathan S, Williams ME, Bloss EB, Stasevich TJ, Speer CM, Nern A, Pfeiffer BD, Hooks BM, Li WP, English BP, Tian T, Henry GL, Macklin JJ, Patel R, Gerfen CR, Zhuang X, Wang Y, Rubin GM, Looger LL (2015) High-performance probes for light and electron microscopy. Nat Methods 12(6):568–576. https://doi.org/10.1038/nmeth.3365
7. Mizuguchi H, Xu Z, Ishii-Watabe A, Uchida E, Hayakawa T (2000) IRES-dependent second gene expression is significantly lower than cap-dependent first gene expression in a bicistronic vector. Mol Ther 1(4):376–382. https://doi.org/10.1006/mthe.2000.0050
8. Wang F, Flanagan J, Su N, Wang LC, Bui S, Nielson A, Wu X, Vo HT, Ma XJ, Luo Y (2012) RNAscope: a novel in situ RNA analysis platform for formalin-fixed, paraffin-embedded tissues. J Mol Diagn 14(1):22–29. https://doi.org/10.1016/j.jmoldx.2011.08.002
9. Levsky JM, Singer RH (2003) Fluorescence in situ hybridization: past, present and future. J Cell Sci 116(Pt 14):2833–2838. https://doi.org/10.1242/jcs.00633
10. Sabatini DD, Bensch K, Barrnett RJ (1963) Cytochemistry and electron microscopy. The preservation of cellular ultrastructure and enzymatic activity by aldehyde fixation. J Cell Biol 17:19–58. https://doi.org/10.1083/jcb.17.1.19
11. Karnovsky MJ (1965) A formaldehyde-glutaraldehyde fixative of high osmolality for use in electron microscopy. J Cell Biol 27:1A–149A
12. Kosaka T, Nagatsu I, Wu JY, Hama K (1986) Use of high concentrations of glutaraldehyde for immunocytochemistry of transmitter-synthesizing enzymes in the central nervous system. Neuroscience 18(4):975–990. https://doi.org/10.1016/0306-4522(86)90112-0
13. Brizzee KR, Ordy JM, Kaack B (1974) Early appearance and regional differences in intraneuronal and extraneuronal lipofuscin accumulation with age in the brain of a nonhuman primate (Macaca mulatta). J Gerontol 29(4):366–381. https://doi.org/10.1093/geronj/29.4.366

14. Schnell SA, Staines WA, Wessendorf MW (1999) Reduction of lipofuscin-like autofluorescence in fluorescently labeled tissue. J Histochem Cytochem 47(6):719–730. https://doi.org/10.1177/002215549904700601
15. Saper CB (2005) An open letter to our readers on the use of antibodies. J Comp Neurol 493(4):477–478. https://doi.org/10.1002/cne.20839
16. Saper CB, Sawchenko PE (2003) Magic peptides, magic antibodies: guidelines for appropriate controls for immunohistochemistry. J Comp Neurol 465(2):161–163. https://doi.org/10.1002/cne.10858
17. Rhodes KJ, Trimmer JS (2006) Antibodies as valuable neuroscience research tools versus reagents of mass distraction. J Neurosci 26(31):8017–8020
18. Hsu SM, Raine L, Fanger H (1981) Use of avidin-biotin-peroxidase complex (ABC) in immunoperoxidase techniques: a comparison between ABC and unlabeled antibody (PAP) procedures. J Histochem Cytochem 29(4):577–580
19. Sternberger LA (1979) The unlabeled antibody (PAP) method, introduction. J Histochem Cytochem 27(12):1657. https://doi.org/10.1177/27.12.392001
20. Galvan A, Hu X, Smith Y, Wichmann T (2012) In vivo optogenetic control of striatal and thalamic neurons in non-human primates. PLoS One 7(11):e50808. https://doi.org/10.1371/journal.pone.0050808
21. Field J, Nikawa J, Broek D, MacDonald B, Rodgers L, Wilson IA, Lerner RA, Wigler M (1988) Purification of a RAS-responsive adenylyl cyclase complex from Saccharomyces cerevisiae by use of an epitope addition method. Mol Cell Biol 8(5):2159–2165

Chapter 16

Considerations for the Use of Viral Vectors in Nonhuman Primates

Martin Bohlen and Sébastien Tremblay

Abstract

The development of new genetic tools has revolutionized our ability to study the functional role of specific neuronal populations and circuits generating behavior. Although this revolution has already taken place in small animal models such as mice, adoption of these techniques has been relatively slow for animals more closely related to humans, such as nonhuman primates. Current challenges include effective delivery to much larger structural targets in the primate brain, cell-type specific transduction, and immunological responses. In this chapter, we will review some of the challenges and considerations that are specific to using these viral technologies in the nonhuman primate brain. Ultimately, these challenges can be met with new advances in surgical technique and gene therapy that will spin out new viral vectors with enhanced features able to compensate for the limitations of current vectors. As the existing challenges are circumvented, this will lead to a revolution in primate neuroscientific research and a greater understanding of the functional role circuits play in complex behaviors relevant to human neurological and psychiatric diseases.

Key words Optogenetics, Chemogenetics, Viral delivery, Viral vectors, Gene therapy, Nonhuman primates, Monkeys

1 Introduction

This chapter covers a range of topics relating to the administration of vectors used to deliver exogenous genes encoding neuronal actuators and sensors to target cellular populations. It covers factors related to the modification of both the viruses and their genomes, routes of delivery, and the validation for proof of efficacy. While the prior chapters in this book relate these topics to all experimental models, this chapter will attempt to directly focus on the maturation of approaches to develop and test genetic tools specifically for use in the nonhuman primate (NHP) model. For a detailed survey of best practices in NHP optogenetics, and for access to the raw data used to generate some of the recommendations made in this chapter, we refer the reader to the following open science resource [1].

Mark A. G. Eldridge and Adriana Galvan (eds.), *Vectorology for Optogenetics and Chemogenetics*,
Neuromethods, vol. 195, https://doi.org/10.1007/978-1-0716-2918-5_16,

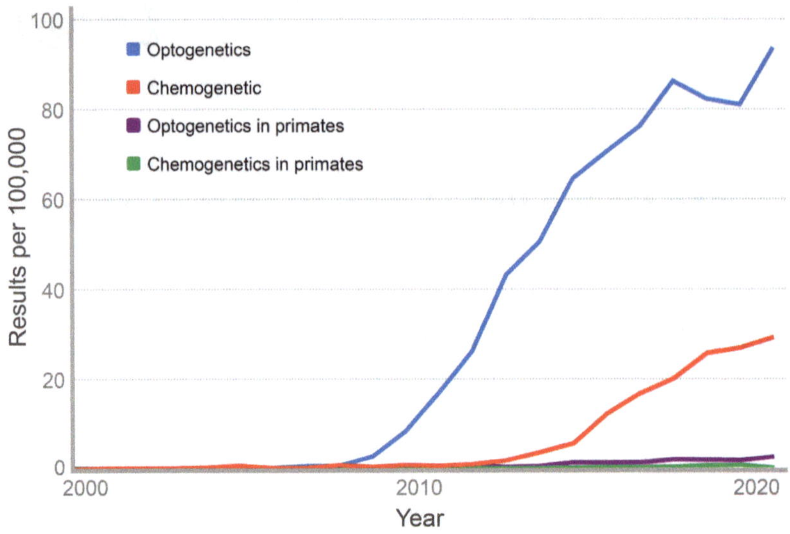

Fig. 1 Proportion of publications per 100,000 citations by year on PubMed, 2000–2021 involving optogenetics (blue line) or chemogenetic (red line), to studies using optogenetics (purple) or chemogenetics (green) in primates. This proportion was calculated by taking the total number of publications discovered for a given year and search term, divided by the total number of publications uploaded to PubMed for the same year then multiplying that number times 100,000. (For more information on how this data was calculated, the reader is referred to: https:/esperr.github.io/pubmed-by-year/about.html Source: https:/esperr.github.io/pubmed-by-year/)

There have been several detailed reviews covering the use of neuronal actuators such as those used in optogenetics and chemogenetics in primate neuroscientific investigations [2–6]. Although the number of publications coming out each year using these neuronal actuators is increasing rapidly, it would appear that there is a lack of global enthusiasm to adopt such techniques in the primate neuroscientific community (Fig. 1). Conversely, a recent sampling of primate labs has revealed that there is, in fact, a large number of primate labs attempting to implement these approaches [1]. The dichotomy between the efficacy of these approaches in phylogenetically lower species like rodents, flies, and fish compared to primates suggests a failure in translation. Further, while the published literature on the matter has shown some behavioral effects in NHPs, these effects tend to be more subtle than those elicited with more established approaches like electrical stimulation or pharmacological activation/inhibition [7]. In contrast, much of rodent research has provided a multitude of convincing behavioral responses to optical or chemogenetic activation or inhibition. Thus, there is still a critical need to refine and optimize these approaches in the primate for neuroscientific research. Much of the refinement will require expertise in the principles of gene therapy and primate neurobiology.

This closing chapter will address some of the important surgical and genetic considerations for the use of viral vectors in NHP experiments specifically. We will highlight technical caveats to avoid and best practices to adopt to maximize the success of future experiments in NHP. More importantly, the value in performing such refinements will extend well beyond the confines of basic neuroscientific research and carry important implications for clinical gene therapies targeting disorders in nearly every excitable cell of the primate body, human and nonhuman alike.

2 Mechanical Considerations in NHP Experiments

2.1 Brain Size and Surgical Approaches

The macaque brain is approximately 200 times the size of a mouse brain, and a similar transformation holds when translating from a macaque to the human brain [8] (Fig. 2). Conventionally, viral injections are made through a single needle, which is used to slowly infuse viral vectors directly into a targeted structure within the brain. Making a single, targeted injection in rodents is often sufficient to achieve neuronal transduction throughout the targeted structure, but is rarely enough to get complete transduction in much larger primate brains. This is believed to be a major culprit for the mixed behavioral effects seen with optogenetics and chemogenetics approaches in NHP.

At present, there are three ways that neuroscientists are attempting to compensate for the larger primate brain. The first approach consists of making multiple, single intraparenchymal infusions throughout the target structure. While this approach is effective at delivering viral vectors to the entirety of the structure, it is time intensive, meaning prolonged sedation times (unless the

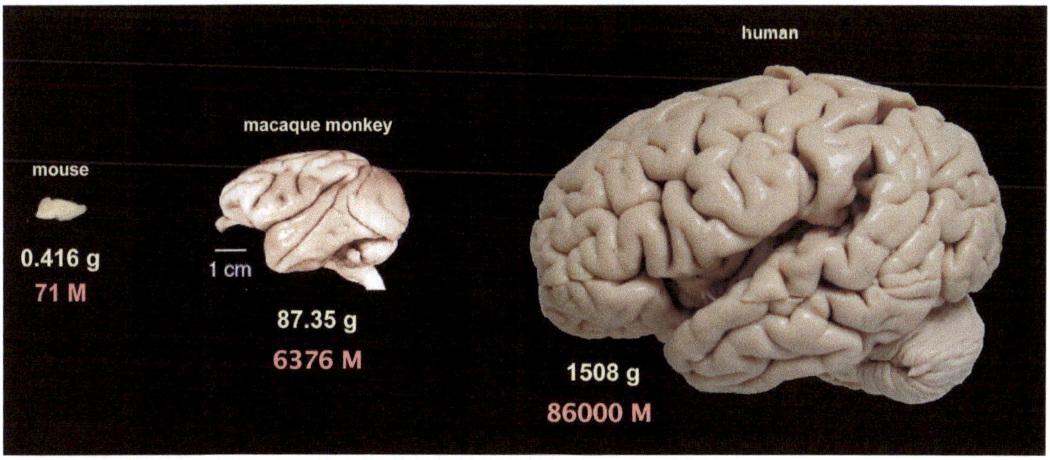

Fig. 2 Relative comparison of brain size in mice, macaque monkeys, and humans. Larger brains require adapted techniques for viral vector injection and actuation. (Image from Herculano-Houzel [8])

injections are done when animals are awake and restrained [9]) and risks for the animal. Namely, passing the needle multiple times through the brain parenchyma increases the propensity for direct damage to the tissue and vasculature. Such trauma, in-turn, promotes local, as well as circulating immunological agents to the injection sites, providing molecular signals for inflammation and the potential fostering of a larger-scale immunological response that can hinder future transduction [10]. Most viral vector experiments in NHP have applied this injection technique, with mixed results [1]. A second, more time-efficient approach is the simultaneous, slow infusion of viral vectors through an array of needles (*see* Chapter 15). This approach allows the simultaneous infusion of viral vectors throughout the structure of interest [11]. Array injections suffer a similar pitfall relating to multiple needles potentially causing local tissue and vascular trauma. Particularly pertaining to the vasculature, with a single needle, it is possible to avoid starting a penetration through a surface vessel, while this may be more difficult with an array of needles. This technique requires an open-brain surgery under sedation, which offers direct visualization of cortical targets and reduces uncertainty about effective targeting. At the moment, this new technology cannot be used in awake animals or to target deeper brain targets, and the array is not yet commercially available.

A third, alternative approach is to perform convection-enhanced delivery (CED; *see* Chapter 12). CED is an approach borrowed from gene therapy where large-scale transduction is achieved through the rapid (~5000 nl/min) delivery of a large volume (~50 μL/site) of viral solution [12–16]. The appeal of CED is that it only requires a single needle penetration and the infusion completes faster than conventional injections because of the high infusion rate (<30 min). A downside of CED is that to prevent reflux of the infused solution along the injection tract, a "reflux-resistant" cannula needs to be used. These larger cannulas (e.g., 670 μm diameter with a 120 μm step on all sides) can cause significant dimpling as they are driven into the brain, which can lead to incomplete penetration when the cortical surface is targeted. Thus, mixing the viral vector with an MRI contrast agent like gadolinium or manganese salt to confirm injection locations in the MRI is highly recommended [13, 16]. Despite these precautions, some investigators have concerns with the use of CED. For one, reflux of the viral solution along the injection tract often leads to transduction in off-target areas (Fig. 3a) [16]. Reflux was also observed by one of the authors (Tremblay) under the microscopic observation of cortical injection sites in about half of the injections attempted (Fig. 3b). This has important implications for chemogenetic approaches that use systemic administration of ligands, or any other approaches that require a restricted transgene expression. These ligands will interact with the chemogenetic receptors

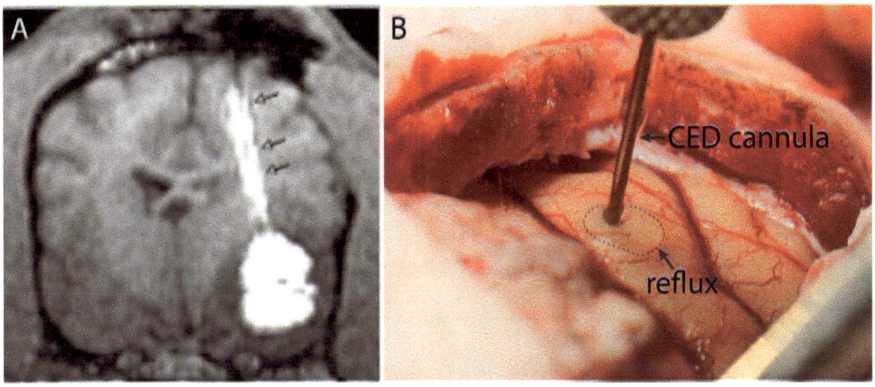

Fig. 3 Reflux of viral solution using convection-enhanced delivery (CED) in the macaque brain. Experiments were pre-approved by the IACUC of the institutions where these were conducted. (**a**) Reflux seen in the MRI scanner using contrast agents (e.g., Gadolinium) mixed with the viral solution (arrows). (**b**) Reflux directly observed using a surgical microscope during open brain surgery in the operating room

wherever they are expressed in the brain, potentially resulting in off-target effects when reflux is present. Second, the high injection pressure used in CED can sometimes lead to damage of the brain tissue, as confirmed through neuropathological examination [14]. Thus, infusion rates are recommended to be kept at or below 5000 nL/min. Third, expensive, commercially available CED cannulas (e.g., MRI Interventions Inc.) have an even larger diameter that causes significant trauma along the penetration tract. Smaller, cheaper, custom-designed CED cannulas built in-lab for less than $100 have worked well in the hands of the authors by combining 450 μm stainless steel sleeves (MacMaster-Carr #8987K73) with 33Ga Hamilton needles. These cannulas can be entirely made out of 304 stainless steel if the injections are conducted outside of the MRI and be connected directly to a Hamilton syringe to avoid the use of tubing in the injection line. Leaks, air bubble infiltration, and clogging when using tubing and fittings in the experimental injection setup are common problems when the needle and the syringe are not directly connected.

Exposing the cortex for the purpose of conducting conventional or CED injections implies performing a surgical opening in the dura mater in NHP. This is not required in rodents where needles are traditionally driven through the much thinner, transparent dura. Although performing an initial dural flap on the NHP brain is feasible using the proper technique and microdissection tools, performing a subsequent second or even third opening of the dura for the purpose of follow-up experiments (optogenetic stimulation or performing additional injections) is much more challenging. If the dura was sewed back in place at the end of the first surgery, the surgeon can expect a significant degree of adhesions between the dura and the underlying meningeal layers upon

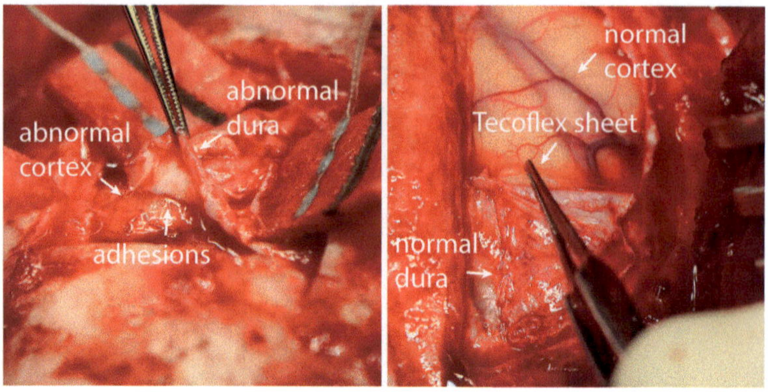

Fig. 4 Reopening of the dura mater in a follow-up surgery. (**a**) Significant adhesions between the dura mater and the underlying cortex is observed at reopening. (**b**) Addition of a Tecoflex layer in between the dura and the cortex during the first surgery prevents dural adhesions with the cortex

reopening (Fig. 4a). The same is true if the dural flap was initially replaced by a layer of dura regeneration matrix (e.g., Durepair, Medtronic #282452) to protect the exposed cortex. These dural replacement products made of collagen invite significant growth of fibrotic tissue that forms tight bonds with the underlying cortex. It becomes practically impossible to re-expose the injection site without causing significant damage to the underlying parenchyma. To circumvent this problem, we recommend inserting a thin layer of biologically inert material (e.g., Tecoflex EG-93A, thickness = 0.005", The Lubrizol Corporation) in between the cortex and the dura before closing the dural flap with sutures. This leads to a significant reduction in adhesions and makes it much easier to reexpose injection sites without any trauma to the brain tissue (Fig. 4b).

2.2 Using Cranial Chambers to Deliver Viruses to the Brain

The use of cranial chambers for an acute neurophysiological recording of single neurons in awake, behaving primates is a widespread practice in NHP neuroscience laboratories. Chambers are surgically secured to the skull to provide long-term access to the brain parenchyma. Through a chamber, it is possible to precisely lower single or multi-electrode probes with the help of a micro-manipulator (e.g., NAN Drive, Kopf Instruments). The same apparatus can be used to lower an injection cannula instead of a recording electrode, which was the first technique adopted by primate researchers to deliver viral vectors carrying opsin genes intraparenchymally to the brain of NHP [17]. Although it is still widely used more than a decade later, this technique presents important challenges that can lead to experimental failure. Here, we will discuss techniques to deliver viral vectors through standard recording chambers, and the alternative "cranial window" technique using artificial dura.

2.2.1 Using Standard Cranial Chambers

Standard cranial chambers, for example, those sold by Crist Instruments Inc., are meant to be installed on top of a circular craniotomy. The dura is usually left intact during this procedure to offer some protection to the otherwise exposed cerebral cortex. As a consequence, the gyral, sulcal, and vascular cortical landmarks are hard to visualize, making it difficult to determine the location of target cortical structures based on those landmarks. We recommend using an MRI-guided neuronavigation system, such as the Brainsight Vet system (Rogue Research Inc.), to plan the positioning of the chamber based on 3D MRI/CT reconstructions of the skull and brain. Less expensive alternatives including specialized MRI-based software for custom implants are also recommended [18, 19]. In addition to identifying target structures in the X-Y plane, depth is also an important variable to test before conducting an injection through a chamber. An overestimation of depth can lead to the injection being placed in the white matter or a deeper structure, and underestimation can lead to diffusion of the viral solution in the meninges where it can mix with the cerebrospinal fluid and travel to distant nontarget sites. To refine the target selection, investigators can conduct extensive electrophysiological "mapping", whereby recording electrodes are lowered at different chamber coordinates and depths to identify a 3D coordinate where neural activity is reliably observed. The mapping is also important to identify a region of interest based on the neurophysiological properties of the neural population (e.g., direction-selective cells in MT). An investigator can then reuse the same 3D coordinates and the same micro-manipulator apparatus to inject the viral solution at the recorded target location.

Data recently collected from many NHP research laboratories indicate that most investigators will wait 4–8 weeks between viral vector injection and experimental stimulation to allow time for the transgene to express (Fig. 5; [1]). During that time, a significant amount of granulation and fibrotic tissue can grow in the chamber on top of the dura. This new tissue can have a significant impact on the trajectory of the neural probe as it descends through the dura to reach its target site. Small angular deviations of the probe's trajectory will be compounded if the target site is deeper within the brain. Moreover, the brain is not perfectly immobile within the cranial cavity. Over days or weeks, the brain tissue underneath the chamber may deviate slightly, leading to potential misalignment with previously defined 3D coordinates [20]. Finally, in the case of cortical injections, the inability to observe the cortex directly within the chamber prevents confirmation of expression based on the epifluorescence of reporter proteins, like eYFP, that are fused to the opsin. Therefore, optogenetic stimulation experiments through cranial chambers have to make the assumption that the transduction was successful, that the brain did not shift with relation to the skull, and that the granulation/fibrotic tissue does not significantly deviate the trajectory of the neural probe.

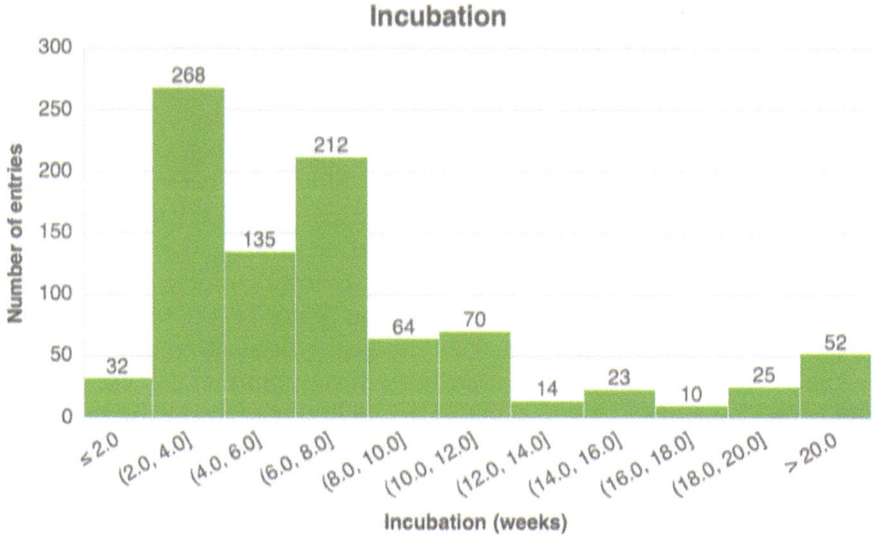

Fig. 5 Distribution of wait times (incubation) between viral vector injection and beginning of experiments in ~1000 injection tests in NHP. (Data from Tremblay et al. [1])

Several of these caveats can be avoided by using some additional precautions. First, to make sure that the injection is targeted to the exact location where the neural population of interest resides, many investigators use an "injectrode", which combines a recording electrode with an injection cannula. Several technical papers detail how to build such devices in the lab [21, 22]. The injectrode allows one to ascertain that the viral solution is delivered within the grey matter at the correct depth by providing an electrophysiological confirmation of the presence of neurons with properties of interest. Second, angular deviations of the neural probe caused by granulation tissue can be limited by performing a dural scrape to remove the granulation tissue before the start of stimulation experiments. Also, regular cleaning of the chamber is needed to avoid infections and may slow down the growth of granulation tissue. Third, it is recommended that, for optogenetic studies, experimenters inject the viral solution over a relatively large area of the target site either by using convection-enhanced delivery or multiple serial injections. This will maximize the chances that transduced neurons will be encountered several weeks after incubation despite possible translation of the cortex and deviations of trajectories inside the chamber. Overall, cranial chambers afford access to a relatively large area of cortical and subcortical tissue (~3 cm^2) to conduct subsequent stimulation, recordings, and inject additional viruses if required. With proper implantation and care, chambers can be maintained for years, allowing for successful long-term experiments in NHPs.

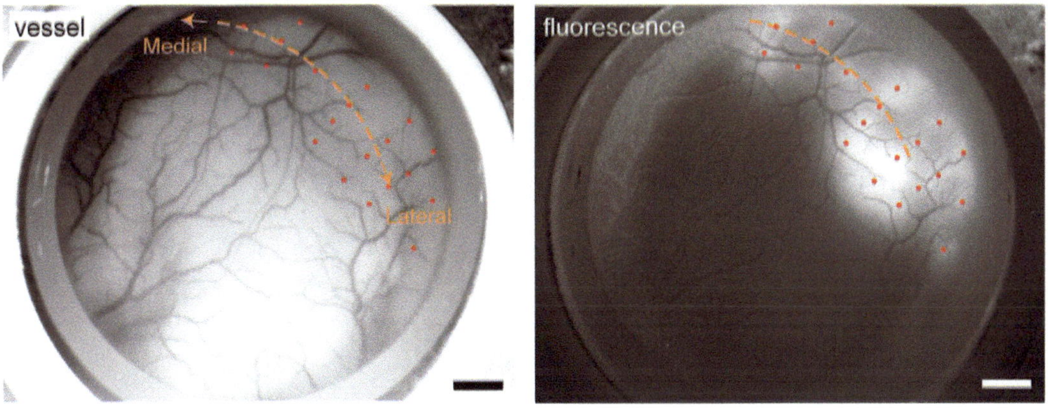

Fig. 6 Cranial windows with transparent artificial dura in NHP. (**a**) Direct visualization of the cortex and its vasculature through a cranial window with the naked eye. (**b**) Visualization of epifluorescence following viral vector injection through the window. Red dots correspond to injection sites. (Image from Nakamichi et al. [151])

2.2.2 The Use of Cranial Windows

The ability to confirm the expression of the transgene in the target tissue before the beginning of experiments is key. As mentioned above, confirmation is difficult to achieve with standard recording chambers where dural and granulation tissue prevents direct observation of epifluorescence from the cortex. While an in vivo fluorescence detector has been proposed for use in chambers, this custom device has not seen widespread adoption in the community [2]. In general, fiber photometry assessments require the insertion of a relatively large fiber into the brain to assess the transgene expression. The downside of this is that each insertion caused tissue damage. Thus, to observe in vivo epifluorescence of the transduced tissue, some investigators have adopted "cranial window" style chambers that use a transparent artificial dura. These windows were originally developed by Ruiz et al. [20] and allow direct visualization of the cortex through the artificial dura (Fig. 6). Using epifluorescence imaging, it is possible to observe the time course of viral expression in the cortex and confirm the success of the transduction before beginning experiments [23]. Investigators have reported that the use of expensive fluorescence stereomicroscopes is not necessary to perform such epifluorescence imaging and that much cheaper fluorescent goggles, such as the ones sold by NightSea Inc. (e.g., Dual Fluorescent Protein Flashlight), do an excellent job (Fig. 7).

Despite these advantages, cranial windows have their drawbacks. For one, they are not commercially available and need to be designed in-lab. Second, they require special surgical skills that include the ability to perform a precise durotomy to insert the artificial dura between the arachnoid layer and the margin of the durotomy. Third, they require a significant amount of maintenance

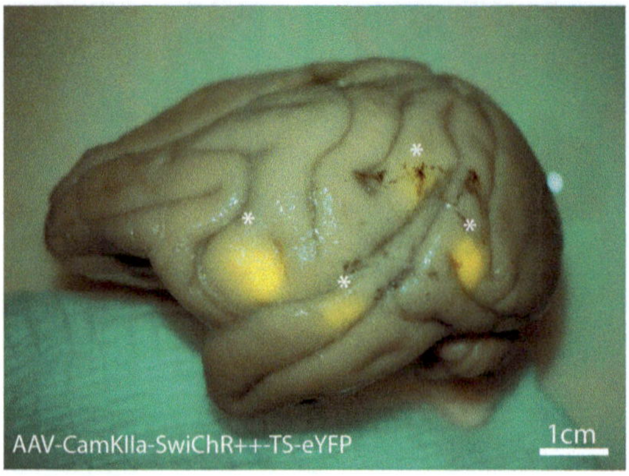

Fig. 7 Epifluorescence seen on a postmortem macaque brain using a fluorescent flashlight designed for marine exploration (NightSea Inc.). Yellow spots (marked by *) represent opsin expression in cortex

to prevent tissue growth that can obscure the window over time [20]. Fourth, the pressure exerted by the artificial dura on the underlying cortex can lead to compression of the cortical tissue, as observed postmortem by some investigators [23]. Despite these caveats, some investigators conducting studies of the cortical surface will opt for a cranial window instead of a standard chamber because of the huge benefit of being able to directly see the tissue that is being investigated. This is of course only possible when the injected tissue is on the cortical surface; the epifluorescence is invisible to the experimenter if the injection targets are deeper. Those windows also afford the opportunity to shine light directly on a large area of the cortex from a light source resting outside the brain. This can be useful for achieving behaviorally meaningful stimulation using optogenetics.

In vivo confirmation of the expression of the transgene is highly valuable, but it may not involve direct observation of epifluorescence in the tissue. Methods for assessing expression using PET imaging are described in Chapter 14. The laboratory of Michael Michaelides at NIDA is developing variants of popular viral constructs that are visible using non-invasive positron emission topography (PET) [24]. For example, by fusing the Ligand Binding Domain of the Estrogen Receptor alpha (ERa-LBD) to effector proteins such as opsins or DREADDs, this method offers a dock for PET-visible ligands, such as (18)F-fluorestradiol, that are used routinely in the clinic [25, 26]. When combined with anatomical scans (e.g., MRI T1), this allows 3D visualization of the expression of the transgene in the brain (Fig. 8), and can be repeated several times to study the time course of the expression. Although this noninvasive imaging technique has not seen widespread adoption yet, it holds great promise both for NHP basic and preclinical research and for eventual clinical applications of these neuromodulatory techniques in human patients.

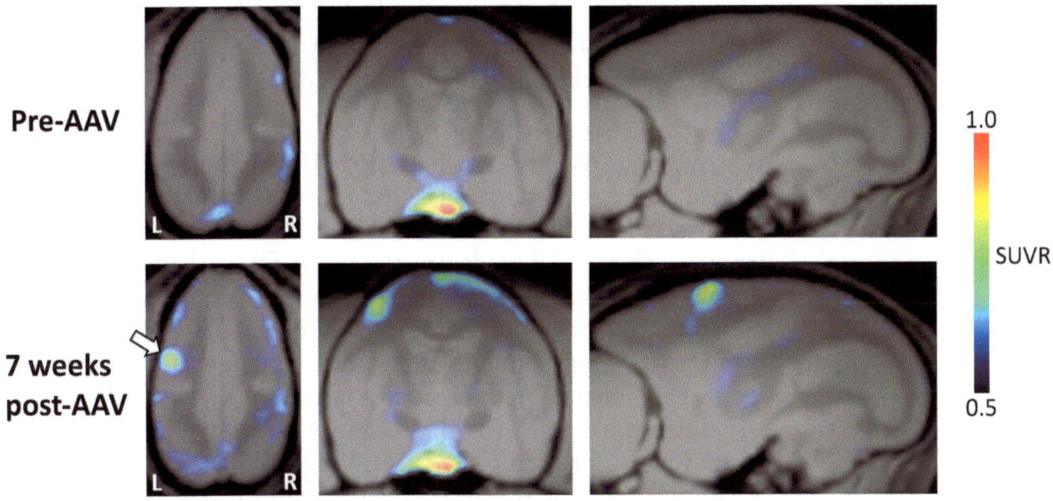

Fig. 8 FES-PET localizes ChRERα opsin in squirrel monkey. Comparison of scans pre-AAV (top) and 7 weeks following injection of AAV-ChRERα (bottom) enables visualization of ChRERα expression in the left motor cortex injection site (indicated by white arrow in bottom left image). Data is shown as average standard uptake value ratios (SUVR) 30–60 min post-FES injection (pituitary used as reference region given endogenous binding of FES observed in pre-AAV scan). (Image courtesy of Dr. Michaelides, NIDA)

3 Viral Considerations in NHP

At present, the most effective approach to deliver exogenous (non-self) genes to targeted cells of NHPs is with viral vectors. This section will describe some of the more common capsids and some of their fundamental features that should be considered, providing a basic framework for the reader.

3.1 Viral Capsid

3.1.1 Adeno-Associated Virus (AAVs)

AAV is a small (~22 nm) non-enveloped dependovirus belonging to the parvovirus family virus with a ~4.7 Kb packaging capacity [27]. AAV was first identified as a contaminant in adenoviral stocks (hence the name, Adeno-*Associated*) [28]. In the wild-type form, the AAV capsid contains a linear, single-stranded DNA genome containing two genes, the left open reading frame contains the *rep* gene while the right contains the *cap* gene, the 5' and 3' ends are flanked by 145 bp inverted terminal repeats (ITRs) [29, 30]. The *rep* gene encodes four replication proteins: Rep78, Rep68, Rep52, and Rep40 (the numbers 78, 68, 52, and 40 designate each protein's molecular weight). The four rep proteins are required for AAV assembly and packaging of the genome into newly synthesized capsids [31]. The cap gene encodes three capsid proteins (Vp1, Vp2, and Vp3) [32]. The AAV capsid is a collection of 60 subunits with a ratio of Vp1:Vp2:Vp3 = 1:1:10 [32]. In the wild, differences in the amino acid sequences of Vp1-3 influence capsid structure, in turn influencing receptor binding, enabling a given capsid to recognize and transduce different cell types. Thus, these subtle structural differences are how AAV serotypes are defined [33]. AAV1-12 were all isolated or detected in different primate species including humans, chimps, baboons, and macaques (rhesus, cynomolgus, and nemestrina) [33].

A Brief Note on AAV Nomenclature
As a vector, AAV1-12 can be stripped of its wild-type genomic payload and, to be functional, the only thing that is required is the inverted terminal repeats (ITRs) on the ends of the new genome. These ITR ends are necessary for packaging the new genome into the, now recombinant, capsid, hence the frequently reported letter "r" immediately in front of the AAV when describing a vector. Another point worth mentioning is that the dominant choice ITR for packaging genomes into rAAVs is the AAV2 ITR. The significance of this is that often in the literature a vector is referred to as, for example, rAAV2/9-CAG-GFP. In this example, "r" means that this is a recombinant capsid where the wild-type genome has been stripped and replaced with a "CAG" promoter and the gene encoding "GFP" (i.e., green fluorescent protein), the "2" refers to the AAV2 ITR that has been used to flank the packaged genome, and the "9" indicates that the genome has been packaged in the AAV9 capsid. Given the frequency with which the AAV2 ITR is used, one can assume that the AAV2 ITR is being used to package a genome of interest, even if that is not reported in the capsid name. Further, it is frequently assumed that the vector is recombinant if one is delivering exogenous genes like an actuator/indicator/anatomical marker, so the "r" is often not reported. Thus, rAAV2/9-CAG-GFP and AAV9-CAG-GFP are functionally considered equivalent. AAV serotype is a predictor of success along several dimensions: transduction efficiency, viral spread from the injection site, cell specificity of transduction in a given neuroanatomical structure, resiliency to immune responses, and pattern of transgene expression.

In primate exogenous actuator research, AAV5 has been used more commonly, but AAV1, AAV2, AAV6, AAV7, AAV9, and AAV5/9, for example, have all been used as well; for review please refer to the following: [2–6]. Comparisons of AAV serotype transduction following primate cortical injections have revealed GFP-positive neurons from AAV9 over 10 mm from their injection sites. AAV2 and 6 diffused, but not nearly as far, while AAV8 showed little spread from the injection site [34]. For the wild-type AAVs, by comparing glial to neuronal transduction in histological sections, injections of different AAV serotypes into the primate brain resulted in the following patterns of neuronal transduction specificity: AAV1 > AAV5 > AAV7 > AAV9 > AAV8 > AAV2 [35, 36]. Markakis et al. [37] showed that, of all tested AAV capsids, AAV2 transduced more neurons than glia, suggesting that capsid structure can provide some degree of cell-type specificity.

3.1.2 Chimeric AAVs

Newly engineered capsids with enhanced or desirable features are increasingly becoming popular. These capsids are most often developed using either rational design or directed evolution (described below). Fundamentally, the ***rational design*** of AAV capsids involves grafting amino acid sequences with known functions from one AAV into the sequences encoding another capsid [38]. This provides a mechanism to harness known, desirable properties from multiple AAV serotypes. For example, a 5 amino acid sequence from AAV1 grafted into AAV2 resulted in enhanced AAV2 receptor binding to produce higher muscle fiber transduction and reduced immunogenicity [39]. In another case, the amino acid sequences from AAV9 encoding the galactose-binding footprint engrafted into AAV2, which has a heparan sulfate-binding footprint, resulted in a chimeric AAV2 that could bind to both receptors and thereby enhanced the chimeric vector's transduction capacity [40]. The alternative approach is to use ***directed evolution*** which is a technique where genes encoding the Cap genes from different AAV serotypes are randomized using error-prone PCR. Through this process, numerous unique, hybrid serotypes are generated which, when combined, form a large library of hybrid AAVs. Capsids are injected, then tissues are harvested, and sequenced to determine which capsids are present in the target tissue/cell type. Through iterative testing, capsids with the most desirable features are selected, while the rest are discarded. Thus, directed evolution is a form of rapid AAV Darwinian evolution [41–44].

Some Examples of Chimeric AAV Capsids

AAV2g9: The development of rAAV2g9 used rational design to merge the galactose-binding footprint from rAAV9 with rAAV2, which contains a heparan sulfate-binding locus. This merger allows rAAV2g9 to reproduce the binding of both rAAV9 and rAAV2 for a gain of function [40]. In rodents, rAAV2g9 features enhanced transduction capacity, extended spread, retrograde transport, and minimal leakage from the central nervous system to mitigate off-target effects [40, 45]. Further, a recent report suggests that two simple arginine amino acid point mutations, R585A and R588A, in the capsid have a robust influence on rAAV2g9 propensity to diffuse from the injection site and provide post-transduction retrograde transport in rats [46]. There are no reported tests of this capsid in the NHP optogenetic database. We tested it in one animal and found that it successfully transduced both neurons and glia at the injection site. We also found evidence for minor retrograde transport in some neuronal populations known to project to the injection site and no obvious differences between the rAAV2g9 and rAAV2g9$_{\text{R585A, R588A}}$ serotypes in their propensity for diffusion from their cortical injection sites (Bohlen, unpublished observation).

AAV-DJ AAV-DJ was an early capsid developed through AAV capsid directed evolution. Its initial development was to escape anti-AAV neutralizing antibody detection and to transduce human liver cells in cell culture. Its sequence contains elements from 8 capsid types with significant fragments from parent capsids AAV2, 8, and 9 [47]. Despite an initial focus on transduction of liver cells, AAV-DJ is efficient at transducing neurons too. For example, in rodents AAV-DJ has been explored for its utility for delivering transgenes to neurons of the ventral tegmental area (VTA) [48], retina [49], striatum [50], spinal cord [51], olfactory bulb [52], and taste cells [53]. According to the NHP optogenetic database, AAV-DJ has only been reportedly tested twice in marmosets [1]. However, there have been three recent reports that have successfully used AAV-DJ to deliver exogenous genes to target neuronal populations. Ninomiya et al. [54] successfully used AAV DJ-TRE-EGFP-eTeNT in combination with AAV2 retro-CMV-rtTAV16 to selectively block ventral premotor cortex neurons projecting to the medial prefrontal cortex. Watanabe et al. [55] infused AAV-DJ-CAG-hChR2(H134R)-EYFP/tdTomato and successfully drove opsin and fluorescent protein expression in neurons of the forelimb region of primary motor cortex in macaque monkeys. Finally, Hasegawa et al. [56] used AAV-DJ-CAG-hM4Di-EGFP-WPRE to reduce activity in the subthalamic nucleus of macaques.

AAV-PHP.B AAV-PHP.B provides an important lesson in the risk of over optimization using directed evolution. AAV-PHP.B is a directly evolved variant from AAV9 that was selected for its high CNS affinity following intravascular administration in a transgenic Cre C57BL/6J mouse line [57, 58]. While extremely effective in some mouse strains [58], the CNS tropisms of PHP.B capsid failed to translate to another inbred strain of mice, BALB/cJ, marmoset, and rhesus monkeys [58–61]. In a subsequent report, intrathecal administration of PHP.B compared to a similar construct, PHP.eB revealed that the intrathecal route improved transduction, and that the PHP.eB capsid was more effective compared to PHP.B at transducing neurons in the primate brain [62]. In another series of experiments, Galvan et al. [63] showed that lateral cerebroventricular injections of AAV9-PHP.B SYN1-EmGFPWPRE in NHPs produced diffuse cortical and subcortical transduction along the rostro-caudal extent of the neuroaxis.

AAV2-retro Recent work using directed evolution has resulted in the development of rAAV2-retro, an evolved AAV variant with the power to transduce and be retrogradely transported to the soma of neurons with projections to the injection site [64]. The authors of this chapter are putting forth a major effort to validate rAAV2-retro in the primate visuomotor circuitry. In anticipation of behavioral experiments using neuronal actuators, we have first sought to anatomically characterize this capsid for its ability to retrogradely deliver exogenous genes to projecting neuronal populations. Thus far we have characterized injections into the macaque frontal eye fields and superior colliculus [65, 66]. First, this effort has shown that rAAV2-retro does provide clear retrograde labeling. For example, left superior colliculus injections resulted in retrogradely labeled neurons in the deep cerebellar nuclei. Left frontal eye field injections resulted in retrogradely labeled neurons in the contralateral frontal eye field. However, several unexpected observations have been made, which may limit the efficacy of rAAV2-retro to specific circuits. Injections in the superior colliculus fail to retrogradely label nigrotectal projecting neurons, a known projection based on conventional tracer experimentation [67]. Similarly, injections into the frontal eye field did not retrogradely label paralaminar mediodorsal thalamocortical projecting neurons, a well-established circuit in the visuomotor system [68, 69]. Further, and despite its naming, rAAV2-retro also produced clear anterograde labeling. For example, tectothalamic terminals are clearly observed in the paralaminar mediodorsal thalamus after rAAV2-retro was injected into the superior colliculus, and corticothalamic axon terminals were observed in paralaminar mediodorsal thalamus after rAAV2-retro was injected into the frontal eye field [65]. The terminal field labeling is very likely to arise predominantly from neurons whose soma is local to the injection site, as neuronal labeling can clearly be observed surrounding the needle tracks and these listed structures are known to receive inputs from the injected structures [66]. Finally, and as discussed below, the promoter portion of the genome can interact with the rAAV2-retro capsid to drive/suppress transgene expression in select retrograde populations [66].

3.1.3 Use of AAVs in Relation to NHP Experimentation

At present, it can be argued that AAVs are the workhorse for delivering transgenes to targeted neuronal populations in the primate model. This is likely a reflection of the commercial availability of this class of viral vectors, their safety profile, how hardy AAV capsids are to different insults (e.g., freeze-thaw cycles, pH changes), and ease of use (most institutions now consider AAVs to be a biosafety level one virus). According to the NHP optogenetic database, in ~84% (974/1163) of cases, investigators reported the use of AAVs. Of those studies 48% (390/974) used AAV5, followed by 22% (216/974 cases) that used AAV9. It is important to highlight that the bias in capsid choice is, in part, likely a reflection of the large neuroanatomical structures that are being targeted—84% (976/1158 cases) are targeting the cortex. At this point, it is also worth making note of two things: First, fundamentally, different capsids are going to be more or less effective depending on the targeted cellular population based on their tropisms [70]. Second, there are newly engineered capsids that are coming online all the time with improved features over parent, wild-type capsids. AAV injection volumes range from 1–1000 μL at rates ranging from 2.3–1200 nl/min [1], the volume has little predictive power in determining a successful outcome. Again, these descriptive statistics likely reflect the aforementioned availability and ease of AAV use.

3.1.4 Lentiviral Vectors

For greater details on lentiviral vectors, the reader is referred to Chapters 1 and 2 of the current text. In terms of the adoption of lentiviral vectors for use in NHPs, these vectors make up 11% (130/1163 cases) of the remaining cases reported to the NHP optogenetic database. As was the case with AAVs, according to the NHP optogenetics database, there is a strong bias toward injections placed in the cerebral cortex 72% (72/130 cases). Cortical injection numbers are followed by injection in the thalamus 16% (21/130), striatum 11% (11/130), and medial septum 1% (1/130), there was no significant effect on successful outcomes based on the targeted structure ($F = 1.107$, $p = 0.349$) [1]. Injection volumes ranged from 1–30 μL, volume has no predictive impact on the experimental outcome. Of those that did report, 96% (103/107) report some sort of successful outcome, while 4% (4/107) report unsuccessful experimental outcomes, and 5 cases fail to report an outcome [1].

3.1.5 Herpes Simplex Viral (HSV) Vectors

For greater details on HSV vectors, the reader is referred to Chapter 3 of the current text. With regards to the adoption of HSV in NHP experimentation, the HSV capsid makes up only 1% (9/1163 cases) of the tested vectors. Despite this, this vector holds promise and warrants further interrogation/optimization in the primate model. HSV-LT-Ef1a has been successfully used in squirrel monkey striatum [71] and, according to records from the NHP optogenetic database, 44% (4/9 cases) were in macaque cortex (V1, M1, and neocortex [72]), the remaining 56% (5/9 cases) were toward motoneuronal transduction following intramuscular injections. Of these cases, only one reported a successful outcome in optogenetic experiments, one weakly successful, and the rest ($n = 7$) either failed or outcome went unreported. Injected volumes for the CNS ranged from 2–12 μL, while muscle injection volumes ranged from 31–100 μL [1].

3.1.6 Rabies Viral Vectors

For greater details on rabies viral vectors, the reader is referred to Chapter 4 of the current text. Only 2% (22/1163) of NHP optogenetic database cases report on the use of rabies as a viral vector in the primate community. The reported cases can be attributed to three locations, V1, the lateral geniculate nucleus, and the subthalamic nucleus. Total volumes injected range from 5–600 μL, with successful transduction reported 62% (13/21 cases) of the time across the spectrum of injection sites, while only 14% (3/21) showed no transduction or 24% (5/21 cases) went unreported [1].

3.1.7 Canine Adenovirus Type 2 (CAV-2)

CAV-2 is a 90 nm nonenveloped icosahedral shell that contains a 32 kbp linear, double-stranded DNA genome with 198 bp ITRs that are required for DNA polymerase-mediated replication [73, 74]. It is particularly appealing as a vector for its high propensity for retrograde transport. Prior work has shown that CAV-2 binds to the coxsackie and adenovirus receptor (CAR) [75–77], once bound, the vector undergoes rapid clathrin-mediated endocytosis before being retrogradely transported to the soma [78]. Given both the genomic payload capacity and its propensity for retrograde transport, it has been used to deliver exogenous genes to the primate striatum in two publications [79, 80] and NHP motoneurons following craniofacial intramuscular injections [81]. According to the NHP optogenetic database, CAV-2 has been tested predominantly in the cortex 51% (19/37 cases) and intramuscular injections 49% (18/37 cases). Of these cases, 55% (17/33) were successful by some metric, 45% (14/33 cases) failed to show any success based on the reported metric, while 3 cases have not been reported. It is worth noting that the high failure rate is biased due to experiments involving intramuscular injections. This is critical for two reasons: first, muscles are highly vascularized. The process of making injections into a muscle causes a small degree of sheering trauma which elicits normal immunological responses to repair the damaged tissue, this also means easy immunological detection of the infused viral vector. One way to potentially evade immunological neutralization is to flood the injection site with capsids to absorb the neutralizing antibodies, allowing some capsids to escape immunological detection and neutralization, known as capsid decoys [82]. The down side of increased volumes is that the larger the payload, the more aggressive the immunological response can become [1, 81]. Thus, there is a balance between having enough virus to get successful transduction and expression of a transgene of interest, while also minimizing immunological detection.

3.2 Viral Promoters and Enhancers

Half the challenge lies with getting the genomic cargo to the target cell and into the nucleus, the viral vector and delivery method serve in this capacity. Once the genome is present in the nucleus, the next challenge is getting the transgenes expressed and the protein appropriately translated and functional. To achieve cell specificity, new promoter technologies with vastly improved cell-type specific expression are on the horizon. At present, transgene expression requires a promoter and/or enhancer element to drive mRNA expression of the exogenous gene. Both promoters and enhancers regulate gene expression, and the line between the two is increasingly being blurred. However, by definition, enhancers require a minimal promoter to be in close proximity to the transcriptional site. The combination is able to strongly express a gene of interest [83–87] and *see* Chapter 9. Finally, it is worth noting that there are a great number of ubiquitous promoters, which are only briefly discussed here.

Ca2+/Calmodulin-Dependent Protein Kinase II (CamKII) CamKII is a 1290 bp [88, 89] neuronal promoter that limits transgene expression primarily to excitatory neurons [4, 36, 90, 91]. According to the NHP optogenetic database, CamKII is the most tested promoter, making up 37% (428/1163 cases) of tested vectors reported in the database. The CamKII promoter has been packaged in several AAV serotypes and lentivirus. Seventy-one percent (813/1145 cases) report a successful outcome (i.e., behavior, physiology, or histology), 11% (128/1145) report a weak outcome, 5% (57/1145) reported a failed outcome, and the remaining 13% were cases that did not report on outcome with the CamKII promoter [1, 7].

Human Synapsin (hSyn) hSyn is an appealing promoter since it constrains transgene expression to transduced neurons and provides stable long-term transgene expression [92]. The hSyn is considered pan-neuronal [93, 94], but it is reportedly biased toward stronger gene expression in excitatory neurons compared to inhibitory ones [90, 93, 95]—though it should be worth noting that there are potential capsid-promoter interactions that may be at play [66]. Particularly, for AAVs, its compact size of 485 bp is appealing since it provides more genomic room for genes encoding actuators and indicators [93]. In the primate community, the hSyn promoter is one of the most widely applied neuron-specific promoters used both in the published literature and in the reports from the NHP optogenetic database (~20% of reported cases). There are striking similarities in hSyn when compared to CamKII; the hSyn promoter has been packaged in several AAV serotypes and lentivirus. 70% (760/1085 cases) report a successful outcome (i.e., behavior, physiology, or histology), 11% (124/1085) report a weak outcome, 5% (55/1085) reported a failed outcome, and remaining 14% were cases that did not report on outcome with the hSyn promoter [1, 7].

Tyrosine hydroxylase (TH) Lerchner et al. [34] found that the lentiviral delivery of a genome driven by the tyrosine hydroxylase promoter was effective at targeting locus coeruleus noradrenergic neurons. In a second important study, Stauffer et al. [96] drove the expression of ChR2 in rhesus midbrain dopaminergic neurons using a two-vector approach. The first was, AAV2/9-rTH-PI-Cre-SV40 which was used to deliver Cre recombinase under a 300 bp 5' fragment of the TH promoter. This constrained Cre expression to dopaminergic neurons. The second vector was,

AAV5-Ef1a-DIO-hChR2(H134R)-EYFP-WPRE-pA, which served as the driver for Cre-recombinase-dependent ChR2 expression. This confines ChR2 expression to dopaminergic neurons that have been dually transduced with both viruses and has a high degree of cell-type specificity (>95% of labeled cells were dopaminergic) [96]. From the NHP optogenetic database, ~7% (84/1163 cases) reported the use of the TH promoter, most of the reported cases packaged the TH promoter in AAV9 with a few reported to have packaged the promoter in AAV5. All AAV9 cases in the database reported some kind of successful outcome, while the AAV5 failed to show any evidence of an effect (i.e., behavior, physiology, or histology) [1, 7].

Dlx5/6 See Chapter 9. De et al. [97] used AAV9–mDlx5/6–ChR2–mCherry to show that the enhancer element, Dlx5/6 could confine ChR2 expression to GABAergic neurons within the visual cortices. The work showed that ChR2 activation of GABAergic neurons resulted in modulation of spiking activity and that this resulted in impaired visual sensitivity in an predictable way based on the location in the primary visual cortex [3, 97–99]. From the NHP optogenetic database ~0.95% (11/1163 cases) reported the use of the Dlx enhancer, all the reported cases were packaged in AAV5, and there were four reports of some sort of positive effect (i.e., behavior, physiology, or histology), the remaining cases were not reported on for efficacy [1, 7].

Parvalbumin (PV) See Chapter 9. *Parvalbumin* is a calcium-binding protein found in a subset of GABAergic neurons. To target this subpopulation of neurons in primates, enhancer elements small enough to be packaged in AAVs have been used with a high degree of anatomically defined specificity at the core of the injection sites [3, 100, 101]. From the NHP optogenetic database ~7% (84/1163 cases) reported the use of the PV promoter [1]. A second enhancer, *E2 is also highly selective for PV neurons*. The Scn1a enhancer sequence limits transgene expression to GABAergic inhibitory VIP and PV-expressing interneurons [100, 102, 103]. E2 is a portion of this enhancer sequence that is reported to have >90% specificity for PV neurons across multiple species, including primates [100].

Glial Fibrillary Acidic Protein (GFAP) Shinohara et al. [104] took the entire GFAP promoter from marmosets and generated progressively shorter variants by making progressively larger deletions from the 5' end. This shortening is required since the packaging capacity of AAV's is only ~4.7 kb [30] and lentivirus is ~10 kb [105]. The result of this effort was the discovery of a 0.3 kb GFAP promoter that maintained a high degree of astrocytic specificity with only a 60% decline in promoter strength. This 0.3 kb promoter was concluded to be an ideal balance between the ability to drive expression while minimizing the genomic burden [104, 106].

L7 In an important first study of its kind, the L7/Pcp2 promoter was used to drive the expression of ChR2 exclusively in Purkinje neurons of the cerebellum [83]. This study provided the first evidence that promoters could limit actuator transgene expression to a specific cell type within the primate brain.

ChAT In a first of its kind, CAV-2-hChAT-GFP was injected into the extraocular muscles (muscles that rotate the eye). While not a perfect answer was achieved due to technical constraints, there was promising evidence that this ChAT promoter drove expression within transduced motoneurons following intramuscular injections in macaques [81]. This early work was followed up with a second publication which showed that CAV-pChAT-GFP successfully drove transgene expression in cholinergic striatal neurons [80]. Both studies used the CAV-2 vector to deliver transgenes [1].

On the horizon is the development of new, cell-type specific targeting elements that will allow for an unparalleled degree of precision. At present, the cutting edge has been the development of the enhancer elements that allow individual cell types and patterns of gene expression to emerge within an organism containing a singular genome [107, 108]. The enhancer-based cell-type specific approaches are showing significant promise for providing unparalleled precision to target cell types in rodents and primates [97, 100, 109–111]. As is always the problem with new technologies, there remains much to be understood [107, 112–115], but such limitations provide ample opportunities for primate neuroscientists to work closely with molecular neurobiologists to validate and refine these elements for the primate model.

Importance of Promoter Choice

Promoter choice is a topic worth deep consideration concerning the experimental goals and how might the promoter influence the outcome. There is much to be learned about the role of the promoter in driving transgene expression and the timelines one might be able to expect for each respective promoter. An important example is a promoter called cytomegalovirus (CMV), which has frequently been used for CNS testing [116]. The CMV promoter is considered ubiquitous and is supposed to provide constitutive expression, however, if one is interested in transduction to cortical pyramidal neurons, gene therapy work has shown that CMV-driven expression is effectively suppressed by cortical pyramidal cells in time [117, 118].

It is worth noting that several other promoters are being used by primate researchers, many of these are not neuronal specific, and therefore likely rely on capsid transduction for cell-type and circuit specificity. This is certainly one of the appeals of rAAV2-retro in the sense that there is cell type specificity in that you are getting projecting neurons and robust transgene expression with constitutive promoters [64–66].

3.3 Interactions Between Capsids, Promoters, and Reporters.

Recent observations have shown that the choice of promoter can interact with the capsid to alter patterns of transduction in unexpected ways. For example, Powell et al. [119] found that AAV9 packaged with the constitutive promoter, chicken β-actin (CBA), and delivered to the striatum of rats, drove transgene expression in striatal neurons, however, a truncated version of the same promoter (CBh), shifted expression to oligodendrocytes. When both viral genomes were tested using AAV2, transgene expression was always detected in striatal neurons. Next, using the CBA promoter in AAV9, small changes to the amino acid sequence of the capsid shifted expression from striatal neurons to oligodendrocytes, while this modification of the CBh promoter shifted expression from oligodendrocytes to neuronal expression [119]. A promoter-capsid interaction was also observed when using rAAV2-retro capsid in the primate model [66]. In this study, patterns of retrogradely transduced neuronal populations depended on whether rAAV2-retro was carrying the hybrid chicken beta actin (CAG) or human synapsin (hSyn) promoters following injections into either the frontal eye field or superior colliculus. For example, injections of rAAV2-retro-hSyn successfully transduced and drove transgene expression in cerebellotectal projecting neurons within the deep cerebellar nuclei, while rAAV2-retro-CAG failed to drive transduction and expression in this same population. These results were made possible because the cell types were morphologically distinguishable (striatal neurons versus oligodendrocytes) [119] and anatomically present in nuclei with known projections to the injected structures [66]. It is highly likely that this promoter-capsid interaction is occurring at the injection site and has been overlooked in studies in which intense reporter expression impairs identification of individual cell types.

In another example, Galvan et al. [120] found that the anatomical marker interfered with the ability for transduced cells to appropriately traffic a chemogenetic protein to the plasma membrane in primates (see Chapter 15). When the authors tested AAV-hSyn-hM4Di-mCherry in monkeys, they discovered that the hM4Di (chemogenetic actuator) protein failed to make it to the plasma membrane. However, if the mCherry (fluorescent protein) was replaced with the HA protein, using AAV-hSyn-HA-hM4Di construct, transduced neurons were successfully able to traffic the actuator protein to the plasma membrane in monkeys.

Both examples highlight the importance of histologically assessing transduction and expression in the neuronal population of interest. More importantly, these examples highlight the fact that it cannot be assumed that changes in the genome (i.e., genomic alterations ranging from single amino acid modifications through large genomic alterations like trading one promotor for another) will not impact patterns of transduction or transgene expression and could very well play a critical role in discrepancies between labs.

3.4 Viral Reporters: Actuators and Indicators

The transgenes that are most commonly expressed in the brain for neuroscientific research can be broken into two broad categories: (1) neuronal actuators and (2) neuronal indicators. While these are discussed independently, it is worth noting that a single vector can deliver both, for example, co-expressing an opsin (i.e., neuronal actuator) with a fluorescent protein (i.e., indicator).

Actuators Actuators are protein receptors that, in the presence of an appropriate stimulus, undergo conformational changes that ultimately allow for the flow of ions and/or induction of second messenger pathways that then impact the physiological state of the affected cell. The two main categories of neuronal actuators used in neuroscientific research are optogenetic and chemogenetic receptors. As a side note, opto- and chemo-genetic actuators are the most popular that are being tested in primates. However, other actuators are being developed, these include the likes of DART (drugs acutely restricted by tethering) [121], luminopsin [122–124], sonogenetic [125], and thermogenetic [126, 127] actuators.

Opsins Opsins are a group of light-sensitive exogenous proteins that, in the presence of the appropriate wavelength of light, undergo confirmations that lead to an alteration in the neuron's electrophysiological state. Opsins have a high spatiotemporal resolution in that the effect is very rapid, but are somewhat limited in the sense that sufficient light must reach the receptors (requiring diffusion through the tissue). At present, optogenetic manipulations are arguably the most published among efforts to use genetic-based approaches in primate animal models; for review see [3–5, 128].

Chemogenetics The second actuator group is chemogenetic receptors, this is a class of receptors that lack any endogenous ligand, but rather are sensitive to artificial ligands delivered experimentally. Once the exogenous ligand is bound, this causes conformational changes to the receptor that result in either a second messenger cascade or direct ion gating, both resulting in a change in the electrophysiological state of the targeted neuronal population. Chemogenetic receptors have a lower temporal resolution, but since the endogenous ligand can cross the blood-brain barrier, it has a high spatial resolution [6, 129–131].

Indicators Neuronal indicators also fall into two general categories; there are the functional biosensors and the anatomical reporters.

Biosensors The most frequently used are Genetically Encoded Calcium Indicators (GECI). GECIs are the synthetic proteins that are a merger between a Green fluorescent protein, a calcium-binding messenger protein—Calmodulin, and the calmodulin's target peptide sequence from myosin light-chain kinase called M13 Protein—that is, GCaMP. This synthetic protein complex combination generates a molecule that is very sensitive to calcium concentration and releases green fluorescence in the presence of calcium and therefore informs the experimenter of neuronal activity [132, 133]. Seidemann et al. [134] and Ju et al. [135] successfully expressed and monitored GCaMP-indicated neuronal activity in the primary visual cortices of awake and behaving macaque monkeys for months. Calcium imaging most certainly holds promise as a noninvasive means for monitoring neuronal activity in the nonhuman primate, but the ability to maintain long-term expression seems to be a potential pitfall that will require some optimization (e.g., codon optimization) with the current generation of GCaMP proteins (Bohlen, unpublished observation). An alternative strategy for circumventing the cessation of GCaMP protein expression, which seems to be effective, is to use utilize the tetracycline-controlled transcriptional activation. This system works by putting the inducible tetracycline promoter in front of the GCaMP gene. In the presence of doxycycline, which can be mixed with the animals' food/water, the tetracycline promoter is turned on, inducing the expression of the GCaMP gene [136]. This arrangement allows for transient expression. The appealing realization is that, if the neuron is able to suppress transgene expression, one can give the neuron a break from expressing the transgene by removing the doxycycline for a few weeks. Then, expression can be reintroduced by again adding the doxorubicin to the animals' diet [137].

Additional biosensors that are coming online include the ability to monitor dopaminergic activity with dLight or GRAB-DA [138–140] and the ability to detect and monitor cholinergic signaling in vivo [141].

Anatomical Reporters The anatomical reporters include the divalent metal transporter gene (DMT1) which provides cellular enhancement on MRI scans, allowing one to assay and monitor neuronal transduction and expression in vivo [142, 143]. Additional anatomical reporters include fluorescent proteins. These are appealing because the experimenter can multiplex to determine what kinds of cells are expressing the transgene that has been delivered (e.g., Stauffer et al. [96]) determining >95% of labeled cells were dopaminergic using virally delivered EYFP under the control of TH, compared to antibody bound TH using a red fluorophore). Along that same line, a method known as, Brainbow, allows individual neurons to take on unique fluorescent identities by randomly expressing different ratios of fluorescent proteins [144, 145]. This approach is fundamental toward disentangling the brain from a connectomic angle [146]. A final category of reporters is not directly detectable, but provides a handle for immunofluorescent or immunohistochemical detection, for example, the hemagglutinin (HA) tag. HA is an epitope derived from the human influenza virus HA protein and can be used to tag proteins [6, 120], *see* Chapter 15.

4 Verification of Gene Transfer Efficacy in NHP

The verification of successful gene delivery to target cells can be investigated at multiple biological levels. One useful classification is to consider separately the anatomical, physiological, and behavioral levels, [1, 7]. The strictest level of verification, the behavioral level, asks whether the transgene has the desired effect at the systemic level, whether through modulation of cognitive or behavioral output, or through assessment of other effectors downstream of the foreign protein's action. The behavioral effect is often the ultimate goal of experiments in awake behaving NHP, and the most challenging one to achieve [147]. It stands to reason that if the behavioral effect is significant that the transduction was also successful. However, details about the safety and specificity of the transduction are difficult to assess at the behavioral level. Thus, gene delivery's success can also be assessed at the physiological level. Physiology is a broad category that includes direct assessment of the behavior of cells (e.g., through electrophysiology) or indirect assessments, such as through fMRI. In vivo electrophysiological confirmation of modulatory effects is the most popular method to confirm the successful expression of opsins in NHP [1]. In optogenetics, this is usually achieved through an optrode that links an optic fiber with a recording microelectrode and allows the isolation of spiking activity during optical stimulation. This technique is powerful because it leverages the extensive experience many NHP labs already have with awake, single-neuron electrophysiology. The same method can be used with chemogenetic approaches following systemic administration of the inducer. This technique offers a direct assessment of neurophysiological effects of stimulation but offers minimal information about the spatial extent of the transduction since neurons are sampled one at a time or in small clusters.

Techniques such as calcium imaging have recently been adopted in some NHP species, mostly in macaques and the smaller marmoset monkey [134, 148]. In combination with 1- or 2-photon excitation of opsins with light, it is possible to observe the effect of the modulation on a large neural population simultaneously in an "all-optical" approach [135]. Although this technique is a very powerful way to reveal the spatial configuration of the controlled population, it is limited to surface imaging of the cortex. Another technique permits imaging of the transduced neurons across the entire brain. Opto-fMRI is a combination of optogenetics and functional imaging using BOLD signals [149]. In this preparation, the NHP is sitting still in the scanner (usually awake and under head restraint) while the light is delivered to the neural population through a recording chamber or window. Using BOLD imaging, one can indirectly assess the modulation effect on large neural populations that span the direct stimulation site and as well as any other brain area. This has been useful to map the indirect effects of optogenetic stimulation on anatomically connected areas, for example [150].

The gold standard to assess transduction efficacy in NHP is the anatomical level (*see* Chapter 15). Using techniques such as immunohistochemistry, immunofluorescence and RNAscope, investigators can precisely image the presence of the protein/RNA across and within cells, and characterize many other parameters of these cells, such as their cellular type [120]. The ultrastructural localization of the protein can be visualized, and signs of pathology, such as somatic aggregation, can be detected. Markers of cells' health can be used to assess mechanical damage to the tissue and toxicity from the injection and the transgene expression. Investigators have used stains such as GFAP, NeuN, Iba1, H&E, Nissl, and MHC-II to image various cellular markers of toxicity, such as gliosis, cell deaths, and microglia reactivity (Fig. 9) [16, 72]. The proportion of cells expressing the transgene can be quantified using histological tools and image analysis software, providing an important piece of information with direct behavioral relevance [2]. For most NHP laboratories, however, it is not feasible to sacrifice several animals for histological examination and optimization of viral expression parameters before beginning a study. Laboratory monkeys are hard to obtain and expensive, and years are invested in their behavioral training. In addition, histological workflows designed for mice histology need to be adapted to the peculiarity of the NHP brain (Fig. 10). However, this does not eliminate the crucial importance of histological examination to confirm the accuracy of expression patterns of transgenes and interpret behavioral effects. There are many experimental pitfalls associated with viral transduction and expression of the transgene, which can only be identified and

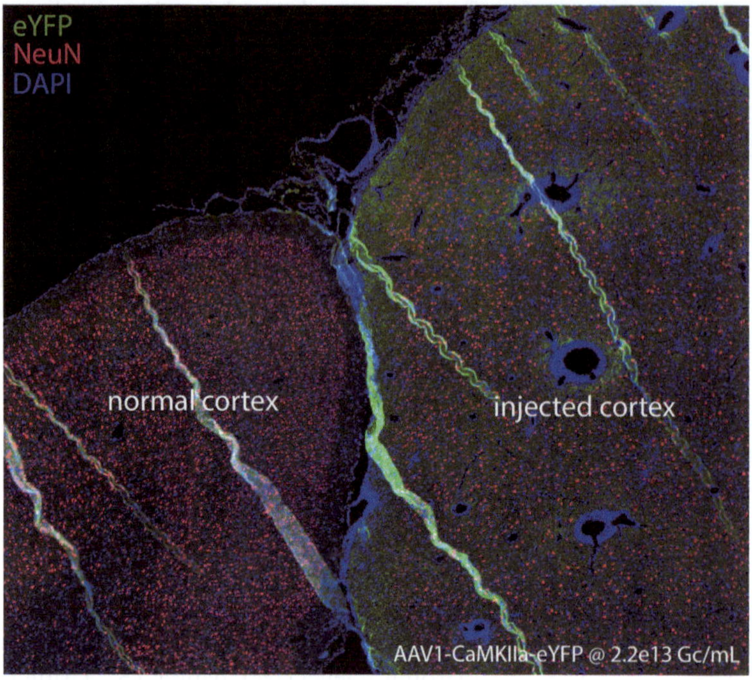

Fig. 9 Neural toxicity following AAV injection in the NHP premotor cortex. Compared to normal cortex, injected cortex exhibits significant neuronal loss (reduced NeuN labeling), perivascular cuffing (seen with DAPI)

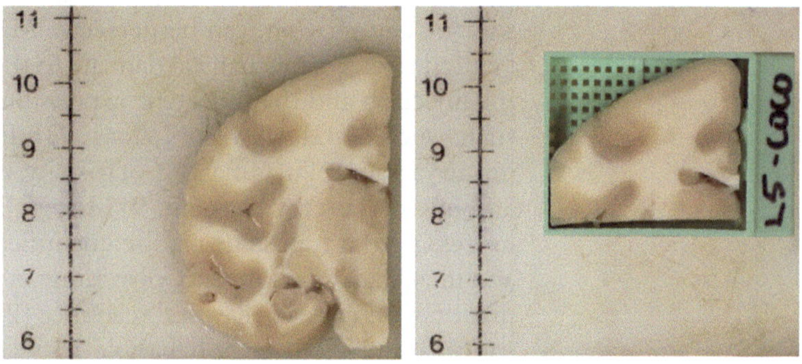

Fig. 10 Size of one macaque brain hemisphere in relation to a 1 × 2in cassette typically used for histological (paraffin-embedded) processing of the mouse brain. Only a small fraction of one hemisphere can be visualized using this histological equipment. The scale on the left is in centimeter

addressed with histological examination to identify the injection site location and the cell types that have been transduced. Open data resources, such as the NHP Optogenetics Open Database, allow investigators to learn from the experience of others and select the best parameters before the beginning of the study, potentially saving significant piloting efforts. We hope such resources will be expanded to include other techniques beyond optogenetics, such as chemogenetic and calcium imaging, and the right incentive structure is put in place so these open resources keep growing over the years.

5 Summary

Though the fields of primate neuronal actuators and indicators are still within their technical infancy, there is much to be excited about and much to be hopeful for as protocols and reagents are refined and optimized for the primate model. As the technology matures and proof of efficacy studies get published, we foresee a rapid adoption of these revolutionary approaches in primate labs akin to the adoption of the tungsten electrode. Importantly, there is also clear clinical relevance not only in the ability to use actuators to model neurological disorders, but also to treat such disorders clinically. Finally, adopting such techniques in the primate model has its challenges in the sense that it is not feasible to test/optimize constructs across several animals, as is often afforded to rodent laboratories. There are, however, ways to circumvent this to some degree. First, it is worth acknowledging that as a community our greatest asset is one another. We posit that the best approach toward adopting viral approaches in a primate lab is to first discuss the experimental goals with other groups conducting such work in NHPs. This is arguably one of the more important outcomes of the NHP optogenetic database. The second, and final point we would like the reader to depart with, if a reagent (i.e., vector/genome) has not been validated in a primate, or the specific neuroanatomical target of interest, there is a fundamental value in performing histological confirmation in one animal prior to performing physiological or behavioral work. This comes at the expense of some time and resources, but can be highly valuable in preventing a major loss of time-spent training an animal where the tested constructs never would have worked in the targeted neuronal population of interest.

References

1. Tremblay S, Acker L, Afraz A, Albaugh DL, Amita H, Andrei AR, Angelucci A, Aschner A, Balan PF, Basso MA, Benvenuti G, Bohlen MO, Caiola MJ, Calcedo R, Cavanaugh J, Chen Y, Chen S, Chernov MM, Clark AM, Dai J, Debes SR, Deisseroth K, Desimone R, Dragoi V, Egger SW, Eldridge MAG, El-Nahal HG, Fabbrini F, Federer F, Fetsch CR, Fortuna MG, Friedman RM, Fujii N, Gail A, Galvan A, Ghosh S, Gieselmann MA, Gulli RA, Hikosaka O, Hosseini EA, Hu X, Huer J, Inoue KI, Janz R, Jazayeri M, Jiang R, Ju N, Kar K, Klein C, Kohn A, Komatsu M, Maeda K, Martinez-Trujillo JC, Matsumoto M, Maunsell JHR, Mendoza-Halliday D, Monosov IE, Muers RS, Nurminen L, Ortiz-Rios M, O'Shea DJ, Palfi S, Petkov CI, Pojoga S, Rajalingham R, Ramakrishnan C, Remington ED, Revsine C, Roe AW, Sabes PN, Saunders RC, Scherberger H, Schmid MC, Schultz W, Seidemann E, Senova YS, Shadlen MN, Sheinberg DL, Siu C, Smith Y, Solomon SS, Sommer MA, Spudich JL, Stauffer WR,

Takada M, Tang S, Thiele A, Treue S, Vanduffel W, Vogels R, Whitmire MP, Wichmann T, Wurtz RH, Xu H, Yazdan-Shahmorad A, Shenoy KV, DiCarlo JJ, Platt ML (2020) An open resource for non-human primate optogenetics. Neuron. https://doi.org/10.1016/j.neuron.2020.09.027

2. Diester I, Kaufman MT, Mogri M, Pashaie R, Goo W, Yizhar O, Ramakrishnan C, Deisseroth K, Shenoy KV (2011) An optogenetic toolbox designed for primates. Nat Neurosci 14(3):387–397. https://doi.org/10.1038/nn.2749

3. El-Shamayleh Y, Horwitz GD (2019) Primate optogenetics: progress and prognosis. Proc Natl Acad Sci U S A. https://doi.org/10.1073/pnas.1902284116

4. Galvan A, Caiola MJ, Albaugh DL (2018) Advances in optogenetic and chemogenetic methods to study brain circuits in non-human primates. J Neural Transm (Vienna) 125(3):547–563. https://doi.org/10.1007/s00702-017-1697-8

5. Galvan A, Stauffer WR, Acker L, El-Shamayleh Y, Inoue KI, Ohayon S, Schmid MC (2017) Nonhuman primate optogenetics: recent advances and future directions. J Neurosci 37(45):10894–10903. https://doi.org/10.1523/JNEUROSCI.1839-17.2017

6. Raper J, Galvan A (2022) Applications of chemogenetics in non-human primates. Curr Opin Pharmacol:102204. https://doi.org/10.1016/j.coph.2022.102204

7. Bliss-Moreau E, Costa VD, Baxter MG (2020) A pragmatic reevaluation of the efficacy of nonhuman primate optogenetics. bioRxiv:20202012201042033l. https://doi.org/10.1101/2020.12.10.420331

8. Herculano-Houzel S (2009) The human brain in numbers: a linearly scaled-up primate brain. Front Hum Neurosci 3:31. https://doi.org/10.3389/neuro.09.031.2009

9. Dai J, Ozden I, Brooks DI, Wagner F, May T, Agha NS, Brush B, Borton D, Nurmikko AV, Sheinberg DL (2015) Modified toolbox for optogenetics in the nonhuman primate. Neurophotonics 2(3):031202. https://doi.org/10.1117/1.NPh.2.3.031202

10. Mendoza SD, El-Shamayleh Y, Horwitz GD (2017) AAV-mediated delivery of optogenetic constructs to the macaque brain triggers humoral immune responses. J Neurophysiol 117(5):2004–2013. https://doi.org/10.1152/jn.00780.2016

11. Fredericks JM, Dash KE, Jaskot EM, Bennett TW, Lerchner W, Dold G, Ide D, Cummins AC, Der Minassian VH, Turchi JN, Richmond BJ, Eldridge MAG (2020) Methods for mechanical delivery of viral vectors into rhesus monkey brain. J Neurosci Methods 339:108730. https://doi.org/10.1016/j.jneumeth.2020.108730

12. Bobo RH, Laske DW, Akbasak A, Morrison PF, Dedrick RL, Oldfield EH (1994) Convection-enhanced delivery of macromolecules in the brain. Proc Natl Acad Sci U S A 91(6):2076–2080. https://doi.org/10.1073/pnas.91.6.2076

13. Khateeb K, Griggs DJ, Sabes PN, Yazdan-Shahmorad A (2019) Convection enhanced delivery of optogenetic Adeno-associated viral vector to the cortex of Rhesus Macaque under guidance of online MRI images. J Vis Exp 147. https://doi.org/10.3791/59232

14. Krauze MT, Saito R, Noble C, Tamas M, Bringas J, Park JW, Berger MS, Bankiewicz K (2005) Reflux-free cannula for convection-enhanced high-speed delivery of therapeutic agents. J Neurosurg 103(5):923–929. https://doi.org/10.3171/jns.2005.103.5.0923

15. Szerlip NJ, Walbridge S, Yang L, Morrison PF, Degen JW, Jarrell ST, Kouri J, Kerr PB, Kotin R, Oldfield EH, Lonser RR (2007) Real-time imaging of convection-enhanced delivery of viruses and virus-sized particles. J Neurosurg 107(3):560–567. https://doi.org/10.3171/JNS-07/09/0560

16. Yazdan-Shahmorad A, Tian N, Kharazia V, Samaranch L, Kells A, Bringas J, He J, Bankiewicz K, Sabes PN (2018) Widespread optogenetic expression in macaque cortex obtained with MR-guided, convection enhanced delivery (CED) of AAV vector to the thalamus. J Neurosci Methods 293:347–358. https://doi.org/10.1016/j.jneumeth.2017.10.009

17. Han X, Qian X, Bernstein JG, Zhou HH, Franzesi GT, Stern P, Bronson RT, Graybiel AM, Desimone R, Boyden ES (2009) Millisecond-timescale optical control of neural dynamics in the nonhuman primate brain. Neuron 62(2):191–198. https://doi.org/10.1016/j.neuron.2009.03.011

18. Miocinovic S, Noecker AM, Maks CB, Butson CR, McIntyre CC (2007) Cicerone: stereotactic neurophysiological recording and deep brain stimulation electrode placement software system. Acta Neurochir Suppl 97(Pt 2):561–567. https://doi.org/10.1007/978-3-211-33081-4_65

19. Chen X, Possel JK, Wacongne C, van Ham AF, Klink PC, Roelfsema PR (2017) 3D printing and modelling of customized

implants and surgical guides for non-human primates. J Neurosci Methods 286:38–55. https://doi.org/10.1016/j.jneumeth.2017.05.013
20. Ruiz O, Lustig BR, Nassi JJ, Cetin A, Reynolds JH, Albright TD, Callaway EM, Stoner GR, Roe AW (2013) Optogenetics through windows on the brain in the nonhuman primate. J Neurophysiol 110(6):1455–1467. https://doi.org/10.1152/jn.00153.2013
21. Kliem MA, Wichmann T (2004) A method to record changes in local neuronal discharge in response to infusion of small drug quantities in awake monkeys. J Neurosci Methods 138(1-2):45–49. https://doi.org/10.1016/j.jneumeth.2004.03.015
22. Noudoost B, Moore T (2011) A reliable microinjectrode system for use in behaving monkeys. J Neurosci Methods 194(2):218–223. https://doi.org/10.1016/j.jneumeth.2010.10.009
23. Yazdan-Shahmorad A, Diaz-Botia C, Hanson TL, Kharazia V, Ledochowitsch P, Maharbiz MM, Sabes PN (2016) A large-scale interface for optogenetic stimulation and recording in nonhuman primates. Neuron 89(5):927–939. https://doi.org/10.1016/j.neuron.2016.01.013
24. Boehm MA, Bonaventura J, Gomez JL, Solis O, Stein EA, Bradberry CW, Michaelides M (2021) Translational PET applications for brain circuit mapping with transgenic neuromodulation tools. Pharmacol Biochem Behav 204:173147. https://doi.org/10.1016/j.pbb.2021.173147
25. Bonaventura J, Eldridge MAG, Hu F, Gomez JL, Sanchez-Soto M, Abramyan AM, Lam S, Boehm MA, Ruiz C, Farrell MR, Moreno A, Galal Faress IM, Andersen N, Lin JY, Moaddel R, Morris PJ, Shi L, Sibley DR, Mahler SV, Nabavi S, Pomper MG, Bonci A, Horti AG, Richmond BJ, Michaelides M (2019) High-potency ligands for DREADD imaging and activation in rodents and monkeys. Nat Commun 10(1):4627. https://doi.org/10.1038/s41467-019-12236-z
26. Hu F, Morris PJ, Bonaventura J, Fan H, Mathews WB, Holt DP, Lam S, Boehm M, Dannals RF, Pomper MG, Michaelides M, Horti AG (2021) (18)F-labeled radiotracers for in vivo imaging of DREADD with positron emission tomography. Eur J Med Chem 213:113047. https://doi.org/10.1016/j.ejmech.2020.113047
27. Grieger JC, Samulski RJ (2005) Adeno-associated virus as a gene therapy vector: vector development, production and clinical applications. Adv Biochem Eng Biotechnol 99:119–145
28. Atchison RW, Casto BC, Hammon WM (1965) Adenovirus-associated defective virus particles. Science 149(3685):754–756. https://doi.org/10.1126/science.149.3685.754
29. Hastie E, Samulski RJ (2015) Adeno-associated virus at 50: a golden anniversary of discovery, research, and gene therapy success--a personal perspective. Hum Gene Ther 26(5):257–265. https://doi.org/10.1089/hum.2015.025
30. Wu Z, Asokan A, Samulski RJ (2006) Adeno-associated virus serotypes: vector toolkit for human gene therapy. Mol Ther 14(3):316–327. https://doi.org/10.1016/j.ymthe.2006.05.009
31. Balakrishnan B, Jayandharan GR (2014) Basic biology of adeno-associated virus (AAV) vectors used in gene therapy. Curr Gene Ther 14(2):86–100. https://doi.org/10.2174/1566523214666140302193709
32. Li C, Samulski RJ (2020) Engineering adeno-associated virus vectors for gene therapy. Nat Rev Genet 21(4):255–272. https://doi.org/10.1038/s41576-019-0205-4
33. Gao G, Vandenberghe LH, Wilson JM (2005) New recombinant serotypes of AAV vectors. Curr Gene Ther 5(3):285–297. https://doi.org/10.2174/1566523054065057
34. Lerchner W, Corgiat B, Der Minassian V, Saunders RC, Richmond BJ (2014) Injection parameters and virus dependent choice of promoters to improve neuron targeting in the nonhuman primate brain. Gene Ther 21(3):233–241. https://doi.org/10.1038/gt.2013.75
35. Gerits A, Vancraeyenest P, Vreysen S, Laramee ME, Michiels A, Gijsbers R, Van den Haute C, Moons L, Debyser Z, Baekelandt V, Arckens L, Vanduffel W (2015) Serotype-dependent transduction efficiencies of recombinant adeno-associated viral vectors in monkey neocortex. Neurophotonics 2(3):031209. https://doi.org/10.1117/1.NPh.2.3.031209
36. Watakabe A, Ohtsuka M, Kinoshita M, Takaji M, Isa K, Mizukami H, Ozawa K, Isa T, Yamamori T (2015) Comparative analyses of adeno-associated viral vector serotypes 1, 2, 5, 8 and 9 in marmoset, mouse and macaque cerebral cortex. Neurosci Res 93:144–157. https://doi.org/10.1016/j.neures.2014.09.002
37. Markakis EA, Vives KP, Bober J, Leichtle S, Leranth C, Beecham J, Elsworth JD, Roth

RH, Samulski RJ, Redmond DE Jr (2010) Comparative transduction efficiency of AAV vector serotypes 1-6 in the substantia nigra and striatum of the primate brain. Mol Ther 18(3):588–593. https://doi.org/10.1038/mt.2009.286

38. Lee EJ, Guenther CM, Suh J (2018) Adeno-Associated Virus (AAV) Vectors: rational design strategies for capsid engineering. Curr Opin Biomed Eng 7:58–63. https://doi.org/10.1016/j.cobme.2018.09.004

39. Bowles DE, McPhee SW, Li C, Gray SJ, Samulski JJ, Camp AS, Li J, Wang B, Monahan PE, Rabinowitz JE, Grieger JC, Govindasamy L, Agbandje-McKenna M, Xiao X, Samulski RJ (2012) Phase 1 gene therapy for Duchenne muscular dystrophy using a translational optimized AAV vector. Mol Ther 20(2):443–455. https://doi.org/10.1038/mt.2011.237

40. Shen S, Horowitz ED, Troupes AN, Brown SM, Pulicherla N, Samulski RJ, Agbandje-McKenna M, Asokan A (2013) Engraftment of a galactose receptor footprint onto adeno-associated viral capsids improves transduction efficiency. J Biol Chem 288(40):28814–28823. https://doi.org/10.1074/jbc.M113.482380

41. Maheshri N, Koerber JT, Kaspar BK, Schaffer DV (2006) Directed evolution of adeno-associated virus yields enhanced gene delivery vectors. Nat Biotechnol 24(2):198–204. https://doi.org/10.1038/nbt1182

42. Marsic D, Zolotukhin S (2016) Altering tropism of rAAV by directed evolution. Methods Mol Biol 1382:151–173. https://doi.org/10.1007/978-1-4939-3271-9_11

43. Nonnenmacher M, van Bakel H, Hajjar RJ, Weber T (2015) High capsid-genome correlation facilitates creation of AAV libraries for directed evolution. Mol Ther 23(4):675–682. https://doi.org/10.1038/mt.2015.3

44. Yang L, Li J, Xiao X (2011) Directed evolution of adeno-associated virus (AAV) as vector for muscle gene therapy. Methods Mol Biol 709:127–139. https://doi.org/10.1007/978-1-61737-982-6_8

45. Murlidharan G, Sakamoto K, Rao L, Corriher T, Wang D, Gao G, Sullivan P, Asokan A (2016) CNS-restricted transduction and CRISPR/Cas9-mediated gene deletion with an engineered AAV vector. Mol Ther Nucleic Acids 5(7):e338. https://doi.org/10.1038/mtna.2016.49

46. Gorbatyuk OS, Warrington KH, Gorbatyuk MS, Zolotukhin I, Lewin AS, Muzyczka N (2019) Biodistribution of adeno-associated virus type 2 with mutations in the capsid that contribute to heparan sulfate proteoglycan binding. Virus Res 274:197771. https://doi.org/10.1016/j.virusres.2019.197771

47. Grimm D, Lee JS, Wang L, Desai T, Akache B, Storm TA, Kay MA (2008) In vitro and in vivo gene therapy vector evolution via multispecies interbreeding and retargeting of adeno-associated viruses. J Virol 82(12):5887–5911. https://doi.org/10.1128/JVI.00254-08

48. Koshimizu Y, Isa K, Kobayashi K, Isa T (2021) Double viral vector technology for selective manipulation of neural pathways with higher level of efficiency and safety. Gene Ther 28(6):339–350. https://doi.org/10.1038/s41434-020-00212-y

49. Katada Y, Kobayashi K, Tsubota K, Kurihara T (2019) Evaluation of AAV-DJ vector for retinal gene therapy. PeerJ 7:e6317. https://doi.org/10.7717/peerj.6317

50. Jang M, Lee SE, Cho IH (2018) Adeno-Associated Viral Vector Serotype DJ-mediated overexpression of N171-82Q-mutant huntingtin in the striatum of Juvenile mice is a new model for Huntington's disease. Front Cell Neurosci 12:157. https://doi.org/10.3389/fncel.2018.00157

51. Candelas M, Reynders A, Arango-Lievano M, Neumayer C, Fruquiere A, Demes E, Hamid J, Lemmers C, Bernat C, Monteil A, Compan V, Laffray S, Inquimbert P, Le Feuvre Y, Zamponi GW, Moqrich A, Bourinet E, Mery PF (2019) Cav3.2 T-type calcium channels shape electrical firing in mouse Lamina II neurons. Sci Rep 9(1):3112. https://doi.org/10.1038/s41598-019-39703-3

52. Kikusui T, Kajita M, Otsuka N, Hattori T, Kumazawa K, Watarai A, Nagasawa M, Inutsuka A, Yamanaka A, Matsuo N, Covington HE 3rd, Mogi K (2018) Sex differences in olfactory-induced neural activation of the amygdala. Behav Brain Res 346:96–104. https://doi.org/10.1016/j.bbr.2017.11.034

53. Taruno A, Kashio M, Sun H, Kobayashi K, Sano H, Nambu A, Marunaka Y (2017) Adeno-associated virus-mediated gene transfer into taste cells in vivo. Chem Senses 42(1):69–78. https://doi.org/10.1093/chemse/bjw101

54. Ninomiya T, Noritake A, Kobayashi K, Isoda M (2020) A causal role for frontal cortico-cortical coordination in social action monitoring. Nat Commun 11(1):5233. https://doi.org/10.1038/s41467-020-19026-y

55. Watanabe H, Sano H, Chiken S, Kobayashi K, Fukata Y, Fukata M, Mushiake H, Nambu A

(2020) Forelimb movements evoked by optogenetic stimulation of the macaque motor cortex. Nat Commun 11(1):3253. https://doi.org/10.1038/s41467-020-16883-5

56. Hasegawa T, Chiken S, Kobayashi K, Nambu A (2022) Subthalamic nucleus stabilizes movements by reducing neural spike variability in monkey basal ganglia. Nat Commun 13(1):2233. https://doi.org/10.1038/s41467-022-29750-2
57. Deverman BE, Pravdo PL, Simpson BP, Kumar SR, Chan KY, Banerjee A, Wu WL, Yang B, Huber N, Pasca SP, Gradinaru V (2016) Cre-dependent selection yields AAV variants for widespread gene transfer to the adult brain. Nat Biotechnol 34(2):204–209. https://doi.org/10.1038/nbt.3440
58. Matsuzaki Y, Tanaka M, Hakoda S, Masuda T, Miyata R, Konno A, Hirai H (2019) Neurotropic properties of AAV-PHP.B are shared among diverse inbred strains of mice. Mol Ther 27(4):700–704. https://doi.org/10.1016/j.ymthe.2019.02.016
59. Matsuzaki Y, Konno A, Mochizuki R, Shinohara Y, Nitta K, Okada Y, Hirai H (2018) Intravenous administration of the adeno-associated virus-PHP.B capsid fails to upregulate transduction efficiency in the marmoset brain. Neurosci Lett 665:182–188. https://doi.org/10.1016/j.neulet.2017.11.049
60. Hordeaux J, Wang Q, Katz N, Buza EL, Bell P, Wilson JM (2018) The neurotropic properties of AAV-PHP.B are limited to C57BL/6J mice. Mol Ther 26(3):664–668. https://doi.org/10.1016/j.ymthe.2018.01.018
61. Liguore WA, Domire JS, Button D, Wang Y, Dufour BD, Srinivasan S, McBride JL (2019) AAV-PHP.B Administration results in a differential pattern of CNS biodistribution in non-human primates compared with mice. Mol Ther 27(11):2018–2037. https://doi.org/10.1016/j.ymthe.2019.07.017
62. Arotcarena ML, Dovero S, Biendon N, Dutheil N, Planche V, Bezard E, Dehay B (2021) Pilot study assessing the impact of intrathecal administration of variants AAV-PHP.B and AAV-PHP.eB on brain transduction in adult rhesus macaques. Front Bioeng Biotechnol 9:762209. https://doi.org/10.3389/fbioe.2021.762209
63. Galvan A, Petkau TL, Hill AM, Korecki AJ, Lu G, Choi D, Rahman K, Simpson EM, Leavitt BR, Smith Y (2021) Intracerebroventricular Administration of AAV9-PHP.B SYN1-EmGFP induces widespread transgene expression in the mouse and monkey central nervous system. Hum Gene Ther 32(11-12):599–615. https://doi.org/10.1089/hum.2020.301
64. Tervo DG, Hwang BY, Viswanathan S, Gaj T, Lavzin M, Ritola KD, Lindo S, Michael S, Kuleshova E, Ojala D, Huang CC, Gerfen CR, Schiller J, Dudman JT, Hantman AW, Looger LL, Schaffer DV, Karpova AY (2016) A designer AAV variant permits efficient retrograde access to projection neurons. Neuron 92(2):372–382. https://doi.org/10.1016/j.neuron.2016.09.021
65. Cushnie AK, El-Nahal HG, Bohlen MO, May PJ, Basso MA, Grimaldi P, Wang MZ, de Velasco EMF, Sommer MA, Heilbronner SR (2020) Using rAAV2-retro in rhesus macaques: promise and caveats for circuit manipulation. J Neurosci Methods 345:108859. https://doi.org/10.1016/j.jneumeth.2020.108859
66. Bohlen MO, McCown TJ, Powell SK, El-Nahal H, Daw T, Basso MA, Sommer MA, Samulski RJ (2020) AAV capsid-promoter interactions in the brain translate from rat to the non-human primate. Hum Gene Ther. https://doi.org/10.1089/hum.2020.196
67. May PJ (2006) The mammalian superior colliculus: laminar structure and connections. Prog Brain Res 151:321–378. https://doi.org/10.1016/S0079-6123(05)51011-2
68. Leichnetz GR (1981) The prefrontal cortico-oculomotor trajectories in the monkey. J Neurol Sci 49(3):387–396
69. Leichnetz GR, Spencer RF, Hardy SG, Astruc J (1981) The prefrontal corticotectal projection in the monkey; an anterograde and retrograde horseradish peroxidase study. Neuroscience 6(6):1023–1041
70. Castle MJ, Turunen HT, Vandenberghe LH, Wolfe JH (2016) Controlling AAV tropism in the nervous system with natural and engineered capsids. Methods Mol Biol 1382:133–149. https://doi.org/10.1007/978-1-4939-3271-9_10
71. Gold SJ, Hoang CV, Potts BW, Porras G, Pioli E, Kim KW, Nadjar A, Qin C, LaHoste GJ, Li Q, Bioulac BH, Waugh JL, Gurevich E, Neve RL, Bezard E (2007) RGS9-2 negatively modulates L-3,4-dihydroxyphenylalanine-induced dyskinesia in experimental Parkinson's disease. J Neurosci 27(52):14338–14348. https://doi.org/10.1523/JNEUROSCI.4223-07.2007
72. O'Shea DJ, Kalanithi P, Ferenczi EA, Hsueh B, Chandrasekaran C, Goo W, Diester I, Ramakrishnan C, Kaufman MT, Ryu SI, Yeom KW, Deisseroth K, Shenoy KV

(2018) Development of an optogenetic toolkit for neural circuit dissection in squirrel monkeys. Sci Rep 8(1):6775. https://doi.org/10.1038/s41598-018-24362-7

73. Del Rio D, Beucher B, Lavigne M, Wehbi A, Gonzalez Dopeso-Reyes I, Saggio I, Kremer EJ (2019) CAV-2 vector development and gene transfer in the central and peripheral nervous systems. Front Mol Neurosci 12:71. https://doi.org/10.3389/fnmol.2019.00071
74. Kremer EJ, Nemerow GR (2015) Adenovirus tales: from the cell surface to the nuclear pore complex. PLoS Pathog 11(6):e1004821. https://doi.org/10.1371/journal.ppat.1004821
75. Loustalot F, Kremer EJ, Salinas S (2016) Membrane dynamics and signaling of the coxsackievirus and adenovirus receptor. Int Rev Cell Mol Biol 322:331–362. https://doi.org/10.1016/bs.ircmb.2015.10.006
76. Soudais C, Boutin S, Hong SS, Chillon M, Danos O, Bergelson JM, Boulanger P, Kremer EJ (2000) Canine adenovirus type 2 attachment and internalization: coxsackievirus-adenovirus receptor, alternative receptors, and an RGD-independent pathway. J Virol 74(22):10639–10649
77. Soudais C, Laplace-Builhe C, Kissa K, Kremer EJ (2001) Preferential transduction of neurons by canine adenovirus vectors and their efficient retrograde transport in vivo. FASEB J 15(12):2283–2285. https://doi.org/10.1096/fj.01-0321fje
78. Salinas S, Bilsland LG, Henaff D, Weston AE, Keriel A, Schiavo G, Kremer EJ (2009) CAR-associated vesicular transport of an adenovirus in motor neuron axons. PLoS Pathog 5(5):e1000442. https://doi.org/10.1371/journal.ppat.1000442
79. Mestre-Frances N, Serratrice N, Gennetier A, Devau G, Cobo S, Trouche SG, Fontes P, Zussy C, De Deurwaerdere P, Salinas S, Mennechet FJ, Dusonchet J, Schneider BL, Saggio I, Kalatzis V, Luquin-Piudo MR, Verdier JM, Kremer EJ (2018) Exogenous LRRK2G2019S induces parkinsonian-like pathology in a nonhuman primate. JCI Insight 3(14). https://doi.org/10.1172/jci.insight.98202
80. Martel AC, Elseedy H, Lavigne M, Scapula J, Ghestem A, Kremer EJ, Esclapez M, Apicella P (2020) Targeted transgene expression in cholinergic interneurons in the monkey striatum using canine adenovirus serotype 2 vectors. Front Mol Neurosci 13:76. https://doi.org/10.3389/fnmol.2020.00076
81. Bohlen MO, El-Nahal HG, Sommer MA (2019) Transduction of craniofacial motoneurons following intramuscular injections of canine adenovirus type-2 (CAV-2) in rhesus macaques. Front Neuroanat 13(84). https://doi.org/10.3389/fnana.2019.00084
82. Mingozzi F, Anguela XM, Pavani G, Chen Y, Davidson RJ, Hui DJ, Yazicioglu M, Elkouby L, Hinderer CJ, Faella A, Howard C, Tai A, Podsakoff GM, Zhou S, Basner-Tschakarjan E, Wright JF, High KA (2013) Overcoming preexisting humoral immunity to AAV using capsid decoys. Sci Transl Med 5(194):194ra192. https://doi.org/10.1126/scitranslmed.3005795
83. El-Shamayleh Y, Kojima Y, Soetedjo R, Horwitz GD (2017) Selective Optogenetic control of purkinje cells in monkey cerebellum. Neuron 95(1):51–62e54. https://doi.org/10.1016/j.neuron.2017.06.002
84. Core LJ, Martins AL, Danko CG, Waters CT, Siepel A, Lis JT (2014) Analysis of nascent RNA identifies a unified architecture of initiation regions at mammalian promoters and enhancers. Nat Genet 46(12):1311–1320. https://doi.org/10.1038/ng.3142
85. Dao LTM, Galindo-Albarran AO, Castro-Mondragon JA, Andrieu-Soler C, Medina-Rivera A, Souaid C, Charbonnier G, Griffon A, Vanhille L, Stephen T, Alomairi J, Martin D, Torres M, Fernandez N, Soler E, van Helden J, Puthier D, Spicuglia S (2017) Genome-wide characterization of mammalian promoters with distal enhancer functions. Nat Genet 49(7):1073–1081. https://doi.org/10.1038/ng.3884
86. Haberle V, Stark A (2018) Eukaryotic core promoters and the functional basis of transcription initiation. Nat Rev Mol Cell Biol 19(10):621–637. https://doi.org/10.1038/s41580-018-0028-8
87. Henriques T, Scruggs BS, Inouye MO, Muse GW, Williams LH, Burkholder AB, Lavender CA, Fargo DC, Adelman K (2018) Widespread transcriptional pausing and elongation control at enhancers. Genes Dev 32(1):26–41. https://doi.org/10.1101/gad.309351.117
88. Yizhar O, Fenno LE, Davidson TJ, Mogri M, Deisseroth K (2011) Optogenetics in neural systems. Neuron 71(1):9–34. https://doi.org/10.1016/j.neuron.2011.06.004
89. Yizhar O, Fenno LE, Prigge M, Schneider F, Davidson TJ, O'Shea DJ, Sohal VS, Goshen I, Finkelstein J, Paz JT, Stehfest K, Fudim R, Ramakrishnan C, Huguenard JR,

Hegemann P, Deisseroth K (2011) Neocortical excitation/inhibition balance in information processing and social dysfunction. Nature 477(7363):171–178. https://doi.org/10.1038/nature10360
90. Yaguchi M, Ohashi Y, Tsubota T, Sato A, Koyano KW, Wang N, Miyashita Y (2013) Characterization of the properties of seven promoters in the motor cortex of rats and monkeys after lentiviral vector-mediated gene transfer. Hum Gene Ther Methods 24(6):333–344. https://doi.org/10.1089/hgtb.2012.238
91. Klein C, Evrard HC, Shapcott KA, Haverkamp S, Logothetis NK, Schmid MC (2016) Cell-targeted optogenetics and electrical microstimulation reveal the primate koniocellular projection to supra-granular visual cortex. Neuron 90(1):143–151. https://doi.org/10.1016/j.neuron.2016.02.036
92. Kugler S, Kilic E, Bahr M (2003) Human synapsin 1 gene promoter confers highly neuron-specific long-term transgene expression from an adenoviral vector in the adult rat brain depending on the transduced area. Gene Ther 10(4):337–347. https://doi.org/10.1038/sj.gt.3301905
93. Nathanson JL, Jappelli R, Scheeff ED, Manning G, Obata K, Brenner S, Callaway EM (2009) Short promoters in Viral Vectors Drive selective expression in mammalian inhibitory neurons, but do not restrict activity to specific inhibitory cell-types. Front Neural Circuits 3:19. https://doi.org/10.3389/neuro.04.019.2009
94. Kugler S, Lingor P, Scholl U, Zolotukhin S, Bahr M (2003) Differential transgene expression in brain cells in vivo and in vitro from AAV-2 vectors with small transcriptional control units. Virology 311(1):89–95. https://doi.org/10.1016/s0042-6822(03)00162-4
95. Hoesche C, Sauerwald A, Veh RW, Krippl B, Kilimann MW (1993) The 5'-flanking region of the rat synapsin I gene directs neuron-specific and developmentally regulated reporter gene expression in transgenic mice. J Biol Chem 268(35):26494–26502
96. Stauffer WR, Lak A, Yang A, Borel M, Paulsen O, Boyden ES, Schultz W (2016) Dopamine neuron-dpecific optogenetic stimulation in rhesus macaques. Cell 166(6):1564–1571. e1566. https://doi.org/10.1016/j.cell.2016.08.024
97. De A, El-Shamayleh Y, Horwitz GD (2020) Fast and reversible neural inactivation in macaque cortex by optogenetic stimulation of GABAergic neurons. Elife 9. https://doi.org/10.7554/eLife.52658
98. Abbas AK, Lichtman AH, Pillai S (2016) Basic immunology : functions and disorders of the immune system, 5th edn. Elsevier, St. Louis
99. Dimidschstein J, Chen Q, Tremblay R, Rogers SL, Saldi GA, Guo L, Xu Q, Liu R, Lu C, Chu J, Grimley JS, Krostag AR, Kaykas A, Avery MC, Rashid MS, Baek M, Jacob AL, Smith GB, Wilson DE, Kosche G, Kruglikov I, Rusielewicz T, Kotak VC, Mowery TM, Anderson SA, Callaway EM, Dasen JS, Fitzpatrick D, Fossati V, Long MA, Noggle S, Reynolds JH, Sanes DH, Rudy B, Feng G, Fishell G (2016) A viral strategy for targeting and manipulating interneurons across vertebrate species. Nat Neurosci 19(12):1743–1749. https://doi.org/10.1038/nn.4430
100. Vormstein-Schneider D, Lin JD, Pelkey KA, Chittajallu R, Guo B, Arias-Garcia MA, Allaway K, Sakopoulos S, Schneider G, Stevenson O, Vergara J, Sharma J, Zhang Q, Franken TP, Smith J, Ibrahim LAKJMA, Sabri E, Huang S, Favuzzi E, Burbridge T, Xu Q, Guo L, Vogel I, Sanchez V, Saldi GA, Gorissen BL, Yuan X, Zaghloul KA, Devinsky O, Sabatini BL, Batista-Brito R, Reynolds J, Feng G, Fu Z, McBain CJ, Fishell G, Dimidschstein J (2020) Viral manipulation of functionally distinct interneurons in mice, non-human primates and humans. Nat Neurosci 23(12):1629–1636. https://doi.org/10.1038/s41593-020-0692-9
101. Mich JK, Graybuck LT, Hess EE, Mahoney JT, Kojima Y, Ding Y, Somasundaram S, Miller JA, Kalmbach BE, Radaelli C, Gore BB, Weed N, Omstead V, Bishaw Y, Shapovalova NV, Martinez RA, Fong O, Yao S, Mortrud M, Chong P, Loftus L, Bertagnolli D, Goldy J, Casper T, Dee N, Opitz-Araya X, Cetin A, Smith KA, Gwinn RP, Cobbs C, Ko AL, Ojemann JG, Keene CD, Silbergeld DL, Sunkin SM, Gradinaru V, Horwitz GD, Zeng H, Tasic B, Lein ES, Ting JT, Levi BP (2021) Functional enhancer elements drive subclass-selective expression from mouse to primate neocortex. Cell Rep 34(13):108754. https://doi.org/10.1016/j.celrep.2021.108754
102. Colasante G, Lignani G, Brusco S, Di Berardino C, Carpenter J, Giannelli S, Valassina N, Bido S, Ricci R, Castoldi V, Marenna S, Church T, Massimino L, Morabito G, Benfenati F, Schorge S, Leocani L, Kullmann DM, Broccoli V

(2020) dCas9-based Scn1a gene activation restores inhibitory interneuron excitability and attenuates seizures in Dravet syndrome mice. Mol Ther 28(1):235–253. https://doi.org/10.1016/j.ymthe.2019.08.018

103. Ogiwara I, Miyamoto H, Morita N, Atapour N, Mazaki E, Inoue I, Takeuchi T, Itohara S, Yanagawa Y, Obata K, Furuichi T, Hensch TK, Yamakawa K (2007) Nav1.1 localizes to axons of parvalbumin-positive inhibitory interneurons: a circuit basis for epileptic seizures in mice carrying an Scn1a gene mutation. J Neurosci 27(22):5903–5914. https://doi.org/10.1523/JNEUROSCI.5270-06.2007

104. Shinohara Y, Konno A, Takahashi N, Matsuzaki Y, Kishi S, Hirai H (2016) Viral vector-based dissection of Marmoset GFAP promoter in Mouse and Marmoset brains. PLoS One 11(8):e0162023. https://doi.org/10.1371/journal.pone.0162023

105. Kalidasan V, Ng WH, Ishola OA, Ravichantar N, Tan JJ, Das KT (2021) A guide in lentiviral vector production for hard-to-transfect cells, using cardiac-derived c-kit expressing cells as a model system. Sci Rep 11(1):19265. https://doi.org/10.1038/s41598-021-98657-7

106. Heffernan KS, Rahman K, Smith Y, Galvan A (2022) Characterization of the GfaABC1D promoter to selectively target astrocytes in the rhesus macaque brain. J Neurosci Methods 372:109530. https://doi.org/10.1016/j.jneumeth.2022.109530

107. Arendt D, Musser JM, Baker CVH, Bergman A, Cepko C, Erwin DH, Pavlicev M, Schlosser G, Widder S, Laubichler MD, Wagner GP (2016) The origin and evolution of cell types. Nat Rev Genet 17(12):744–757. https://doi.org/10.1038/nrg.2016.127

108. Hrvatin S, Tzeng CP, Nagy MA, Stroud H, Koutsioumpa C, Wilcox OF, Assad EG, Green J, Harvey CD, Griffith EC, Greenberg ME (2019) A scalable platform for the development of cell-type-specific viral drivers. Elife 8. https://doi.org/10.7554/eLife.48089

109. Blankvoort S, Descamps LAL, Kentros C (2020) Enhancer-Driven Gene Expression (EDGE) enables the generation of cell type specific tools for the analysis of neural circuits. Neurosci Res 152:78–86. https://doi.org/10.1016/j.neures.2020.01.009

110. Blankvoort S, Witter MP, Noonan J, Cotney J, Kentros C (2018) Marked diversity of unique cortical enhancers enables neuron-specific tools by enhancer-driven gene expression. Curr Biol 28(13):2103–2114 e2105. https://doi.org/10.1016/j.cub.2018.05.015

111. Hardison RC, Taylor J (2012) Genomic approaches towards finding cis-regulatory modules in animals. Nat Rev Genet 13(7): 469–483. https://doi.org/10.1038/nrg3242

112. Arnold CD, Gerlach D, Stelzer C, Boryn LM, Rath M, Stark A (2013) Genome-wide quantitative enhancer activity maps identified by STARR-seq. Science 339(6123):1074–1077. https://doi.org/10.1126/science.1232542

113. Farley EK, Olson KM, Levine MS (2015) Regulatory principles governing tissue specificity of developmental enhancers. Cold Spring Harb Symp Quant Biol 80:27–32. https://doi.org/10.1101/sqb.2015.80.027227

114. Kittelmann S, Preger-Ben Noon E, McGregor AP, Frankel N (2021) A complex gene regulatory architecture underlies the development and evolution of cuticle morphology in Drosophila. Curr Opin Genet Dev 69:21–27. https://doi.org/10.1016/j.gde.2021.01.003

115. Yue F, Cheng Y, Breschi A, Vierstra J, Wu W, Ryba T, Sandstrom R, Ma Z, Davis C, Pope BD, Shen Y, Pervouchine DD, Djebali S, Thurman RE, Kaul R, Rynes E, Kirilusha A, Marinov GK, Williams BA, Trout D, Amrhein H, Fisher-Aylor K, Antoshechkin I, DeSalvo G, See LH, Fastuca M, Drenkow J, Zaleski C, Dobin A, Prieto P, Lagarde J, Bussotti G, Tanzer A, Denas O, Li K, Bender MA, Zhang M, Byron R, Groudine MT, McCleary D, Pham L, Ye Z, Kuan S, Edsall L, Wu YC, Rasmussen MD, Bansal MS, Kellis M, Keller CA, Morrissey CS, Mishra T, Jain D, Dogan N, Harris RS, Cayting P, Kawli T, Boyle AP, Euskirchen G, Kundaje A, Lin S, Lin Y, Jansen C, Malladi VS, Cline MS, Erickson DT, Kirkup VM, Learned K, Sloan CA, Rosenbloom KR, Lacerda de Sousa B, Beal K, Pignatelli M, Flicek P, Lian J, Kahveci T, Lee D, Kent WJ, Ramalho Santos M, Herrero J, Notredame C, Johnson A, Vong S, Lee K, Bates D, Neri F, Diegel M, Canfield T, Sabo PJ, Wilken MS, Reh TA, Giste E, Shafer A, Kutyavin T, Haugen E, Dunn D, Reynolds AP, Neph S, Humbert R, Hansen RS, De Bruijn M, Selleri L, Rudensky A, Josefowicz S, Samstein R, Eichler EE, Orkin SH, Levasseur D, Papayannopoulou T, Chang KH, Skoultchi A, Gosh S, Disteche C,

Treuting P, Wang Y, Weiss MJ, Blobel GA, Cao X, Zhong S, Wang T, Good PJ, Lowdon RF, Adams LB, Zhou XQ, Pazin MJ, Feingold EA, Wold B, Taylor J, Mortazavi A, Weissman SM, Stamatoyannopoulos JA, Snyder MP, Guigo R, Gingeras TR, Gilbert DM, Hardison RC, Beer MA, Ren B, Mouse EC (2014) A comparative encyclopedia of DNA elements in the mouse genome. Nature 515(7527):355–364. https://doi.org/10.1038/nature13992

116. Bartlett JS, Samulski RJ, McCown TJ (1998) Selective and rapid uptake of adeno-associated virus type 2 in brain. Hum Gene Ther 9(8):1181–1186. https://doi.org/10.1089/hum.1998.9.8-1181

117. Gray SJ, Foti SB, Schwartz JW, Bachaboina L, Taylor-Blake B, Coleman J, Ehlers MD, Zylka MJ, McCown TJ, Samulski RJ (2011) Optimizing promoters for recombinant adeno-associated virus-mediated gene expression in the peripheral and central nervous system using self-complementary vectors. Hum Gene Ther 22(9):1143–1153. https://doi.org/10.1089/hum.2010.245

118. McCown TJ, Xiao X, Li J, Breese GR, Samulski RJ (1996) Differential and persistent expression patterns of CNS gene transfer by an adeno-associated virus (AAV) vector. Brain Res 713(1-2):99–107

119. Powell SK, Samulski RJ, McCown TJ (2020) AAV capsid-promoter interactions determine CNS cell-selective gene expression in vivo. Mol Ther. https://doi.org/10.1016/j.ymthe.2020.03.007

120. Galvan A, Raper J, Hu X, Pare JF, Bonaventura J, Richie CT, Michaelides M, Mueller SAL, Roseboom PH, Oler JA, Kalin NH, Hall RA, Smith Y (2019) Ultrastructural localization of DREADDs in monkeys. Eur J Neurosci. https://doi.org/10.1111/ejn.14429

121. Shields BC, Kahuno E, Kim C, Apostolides PF, Brown J, Lindo S, Mensh BD, Dudman JT, Lavis LD, Tadross MR (2017) Deconstructing behavioral neuropharmacology with cellular specificity. Science 356(6333). https://doi.org/10.1126/science.aaj2161

122. Berglund K, Clissold K, Li HE, Wen L, Park SY, Gleixner J, Klein ME, Lu D, Barter JW, Rossi MA, Augustine GJ, Yin HH, Hochgeschwender U (2016) Luminopsins integrate opto- and chemogenetics by using physical and biological light sources for opsin activation. Proc Natl Acad Sci U S A 113(3):E358–E367. https://doi.org/10.1073/pnas.1510899113

123. Berglund K, Fernandez AM, Gutekunst CN, Hochgeschwender U, Gross RE (2019) Step-function luminopsins for bimodal prolonged neuromodulation. J Neurosci Res. https://doi.org/10.1002/jnr.24424

124. Berglund K, Gross RE (2019) Opto-chemogenetics with luminopsins: a novel avenue for targeted control of neuronal activity. J Neurosci Res. https://doi.org/10.1002/jnr.24473

125. Yang Y, Pacia CP, Ye D, Zhu L, Baek H, Yue Y, Yuan J, Miller MJ, Cui J, Culver JP, Bruchas MR, Chen H (2020) Sonogenetics for noninvasive and cellular-level neuromodulation in rodent brain. bioRxiv:2020200120289199l0. https://doi.org/10.1101/2020.01.28.919910

126. Ermakova YG, Lanin AA, Fedotov IV, Roshchin M, Kelmanson IV, Kulik D, Bogdanova YA, Shokhina AG, Bilan DS, Staroverov DB, Balaban PM, Fedotov AB, Sidorov-Biryukov DA, Nikitin ES, Zheltikov AM, Belousov VV (2017) Thermogenetic neurostimulation with single-cell resolution. Nat Commun 8:15362. https://doi.org/10.1038/ncomms15362

127. Bernstein JG, Garrity PA, Boyden ES (2012) Optogenetics and thermogenetics: technologies for controlling the activity of targeted cells within intact neural circuits. Curr Opin Neurobiol 22(1):61–71. https://doi.org/10.1016/j.conb.2011.10.023

128. Smith Y, Galvan A (2018) Non-human primate research of basal ganglia and movement disorders: advances and challenges. J Neural Transm (Vienna) 125(3):275–278. https://doi.org/10.1007/s00702-018-1849-5

129. Sternson SM, Roth BL (2014) Chemogenetic tools to interrogate brain functions. Annu Rev Neurosci 37:387–407. https://doi.org/10.1146/annurev-neuro-071013-014048

130. Atasoy D, Sternson SM (2018) Chemogenetic tools for causal cellular and neuronal biology. Physiol Rev 98(1):391–418. https://doi.org/10.1152/physrev.00009.2017

131. Roth BL (2016) DREADDs for Neuroscientists. Neuron 89(4):683–694. https://doi.org/10.1016/j.neuron.2016.01.040

132. Barnett LM, Hughes TE, Drobizhev M (2017) Deciphering the molecular mechanism responsible for GCaMP6m's Ca2+-dependent change in fluorescence. PLoS One 12(2):e0170934. https://doi.org/10.1371/journal.pone.0170934

133. Nakai J, Ohkura M, Imoto K (2001) A high signal-to-noise Ca(2+) probe composed of a

single green fluorescent protein. Nat Biotechnol 19(2):137–141. https://doi.org/10.1038/84397

134. Seidemann E, Chen Y, Bai Y, Chen SC, Mehta P, Kajs BL, Geisler WS, Zemelman BV (2016) Calcium imaging with genetically encoded indicators in behaving primates. Elife 5. https://doi.org/10.7554/eLife.16178

135. Ju N, Jiang R, Macknik SL, Martinez-Conde-S, Tang S (2018) Long-term all-optical interrogation of cortical neurons in awake-behaving nonhuman primates. PLoS Biol 16(8):e2005839. https://doi.org/10.1371/journal.pbio.2005839

136. Bollimunta A, Bogadhi AR, Krauzlis RJ (2018) Comparing frontal eye field and superior colliculus contributions to covert spatial attention. Nat Commun 9(1):3553. https://doi.org/10.1038/s41467-018-06042-2

137. Kinoshita M, Matsui R, Kato S, Hasegawa T, Kasahara H, Isa K, Watakabe A, Yamamori T, Nishimura Y, Alstermark B, Watanabe D, Kobayashi K, Isa T (2012) Genetic dissection of the circuit for hand dexterity in primates. Nature 487(7406):235–238. https://doi.org/10.1038/nature11206

138. Sun F, Zhou J, Dai B, Qian T, Zeng J, Li X, Zhuo Y, Zhang Y, Wang Y, Qian C, Tan K, Feng J, Dong H, Lin D, Cui G, Li Y (2020) Next-generation GRAB sensors for monitoring dopaminergic activity in vivo. Nat Methods 17(11):1156–1166. https://doi.org/10.1038/s41592-020-00981-9

139. Patriarchi T, Cho JR, Merten K, Howe MW, Marley A, Xiong WH, Folk RW, Broussard GJ, Liang R, Jang MJ, Zhong H, Dombeck D, von Zastrow M, Nimmerjahn A, Gradinaru V, Williams JT, Tian L (2018) Ultrafast neuronal imaging of dopamine dynamics with designed genetically encoded sensors. Science 360(6396). https://doi.org/10.1126/science.aat4422

140. Sun F, Zeng J, Jing M, Zhou J, Feng J, Owen SF, Luo Y, Li F, Wang H, Yamaguchi T, Yong Z, Gao Y, Peng W, Wang L, Zhang S, Du J, Lin D, Xu M, Kreitzer AC, Cui G, Li Y (2018) A genetically encoded fluorescent sensor enables rapid and specific detection of dopamine in flies, fish, and mice. Cell 174(2):481–496 e419. https://doi.org/10.1016/j.cell.2018.06.042

141. Borden PM, Zhang P, Shivange AV, Marvin JS, Cichon J, Dan C, Podgorski K, Figueiredo A, Novak O, Tanimoto M, Shigetomi E, Lobas MA, Kim H, Zhu PK, Zhang Y, Zheng WS, Fan C, Wang G, Xiang B, Gan L, Zhang G-X, Guo K, Lin L, Cai Y, Yee AG, Aggarwal A, Ford CP, Rees DC, Dietrich D, Khakh BS, Dittman JS, Gan W-B, Koyama M, Jayaraman V, Cheer JF, Lester HA, Zhu JJ, Looger LL (2020) A fast genetically encoded fluorescent sensor for faithful in vivo acetylcholine detection in mice, fish, worms and flies. bioRxiv:2020.2002.2007.939504. https://doi.org/10.1101/2020.02.07.939504

142. Bartelle BB, Szulc KU, Suero-Abreu GA, Rodriguez JJ, Turnbull DH (2013) Divalent metal transporter, DMT1: a novel MRI reporter protein. Magn Reson Med 70(3):842–850. https://doi.org/10.1002/mrm.24509

143. Patrick PS, Hammersley J, Loizou L, Kettunen MI, Rodrigues TB, Hu DE, Tee SS, Hesketh R, Lyons SK, Soloviev D, Lewis DY, Aime S, Fulton SM, Brindle KM (2014) Dual-modality gene reporter for in vivo imaging. Proc Natl Acad Sci U S A 111(1):415–420. https://doi.org/10.1073/pnas.1319000111

144. Livet J, Weissman TA, Kang H, Draft RW, Lu J, Bennis RA, Sanes JR, Lichtman JW (2007) Transgenic strategies for combinatorial expression of fluorescent proteins in the nervous system. Nature 450(7166):56–62. https://doi.org/10.1038/nature06293

145. Cai D, Cohen KB, Luo T, Lichtman JW, Sanes JR (2013) Improved tools for the Brainbow toolbox. Nat Methods 10(6):540–547. https://doi.org/10.1038/nmeth.2450

146. Lichtman JW, Livet J, Sanes JR (2008) A technicolour approach to the connectome. Nat Rev Neurosci 9(6):417–422. https://doi.org/10.1038/nrn2391

147. Deng C, Yuan H, Dai J (2018) Behavioral manipulation by optogenetics in the nonhuman primate. Neuroscientist 24(5):526–539. https://doi.org/10.1177/1073858417728459

148. Macknik SL, Alexander RG, Caballero O, Chanovas J, Nielsen KJ, Nishimura N, Schaffer CB, Slovin H, Babayoff A, Barak R, Tang S, Ju N, Yazdan-Shahmorad A, Alonso JM, Malinskiy E, Martinez-Conde S (2019) Advanced circuit and cellular imaging methods in nonhuman primates. J Neurosci 39(42):8267–8274. https://doi.org/10.1523/JNEUROSCI.1168-19.2019

149. Gerits A, Farivar R, Rosen BR, Wald LL, Boyden ES, Vanduffel W (2012) Optogenetically induced behavioral and functional network changes in primates. Curr Biol 22(18):

1722–1726. https://doi.org/10.1016/j.cub.2012.07.023

150. Ohayon S, Grimaldi P, Schweers N, Tsao DY (2013) Saccade modulation by optical and electrical stimulation in the macaque frontal eye field. J Neurosci 33(42):16684–16697. https://doi.org/10.1523/JNEUROSCI.2675-13.2013

151. Nakamichi Y, Okubo K, Sato T, Hashimoto M, Tanifuji M (2019) Optical intrinsic signal imaging with optogenetics reveals functional cortico-cortical connectivity at the columnar level in living macaques. Sci Rep 9(1):6466. https://doi.org/10.1038/s41598-019-42923-2

INDEX

Mark A. G. Eldridge and Adriana Galvan (eds.), *Vectorology for Optogenetics and Chemogenetics*, Neuromethods, vol. 195, https://doi.org/10.1007/978-1-0716-2918-5,

R

S

T

V

MIX
Papier aus verantwortungsvollen Quellen
Paper from responsible sources
FSC® C105338

If you have any concerns about our products,
you can contact us on
ProductSafety@springernature.com

In case Publisher is established outside the EU,
the EU authorized representative is:
Springer Nature Customer Service Center GmbH
Europaplatz 3, 69115 Heidelberg, Germany

Printed by Libri Plureos GmbH
in Hamburg, Germany